IEE Electromagnetic Waves Series 21

Series Editors: Professors P. J. B. Clarricoats,
E. D. R. Shearman and J. R. Wait

Waveguide Handbook

Previous volumes in this series

Volume 1 Geometrical theory of diffraction for electromagnetic waves
Graeme L. James
Volume 2 Electromagnetic waves and curved structures
Leonard Lewin, David C. Chang and Edward F. Kuester
Volume 3 Microwave homodyne systems
Ray J. King
Volume 4 Radio direction-finding
P. J. D. Gething
Volume 5 ELF communications antennas
Michael L. Burrows
Volume 6 Waveguide tapers, transitions and couplers
F. Sporleder and H. G. Unger
Volume 7 Reflector antenna analysis and design
P. J. Wood
Volume 8 Effects of the troposphere on radio communications
Martin P. M. Hall
Volume 9 Schumann resonances in the earth-ionosphere cavity
P. V. Bliokh, A. P. Nikolaenko and Y. F. Filippov
Volume 10 Aperture antennas and diffraction theory
E. V. Jull
Volume 11 Adaptive array principles
J. E. Hudson
Volume 12 Microstrip antenna theory and design
J. R. James, P. S. Hall and C. Wood
Volume 13 Energy in electromagnetism
H. G. Booker
Volume 14 Leaky feeders and subsurface radio communications
P. Delogne
Volume 15 The handbook of antenna design volume 1
Editors: A. W. Rudge, K. Milne, A. D. Olver, P. Knight
Volume 16 The handbook of antenna design volume 2
Editors: A. W. Rudge, K. Milne, A. D. Olver, P. Knight
Volume 17 Surveillance radar performance prediction
P. Rohan
Volume 18 Corrugated horns for microwave antennas
P. J. B. Clarricoats and A. D. Olver
Volume 19 Microwave antenna theory and design
S. Silver
Volume 20 Advances in radar techniques
J. Clarke

Waveguide Handbook

N. Marcuvitz

Peter Peregrinus Ltd
On behalf of The Institution of Electrical Engineers

Published by Peter Peregrinus Ltd., London, UK

© 1986 Errata and preface to this reprint, Peter Peregrinus Ltd

This book was first published in 1951 by the McGraw-Hill Book Company Inc.
ISBN 0 86341 058 8

Printed in England by Short Run Press Ltd., Exeter

Foreword

THE tremendous research and development effort that went into the development of radar and related techniques during World War II resulted not only in hundreds of radar sets for military (and some for possible peacetime) use but also in a great body of information and new techniques in the electronics and high-frequency fields. Because this basic material may be of great value to science and engineering, it seemed most important to publish it as soon as security permitted.

The Radiation Laboratory of MIT, which operated under the supervision of the National Defense Research Committee, undertook the great task of preparing these volumes. The work described herein, however, is the collective result of work done at many laboratories, Army, Navy, university, and industrial, both in this country and in England, Canada, and other Dominions.

The Radiation Laboratory, once its proposals were approved and finances provided by the Office of Scientific Research and Development, chose Louis N. Ridenour as Editor-in-Chief to lead and direct the entire project. An editorial staff was then selected of those best qualified for this type of task. Finally the authors for the various volumes or chapters or sections were chosen from among those experts who were intimately familiar with the various fields and who were able and willing to write the summaries of them. This entire staff agreed to remain at work at MIT for six months or more after the work of the Radiation Laboratory was complete. These volumes stand as a monument to this group.

These volumes serve as a memorial to the unnamed hundreds and thousands of scientists, engineers, and others who actually carried on the research, development, and engineering work the results of which are herein described. There were so many involved in this work and they worked so closely together even though often in widely separated laboratories that it is impossible to name or even to know those who contributed to a particular idea or development. Only certain ones who wrote reports or articles have even been mentioned. But to all those who contributed in any way to this great cooperative development enterprise, both in this country and in England, these volumes are dedicated.

L. A. DuBridge

Preface

THIS book endeavors to present the salient features in the reformulation of microwave field problems as microwave network problems. The problems treated are the class of electromagnetic "boundary value" or "diffraction" problems descriptive of the scattering properties of discontinuities in waveguides. Their reformulation as network problems permits such properties to be calculated in a conventional network manner from equivalent microwave networks composed of transmission lines and lumped constant circuits. A knowledge of the values of the equivalent network parameters is a necessary prerequisite to quantitative calculations. The theoretical evaluation of microwave network parameters entails in general the solution of three-dimensional boundary-value problems and hence belongs properly in the domain of electromagnetic field theory. In contrast, the network calculations of power distribution, frequency response, resonance properties, etc., characteristic of the "far-field" behavior in microwave structures, involve mostly algebraic problems and hence may be said to belong in the domain of microwave network theory. The independence of the roles played by microwave field and network theories is to be emphasized; it has a counterpart in conventional low-frequency electrical theory and accounts in no small measure for the far-reaching development of the network point of view both at microwave and low frequencies.

In the years 1942 to 1946 a rather intensive and systematic exploitation of both the field and network aspects of microwave problems was carried out at the Radiation Laboratory of MIT by a group of workers among whom J. Schwinger played a dominant role. By means of an integral-equation formulation of field problems, Schwinger pointed the way both in the setting up and solving of a wide variety of microwave problems. These developments resulted in a rigorous and general theory of microwave structures in which conventional low-frequency electrical theory appeared as a special case. As is to be expected, the presentation of the results of these developments involves the work of many individuals both in this country and abroad, as well as much material which is now more or less standard in mathematical and engineering literature. Unfortunately, it has not been possible to document adequately these sources in the present edition. It is hoped that these and other omissions will be remedied in a subsequent and more up-to-date edition.

Although the primary aim of this book is to present the equivalent-circuit parameters for a large number of microwave structures, a brief but coherent account of the fundamental concepts necessary for their proper utilization is included. Thus there is summarized in the first three chapters both the field and network theoretic considerations necessary for the derivation and utilization of the basic transmission line—equivalent-circuit formalism. The mode concept and transmission-line formulation of the field equations are introduced in Chapter 1. This chapter contains an engineering treatment of the transmission-line theory necessary for the description of propagating and nonpropagating modes in the more important types of uniform and nonuniform waveguides. The field-structure, propagation, attenuation, etc., characteristics of the transmission-line modes so described are compiled in Chapter 2, with both quantitative and pictorial detail. The elements of microwave-network theory required for the analysis, representation, and measurement of the equivalent circuits for N-terminal microwave structures are outlined in Chapter 3; also contained in this chapter is a sketch of some of the field theoretic methods employed in the derivation of the equivalent-circuit parameters reported in Chapters 4 to 8. Although most of the above material is written for the impedance-minded microwave engineer, some of the sections should be of interest to the applied mathematician.

The remaining chapters contain a compilation of the equivalent-circuit parameters for a variety of nondissipative N-terminal microwave structures. These results are presented usually both analytically and graphically in individual sections having an intentionally concise format to avoid repetition. Since the analytical formulae are frequently cumbersome to evaluate, care has been taken to achieve a reasonable degree of accuracy in the graphical plots. In Chapter 4 a number of two-terminal structures, such as beyond-cutoff and radiative waveguide terminations, are treated. Obstacle and aperture discontinuities in waveguides, gratings in free space, etc., are among the four-terminal structures described in Chapter 5. Chapter 6 deals with six-terminal microwave structures and contains the equivalent-circuit parameters for a number of E- and H-plane T- and Y-junctions, bifurcations, etc. Several eight-terminal structures are treated in Chapter 7. Chapter 8 contains the circuit description of a number of typical composite microwave structures: dielectric-filled guides, thick apertures, etc. In contrast to the relatively complicated field calculations employed to obtain the previous results, only simple microwave network calculations are required to find the circuit parameters and properties of these composite structures.

The equivalent-circuit results in the various sections of Chapters 4 to 8 involve the expenditure of considerable time and effort on the part of many workers—often not at all commensurate with the space devoted to the presentation of these results. Each section usually represents the contributions of many individuals who unfortunately are not acknowledged in each instance. In addition to J. Schwinger, the following helped with direct theoretical contributions to these sections:

J. F. Carlson, A. E. Heins, H. A. Levine
P. M. Marcus, and D. S. Saxon.

Indirect contributions were made by H. A. Bethe, N. H. Frank, and R. M. Whitmer. The efforts of Levine and Marcus, who remained with the office of publications until its close in 1946, are particularly acknowledged; the latter correlated all the tabulated work reported in the appendix. The continued interest and criticism of Levine and Schwinger since the close of the laboratory are greatly appreciated. Although a great deal of experimental work on the measurement of equivalent-circuit parameters was carried out, only that part which is not covered by or in agreement with theory is included in Chapters 4 to 8. The work of W. H. Pickering *et al.,* of California Institute of Technology, and of C. G. and D. D. Montgomery should be cited in this connection.

A considerable amount of technical assistance was rendered by many others. Mrs. A. Marcus did most of the work on the mode plots presented in Chapter 2. C. W. Zabel correlated some of the theoretical and experimental data on T sections in Chapter 6. Most of the numerical computations were carried out under the direction of A. E. Heins by M. Karakashian, R. Krock, D. Perkins, B. Siegle, and others. Finally, the valuable editorial assistance, planning, and criticism of H. M. James in the initial stages of preparation of this book should be mentioned.

Although conceived at the Radiation Laboratory of MIT, the greater part of this book was written in the years subsequent to its close while the author was a staff member of the Polytechnic Institute of Brooklyn. The author wishes to thank Professor E. Weber, Director of Microwave Research Institute at the Polytechnic Institute, for use of the technical and clerical facilities of the laboratory in the preparation of this book; also various members of the Institute for their criticism and proofreading of many sections of this book; and lastly his wife, Muriel, for her continuous help and encouragement.

N. Marcuvitz

Brooklyn, N.Y.
September, 1950

Preface to 1986 Edition

THIRTY-FIVE years have elapsed since the first printing of the Waveguide Handbook. Public domain editions by various publishers have appeared in the interim, but the current "out of print" status of the Handbook has been a cause of lament by interested students and microwave researchers. The decision by the Institution of Electrical Engineers to undertake a new printing remedies this condition and comes as a personal pleasure and tribute to the efforts of the individuals acknowledged in the original preface. The new edition also provides an opportunity to correct a partial list of typographical and substantive errors pointed out over the past years by colleagues.

N. MARCUVITZ

POLYTECHNIC INSTITUTE OF NEW YORK
December, 1985

Errata

Page	Reads	Should read
12, 4th line from bottom	Eqn(21)	Eqn(22)
30, 4th line after Eqn(62):	implies, in general,	does not imply
6th line after Eqn(62):	must consequently be effected on a scalar basis	can be effected in terms of vector modes with $\mathbf{e} \neq \mathbf{h} \times \mathbf{r_o}$
35, Fig.1.10b legend:	zero	infinite
37, Fig.1.11b legend:	zero	infinite
40, line 2:	et(x,y)	ct(x,y)
52, Eqn(97a)	J_n	$\hat{J}_n$
Eqn(97c)	J_o	$\hat{J}_o$
47, line 1:	Fig.1.8	Fig.1.8b
53, end of first paragraph	are not as yet	are
63, top figure labelled H_{10} 3rd and 4th columns of arrows:		should be reversed i.e. ▶ ···· ◀ ▶ ···· ◀ ▶ ···· ◀ ▶ ···· ◀
72, Eqn(31):	V'_{00} I'_{00}	$-V'_{00}$ $-I'_{00}$
73, Eqn(32):	V I	$-V$ $-I$
79, Eqn at top figure:	1.873	1.946
88, 2nd line after Eqn(63):	falls normally	is incident
104, line 5:	$Z_{mn} = Z_{nm}$ and $\mathrm{Re}(Z_{mn}) = 0$	$Z_{mn} = Z_{nm}$ for reciprocal and $\mathrm{Re}(Z_{mn}) = 0$ for nondissipative structures
108, Eqn(10):	$S_{mn} = S_{nm}$.	$S_{mn} = S_{nm}$ (for reciprocal structures)
111, Fig.3.3 labels:	(c) (b)	(b), (c)
115, Fig.3.7 legend:	H-plane	E-plane
126, line 2:	the shunt	the ± shunt

xii

Page	Reads	Should read
155, Eqn(88a):	$E(y)$ $H(y)$	$E(y,z)$ $H(y,z)$
155, Eqn(88b):	$E_n(y)$ $H_n(y)$	$E_n(y,z)$ $H_n(y,z)$
178, Eqn(1):	$\dfrac{4b}{\lambda} \ln \dfrac{a}{b} ($	$\left(\dfrac{4b}{\lambda} \ln \dfrac{a}{b}\right)($
182, Fig.4.6-5	$\dfrac{b}{\lambda^1}$	$\dfrac{b}{\lambda'}$
184, Eqn(2a):	$N_o(x)$	$-N_o(x)$
196, line above Eqn(1):	where	where ($\gamma = 1.781$)
201, line above Eqn(1):	by	by ($\gamma = 1.781$)
208, line above Eqn(1):	where	where ($\gamma = 1.781$)
208, line 3 in Eqn(1):	l	1
216, 7 lines from bottom:	range of validity	range of validity, except for $a \approx b$,
224, line 3-4 of *Restrictions:*	an incident lowest mode	*two* incident lowest modes
238, line after Eqn(1b):	$\tan^3$	$9 \tan^3$
line 3 after Eqn(1b):	bd	bd^2
239, line 8	Σ	$\acute{\Sigma}$
246, line 4 in Experimental Results:	dotted	solid
line 7 in Experimental Results:	solid	dotted
256, Eqn after Eqn(6b):	$\dfrac{l}{d'}$	$\dfrac{1}{d'}$
257, Eqn(7)	4_a	$4a$
Eqn(8):	$\left(\dfrac{\pi D_1}{a}\right)^2$	$\left(\dfrac{\pi D_1}{a}\right)^4$
266, Eqn(1a) + (1b):	$\dfrac{a}{2\lambda g} \csc^2 \dfrac{\pi x}{a}$	$\left(\dfrac{a}{2\lambda g} \csc^2 \dfrac{\pi x}{a}\right)$
266, 3rd line after Eqn(2b):	$\sin^2 \dfrac{n\pi x}{a} [\ldots\ldots]$	$[\ldots\ldots] \dfrac{\sin^2 n\pi x}{a}$
289, Fig.5.22-1 Side View:	d' d	d d'
Transpose	↑ E ↕Θ - - - - ⇀	↑ E ↕Θ - - - - ⇀

xii

Page	Reads	Should read
339, Fig. 6.1-3 Side View:	T_2' should be below line	$\underset{T_2'}{\llcorner\rule{2cm}{0.4pt}}$
339, Numerical Results —line 3	$\dfrac{b}{\lambda g}$	$\dfrac{2b}{\lambda g}$
Experimental Results —line 3	$\dfrac{b}{\lambda g}$	$\dfrac{2b}{\lambda g}$
340, Fig.6.1-4	$\dfrac{b}{\lambda g} = \dfrac{b}{b}$	$\dfrac{2b}{\lambda g} = \dfrac{b'}{b}$
341, Fig.6.1-5	$\dfrac{b}{\lambda g} =$	$\dfrac{2b}{\lambda g} =$
342, Fig.6.1-6	$\dfrac{b}{\lambda g} =$	$\dfrac{2b}{\lambda g} =$
343, Fig.6.1-7	$\dfrac{b}{\lambda g} =$	$\dfrac{2b}{\lambda g} =$
344, Fig.6.1-8	$\dfrac{b}{\lambda g} =$	$\dfrac{2b}{\lambda g} =$
345, Fig.6.1-9	$\dfrac{b}{\lambda g} =$	$\dfrac{2b}{\lambda g} =$
346, Fig.6.1-10	$\dfrac{b}{\lambda g} =$	$\dfrac{2b}{\lambda g} =$
347, Fig.6.1-11	$\dfrac{b}{\lambda g} =$	$\dfrac{2b}{\lambda g} =$
348, Fig.6.1-12	$\dfrac{b}{\lambda g} =$	$\dfrac{2b}{\lambda g} =$
349, Fig.6.1-13	$\dfrac{b}{\lambda g} =$	$\dfrac{2b}{\lambda g} =$
350, Fig.6.1-14	$\dfrac{b}{\lambda g} =$	$\dfrac{2b}{\lambda g} =$

Should corrections other than the above be noticed, the publishers and Series Editors of the IEE's Electromagnetic Waves Series would be grateful if they could be communicated to them so that the errata sheet can be updated in future editions, thus benefiting the next generation of research workers.

Contents

FOREWORD v

PREFACE vii

CHAP. 1. TRANSMISSION LINES 1

    1·1. Waveguides as Transmission Lines 1
    1·2. Field Representation in Uniform Waveguides 3
    1·3. Uniform Transmission Lines. Impedance Descriptions 7
    1·4. Uniform Transmission Lines. Scattering Descriptions 13
    1·5. Interrelations among Uniform Transmission-line Descriptions . . 16
    1·6. Uniform Transmission Lines with Complex Parameters 17
        (a) Waveguides with dissipation
        (b) Waveguides beyond cutoff
    1·7. Field Representation in Nonuniform Radial Waveguides 29
    1·8. Field Representation in Nonuniform Spherical Waveguides . . . 47

CHAP. 2. TRANSMISSION-LINE MODES 55

    2·1. Mode Characteristics 55
    2·2. Rectangular Waveguides 56
        (a) E-modes
        (b) H-modes
        (c) Modes in a Parallel Plate Guide
    2·3. Circular Waveguides 66
        (a) E-modes
        (b) H-modes
    2·4. Coaxial Waveguides 72
        (a) E-modes
        (b) H-modes
    2·5. Elliptical Waveguides 79
    2·6. Space as a Uniform Waveguide 84
        (a) Fields in free space
        (b) Field in the vicinity of gratings
    2·7. Radial Waveguides 89
        (a) Cylindrical cross-sections
        (b) Cylindrical sector cross-sections
    2·8. Spherical Waveguides 96
        (a) Fields in free space
        (b) Conical waveguides

CHAP. 3. MICROWAVE NETWORKS 101

    3·1. Representation of Waveguide Discontinuities 101
        (a) Impedance representation
        (b) Admittance representation
        (c) Scattering representation

3·2.	Equivalent Circuits for Waveguide Discontinuities.	108
3·3.	Equivalent Representations of Microwave Networks	117
3·4.	Measurement of Network Parameters	130
3·5.	Theoretical Determination of Circuit Parameters	138

Chap. 4. TWO-TERMINAL STRUCTURES. 168

Lines Terminating in Guides Beyond Cutoff 168

4·1. Change of Cross Section, H-plane 168
    (a) Symmetrical case
    (b) Asymmetrical case
4·2. Bifurcation of a Rectangular Guide, H-plane 172
4·3. Coupling of a Coaxial Line to a Circular Guide 174
4·4. Rectangular to Circular Change in Cross Section. 176
4·5. Termination of a Coaxial Line by a Capacitive Gap 178

Lines Radiating into Space 179

4·6a. Parallel-plate Guide into Space, E-plane 179
4·6b. Rectangular Guide into Bounded Space, E-plane 183
4·7a. Parallel-plate Guide Radiating into Space, E-plane. 183
4·7b. Rectangular Guide Radiating into Bounded Half Space, E-plane 184
4·8. Parallel-plate Guide into Space, H-plane 186
4·9. Parallel-plate Guide Radiating into Half Space, H-plane . . . 187
4·10. Apertures in Rectangular Guide. 193
    (a) Rectangular apertures
    (b) Circular apertures
4·11. Array of Semi-infinite Planes, H-plane 195
4·12. Radiation from a Circular Guide, E_{01}-mode. 196
4·13. Radiation from a Circular Guide, H_{01}-mode. 201
4·14. Radiation from a Circular Guide, H_{11}-mode. 206
4·15. Coaxial Line with Infinite-center Conductor 208
4·16. Coaxial Line Radiating into Semi-infinite Space 213

Chap. 5. FOUR-TERMINAL STRUCTURES 217

Structures with Zero Thickness. 218

5·1. Capacitive Obstacles and Windows in Rectangular Guide . . . 218
    (a) Window formed by two obstacles
    (b) Window formed by one obstacle
    (c) Symmetrical obstacle
5·2. Inductive Obstacles and Windows in Rectangular Guide . . . 221
    (a) Symmetrical window
    (b) Asymmetrical window
    (c) Symmetrical obstacle
5·3. Capacitive Windows in Coaxial Guide 229
    (a) Disk on inner conductor
    (b) Disk on outer conductor
5·4. Circular and Elliptical Apertures in Rectangular Guide. . . . 238
    (a) Centered circular aperture
    (b) Small elliptical or circular aperture
5·5. Elliptical and Circular Apertures in Circular Guide 243
5·6. Small Elliptical and Circular Apertures in Coaxial Guide. . . 246

5·7. Annular Window in Circular Guide 247
5·8. Annular Obstacles in Circular Guide. 249
STRUCTURES WITH FINITE THICKNESS 249
5·9. Capacitive Obstacles of Finite Thickness 249
    (a) Window formed by two obstacles
    (b) Window formed by one obstacle
5·10. Inductive Windows of Finite Thickness. 253
    (a) Symmetrical window
    (b) Asymmetrical window
5·11. Solid Inductive Post in Rectangular Guide 257
    (a) Off-centered post
    (b) Centered post
    (c) Noncircular posts
5·12. Dielectric Posts in Rectangular Guide 266
5·13. Capacitive Post in Rectangular Guide 268
5·14. Post of Variable Height in Rectangular Guide. 271
5·15. Spherical Dent in Rectangular Guide. 274
5·16. Circular Obstacle of Finite Thickness in Rectangular Guide. . . 274
5·17. Resonant Ring in Circular Guide 275
GRATINGS AND ARRAYS IN FREE SPACE 280
5·18. Capacitive Strips 280
5·19. Inductive Strips. 284
5·20. Capacitive Posts. 285
5·21. Inductive Posts. 286
5·22. Array of Semi-infinite Planes, E-plane 289
5·23. Array of Semi-infinite Planes, H-plane 292
ASYMMETRIC STRUCTURES; COUPLING OF TWO GUIDES 296
5·24. Junction of Two Rectangular Guides, H-plane. 296
    (a) Symmetrical case
    (b) Asymmetrical case
5·25. Bifurcation of a Rectangular Guide, H-plane 302
5·26. Change in Height of Rectangular Guide 307
    (a) Symmetrical case
    (b) Asymmetrical case
5·27. Change in Radius of Coaxial Guide 310
    (a) Equal outer radii
    (b) Equal inner radii
5·28. E-plane Corners. 312
    (a) Right-angle bends
    (b) Arbitrary angle bends
5·29. H-plane Corners. 318
    (a) Right-angle bends
    (b) Arbitrary angle H-plane corners
5·30. Junction of a Rectangular and a Radial Guide, E-plane. 322
5·31. Coupling of a Coaxial and a Circular Guide. 323
5·32. Coupling of Rectangular and Circular Guide 324
5·33. Aperture Coupling of Two Guides. 329
    (a) Junction of two rectangular guides
    (b) Junction of two circular guides
    (c) Junction of two coaxial guides
    (d) Junction of a rectangular and circular guide

CONTENTS

- 5·34. Circular Bends, E-plane 333
- 5·35. Circular Bends, H-plane 334

Chap. 6. SIX-TERMINAL STRUCTURES 336

- 6·1. Open T-junction, E-plane. 337
- 6·2. Slit-coupled T-junctions in Rectangular Guide, E-plane. 339
- 6·3. 120° Y-junction, E-plane 352
- 6·4. E-plane Bifurcation 353
- 6·5. Open T-junction, H-plane. 355
- 6·6. Slit-coupled T-junction in Rectangular Guide, H-plane 360
- 6·7. 120° Y-junction, H-plane. 362
- 6·8. Aperture-coupled T-junctions, E-plane. 363
  - (a) Rectangular stub guide
  - (b) Circular stub guide
- 6·9. Aperture-coupled T-junction in Rectangular Guide, H-plane. . . 366
  - (a) Rectangular stub guide
  - (b) Circular stub guide
- 6·10. Aperture-coupled T-junction in Coaxial Guide. 368
- 6·11. Bifurcation of a Coaxial Line 369

Chap. 7. EIGHT-TERMINAL STRUCTURES. 373

- 7·1. Slit Coupling of Rectangular Guides, E-plane. 373
- 7·2. Small-aperture Coupling of Rectangular Guides, E-plane. . . . 375
- 7·3. Aperture Coupling of Coaxial Guides. 377
- 7·4. Slit Coupling of Rectangular Guides, H-plane. 378
- 7·5. Aperture Coupling of Rectangular Guides, H-plane 379
- 7·6. 0° Y-junction, E-plane. 380
- 7·7. 0° Y-junction, H-plane. 383
- 7·8. Magic-T (Hybrid) Junction. 386

Chap. 8. COMPOSITE STRUCTURES. 387

Propagation in Composite Guides 387

- 8·1. Rectangular Guide with Dielectric Slabs Parallel to E 388
- 8·2. Rectangular Guide with Dielectric Slabs Perpendicular to E . . 391
- 8·3. Circular Guide with Dielectric Cylinders 393
- 8·4. Coaxial Guide with Dielectric Cylinders 396
- 8·5. Rectangular Guide with "Nonradiating" Slit 397
- 8·6. Rectangular Guides with Ridges. 399
- 8·7. Rectangular Guide with Resistive Strip. 402

Thickness Effects 404

- 8·8. Capacitive Obstacles of Large Thickness 404
- 8·9. Inductive Obstacles of Large Thickness. 407
  - (a) Window formed by two obstacles
  - (b) Window formed by one obstacle
- 8·10. Thick Circular Window. 408
- 8·11. E-plane T with Slit Coupling 412

APPENDIX 415

GLOSSARY 421

INDEX 423

CHAPTER 1

TRANSMISSION LINES

1·1. Waveguides as Transmission Lines.—The determination of the electromagnetic fields within any region is dependent upon one's ability to solve explicitly the Maxwell field equations in a coordinate system appropriate to the region. Complete solutions of the field equations, or equivalently of the wave equation, are known for only relatively few types of regions. Such regions may be classified as either uniform or nonuniform. Uniform regions are characterized by the fact that cross sections transverse to a given symmetry, or propagation, direction are almost everywhere identical with one another in both size and shape. Nonuniform regions are likewise characterized by a symmetry, or propagation, direction but the transverse cross sections are similar to rather than identical with one another.

Examples of uniform regions are provided by regions cylindrical about the symmetry direction and having planar cross sections with rectangular, circular, etc., peripheries. Regions not cylindrical about the symmetry direction and having nonplanar cross sections of cylindrical, spherical, etc., shapes furnish examples of nonuniform regions (*cf.* Secs. 1·7 and 1·8). In either case the cross sections may or may not be limited by metallic boundaries. Within such regions the electromagnetic field may be represented as a superposition of an infinite number of standard functions that form a mathematically complete set. These complete sets of functional solutions are classical and have been employed in the mathematical literature for some time. However, in recent years the extensive use of ultrahigh frequencies has made it desirable to reformulate these mathematical solutions in engineering terms. It is with this reformulation that the present chapter will be concerned.

The mathematical representation of the electromagnetic field within a uniform or nonuniform region is in the form of a superposition of an infinite number of modes or wave types. The electric and magnetic field components of each mode are factorable into form functions, depending only on the cross-sectional coordinates transverse to the direction of propagation, and into amplitude functions, depending only on the coordinate in the propagation direction. The transverse functional form of each mode is dependent upon the cross-sectional shape of the given region and, save for the amplitude factor, is identical at every cross section. As a result the amplitudes of a mode completely characterize the mode at every cross section. The variation of each amplitude along the propagation direction is given implicitly as a solution of a one-dimensional wave or transmission-line equation. According to the mode in question the wave amplitudes may be either propagating or attenuating along the transmission direction.

In many regions of practical importance, as, for example, in waveguides, the dimensions and field excitation are such that only one mode is capable of propagation. As a result the electromagnetic field almost everywhere is characterized completely by the amplitudes of this one dominant wave type. Because of the transmission-line behavior of the mode amplitudes it is suggestive to define the amplitudes that measure the transverse electric and magnetic field intensities of this dominant mode as voltage and current, respectively. It is thereby implied that the electromagnetic fields may be described almost everywhere in terms of the voltage and current on an appropriate transmission line. This transmission line completely characterizes the behavior of the dominant mode everywhere in the waveguide. The knowledge of the real characteristic impedance and wave number of the transmission line then permits one to describe rigorously the propagation of this dominant mode in familiar impedance terms.

The impedance description may be extended to describe the behavior of the nonpropagating or higher modes that are present in the vicinity of cross-sectional discontinuities. Mode voltages and currents are introduced as measures of the amplitudes of the transverse electric and magnetic field intensities of each of the higher modes. Thus, as before, each of the higher modes is represented by a transmission line but now the associated characteristic impedance is reactive and the wave number imaginary, i.e., attenuating. In this manner the complete description of the electromagnetic field in a waveguide may be represented in terms of the behavior of the voltages and currents on an infinite number of transmission lines. The quantitative use of such a representation in a given waveguide geometry presupposes the ability to determine explicitly the following:

1. The transverse functional form of each mode in the waveguide cross section.
2. The transmission-line equations for the mode amplitudes together with the values of the mode characteristic impedance and propagation constant for each mode.
3. Expressions for the field components in terms of the amplitudes and functional form of the modes.

The above-described impedance or transmission-line reformulation of the electromagnetic field will be carried out for a number of practical uniform and nonuniform waveguides.

1·2. Field Representation in Uniform Waveguides.—By far the largest class of waveguide regions is the uniform type represented in Fig. 1·1. Such regions are cylindrical and have, in general, an arbitrary cross section that is generated by a straight line moving parallel to the symmetry or transmission direction, the latter being characterized by the unit vector z_0. In many practical waveguides the cross sectional geometry is described by a coordinate system appropriate to the boundary

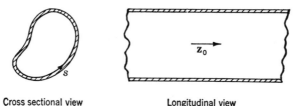

Cross sectional view   Longitudinal view

Fig. 1·1.—Uniform waveguide of arbitrary cross section.

curves although this is not a necessary requirement. Since the transmission-line description of the electromagnetic field within *uniform* guides is independent of the particular form of coordinate system employed to describe the cross section, no reference to cross-sectional coordinates will be made in this section. Special coordinate systems appropriate to rectangular, circular, and elliptical cross sections, etc., will be considered in Chap. 2. To stress the independence of the transmission-line description upon the cross-sectional coordinate system an invariant transverse vector formulation of the Maxwell field equations will be employed in the following. This form of the field equations is obtained by elimination of the field components along the transmission, or z, direction and can be written, for the steady state of angular frequency ω, as

$$\left. \begin{array}{l} \dfrac{\partial \mathbf{E}_t}{\partial z} = -jk\zeta(\varepsilon + \dfrac{1}{k^2} \boldsymbol{\nabla}_t \boldsymbol{\nabla}_t) \cdot (\mathbf{H}_t \times \mathbf{z}_0), \\[2mm] \dfrac{\partial \mathbf{H}_t}{\partial z} = -jk\eta(\varepsilon + \dfrac{1}{k^2} \boldsymbol{\nabla}_t \boldsymbol{\nabla}_t) \cdot (\mathbf{z}_0 \times \mathbf{E}_t). \end{array} \right\} \quad (1)$$

Vector notation is employed with the following meanings for the symbols:

$\mathbf{E}_t = \mathbf{E}_t(x,y,z)$ = the rms electric-field intensity transverse to the z-axis.

$\mathbf{H}_t = \mathbf{H}_t(x,y,z)$ = the rms magnetic-field intensity transverse to the z-axis.

ζ = intrinsic impedance of the medium = $1/\eta = \sqrt{\mu/\epsilon}$

k = propagation constant in medium = $\omega \sqrt{\mu\epsilon} = 2\pi/\lambda$

$\boldsymbol{\nabla}_t$ = gradient operator transverse to z-axis[1] = $\boldsymbol{\nabla} - \mathbf{z}_0 \dfrac{\partial}{\partial z}$

$\boldsymbol{\varepsilon}$ = unit dyadic defined such that $\boldsymbol{\varepsilon} \cdot \mathbf{A} = \mathbf{A} \cdot \boldsymbol{\varepsilon} = \mathbf{A}$

The time variation of the field is assumed to be exp $(+j\omega t)$. The z components of the electric and magnetic fields follow from the transverse components by the relations

$$\left. \begin{array}{l} jk\eta E_z = \boldsymbol{\nabla}_t \cdot (\mathbf{H}_t \times \mathbf{z}_0), \\ jk\zeta H_z = \boldsymbol{\nabla}_t \cdot (\mathbf{z}_0 \times \mathbf{E}_t). \end{array} \right\} \quad (2)$$

Equations (1) and (2), which are fully equivalent to the Maxwell equations, make evident in transmission-line guise the separate dependence of the field on the cross-sectional coordinates and on the longitudinal coordinate z. The cross-sectional dependence may be integrated out of Eqs. (1) by means of a suitable set of vector orthogonal functions. Functions such that the result of the operation $\boldsymbol{\nabla}_t \boldsymbol{\nabla}_t \cdot$ on a function is proportional to the function itself are of the desired type provided they satisfy, in addition, appropriate conditions on the boundary curve or curves s of the cross section. Such vector functions are known to be of two types: the E-mode functions $\mathbf{e}_i'$ defined by

where
$$\left. \begin{array}{l} \mathbf{e}_i' = -\boldsymbol{\nabla}_t \Phi_i, \\ \mathbf{h}_i' = \mathbf{z}_0 \times \mathbf{e}_i', \end{array} \right\} \quad (3a)$$

$$\left. \begin{array}{l} \nabla_t^2 \Phi_i + k_{ci}'^2 \Phi_i = 0 \\ \Phi_i = 0 \text{ on } s \text{ if } k_{ci}' \neq 0 \\ \dfrac{\partial \Phi_i}{\partial s} = 0 \text{ on } s \text{ if } k_{ci}' = 0;* \end{array} \right\} \quad (3b)$$

and the H-mode functions $\mathbf{e}_i''$ defined by

where
$$\left. \begin{array}{l} \mathbf{e}_i'' = \mathbf{z}_0 \times \boldsymbol{\nabla}_t \Psi_i, \\ \mathbf{h}_i'' = \mathbf{z}_0 \times \mathbf{e}_i'', \end{array} \right\} \quad (4a)$$

$$\left. \begin{array}{l} \nabla_t^2 \Psi_i + k_{ci}''^2 \Psi_i = 0, \\ \dfrac{\partial \Psi_i}{\partial \nu} = 0 \text{ on } s, \end{array} \right\} \quad (4b)$$

---

[1] For a cross-section defined by a rectangular xy coordinate system

$$\boldsymbol{\nabla}_t = \mathbf{x}_0 \dfrac{\partial}{\partial x} + \mathbf{y}_0 \dfrac{\partial}{\partial y},$$

where $\mathbf{x}_0$ and $\mathbf{y}_0$ are unit vectors in the x and y directions.

* The case $k_{ci}' = 0$ arises in multiply connected cross sections such as those encountered in coaxial waveguides. The vanishing of the tangential derivative of Φ_i on s implies that Φ_i is a constant on each periphery.

where i denotes a double index mn and ν is the outward normal to s in the cross-section plane. For the sake of simplicity, the explicit dependence of $\mathbf{e}'_i$, $\mathbf{e}''_i$, Φ_i, and Ψ_i on the cross-sectional coordinates has been omitted in the writing of the equations. The constants k'_{ci} and k''_{ci} are defined as the cutoff wave numbers or eigenvalues associated with the guide cross section. Explicit expressions for the mode functions and cutoff wave numbers of several waveguide cross sections are presented in Chap. 2.

The functions $\mathbf{e}_i$ possess the vector orthogonality properties

$$\iint \mathbf{e}'_i \cdot \mathbf{e}'_j \, dS = \iint \mathbf{e}''_i \cdot \mathbf{e}''_j \, dS = \begin{cases} 1 \text{ for } i = j \\ 0 \text{ for } i \neq j \end{cases} \\ \iint \mathbf{e}'_i \cdot \mathbf{e}''_j \, dS = 0, \tag{5}$$

with the integration extended over the entire guide cross section. The product $\mathbf{e}_i \cdot \mathbf{e}_j$ is a simple scalar product or an Hermitian (i.e., complex conjugate) product depending on whether or not the mode vectors are real or complex.

The transverse electric and magnetic fields can be expressed in terms of the above-defined orthogonal functions by means of the representation

$$\mathbf{E}_t = \sum_i V'_i(z)\mathbf{e}'_i + \sum_i V''_i(z)\mathbf{e}''_i, \\ \mathbf{H}_t = \sum_i I'_i(z)\mathbf{h}'_i + \sum_i I''_i(z)\mathbf{h}''_i, \tag{6a}$$

and inversely the amplitudes V_i and I_i can be expressed in terms of the fields as

$$V'_i = \iint \mathbf{E}_t \cdot \mathbf{e}'_i \, dS, \qquad V''_i = \iint \mathbf{E}_t \cdot \mathbf{e}''_i \, dS, \\ I'_i = \iint \mathbf{H}_t \cdot \mathbf{h}'_i \, dS, \qquad I''_i = \iint \mathbf{H}_t \cdot \mathbf{h}''_i \, dS. \tag{6b}$$

The longitudinal field components then follow from Eqs. (2), (3), (4), and (6a) as

$$jk\eta E_z = \sum_i I'_i(z) k'^{2}_{ci} \Phi_i, \\ jk\zeta H_z = \sum_i V''_i(z) k''^{2}_{ci} \Psi_i. \tag{6c}$$

In view of the orthogonality properties (5) and the representation (6a), the total average power flow along the guide at z and in the $\mathbf{z}_0$ direction is,

$$P_z = \text{Re}\left(\iint \mathbf{E}_t \times \mathbf{H}_t^* \cdot \mathbf{z}_0 \, dS\right) = \text{Re}\left(\sum_i V_i' I_i'^* + \sum_i V_i'' I_i''^*\right), \quad (7)$$

where all quantities are rms and the asterisk denotes the complex conjugate.

For uniform guides possessing no discontinuities within the guide cross section or on the guide walls the substitution of Eqs. (6a) transforms Eqs. (1) into an infinite set of equations of the type

$$\left. \begin{array}{l} \dfrac{dV_i}{dz} = -j\kappa_i Z_i I_i, \\[4pt] \dfrac{dI_i}{dz} = -j\kappa_i Y_i V_i, \end{array} \right\} \quad (8)$$

which define the variation with z of the mode amplitudes V_i and I_i. The superscript distinguishing the mode type has been omitted, since the equations are of the same form for both modes. The parameters κ_i and Z_i are however of different form; for E-modes

$$\kappa_i' = \sqrt{k^2 - k_{ci}'^2}, \qquad Z_i' = \zeta \frac{\kappa_i'}{k} = \frac{\kappa_i'}{\omega \epsilon}; \quad (9a)$$

for H-modes

$$\kappa_i'' = \sqrt{k^2 - k_{ci}''^2}, \qquad Z_i'' = \zeta \frac{k}{\kappa_i''} = \frac{\omega \mu}{\kappa_i''}. \quad (9b)$$

Equations (8) are of standard transmission-line form. They constitute the basis for the definition of the amplitudes V_i as mode voltages, of the amplitudes I_i as mode currents, and concomitantly of the parameters κ_i and Z_i as the mode propagation constant and mode characteristic impedance, respectively. The functional dependence of the parameters κ_i and Z on the cross-sectional dimensions is given in Chap. 2 for several waveguides of practical importance.

The field representation given by Eqs. (6a) and (8) provides a general solution of the field equations that is particularly appropriate for the description of the guide fields in the vicinity of transverse discontinuities—such as apertures in transverse plates of zero thickness, or changes of cross section. The field representation given in Eqs. (6a) is likewise applicable to the description of longitudinal discontinuities—such as obstacles of finite thickness or apertures in the guide walls. However, as is evident on substitution of Eqs. (6a) into Eqs. (1), the transmission-line equations (8) for the determination of the voltage and current amplitudes must be modified to take into account the presence of longitudinal discontinuities within the cross section. This modification results in the addition of z-dependent "generator" voltage and current terms to the right-hand members of Eqs. (8). The determination of the mode amplitude for the case of longitudinal discontinuities is thus somewhat more complicated than for the case of transverse discontinuities. Both cases, however, constitute more or less conventional transmission-line problems.

1·3. Uniform Transmission Lines.—As shown in Sec. 1·2 the representation of the electric and magnetic fields within an arbitrary but uniform waveguide (*cf.* Fig. 1·1) can be reformulated into an engineering description in terms of an infinite number of mode voltages and currents. The variation of each mode voltage and current along the guide axis is described in terms of the corresponding variation of voltage and current along an appropriate transmission line. The description of the entire field within the guide is thereby reduced to the description of the electrical behavior on an infinite set of transmission lines. In this section two distinctive ways of describing the electrical behavior on a transmission line will be sketched: (1) the impedance (admittance) description, (2) the scattering (reflection and transmission coefficient) description.

The transmission-line description of a waveguide mode is based on the fact, noted in the preceding section, that the transverse electric field $\mathbf{E}_t$ and transverse magnetic field $\mathbf{H}_t$ of *each* mode can be expressed as

$$\left. \begin{array}{l} \mathbf{E}_t(x,y,z) = V(z)\mathbf{e}(x,y), \\ \mathbf{H}_t(x,y,z) = I(z)\mathbf{h}(x,y), \end{array} \right\} \quad (10)$$

where $\mathbf{e}(x,y)$ and $\mathbf{h}(x,y)$ are vector functions indicative of the cross-sectional form of the mode fields, and $V(z)$ and $I(z)$ are voltage and current functions that measure the rms amplitudes of the transverse electric and magnetic fields at any point z along the direction of propagation. As a consequence of the Maxwell field equations (*cf.* Sec. 1·2) the voltage and current are found to obey transmission-line equations of the form

$$\left. \begin{array}{l} \dfrac{dV}{dz} = -j\kappa ZI, \\ \dfrac{dI}{dz} = -j\kappa YV, \end{array} \right\} \quad (11)$$

where, for a medium of uniform dielectric constant and permeability,

$$\kappa = \sqrt{k^2 - k_c^2}$$

$$Z = \frac{1}{Y} \left\{ \begin{array}{l} = \zeta \dfrac{k}{\kappa} = \sqrt{\dfrac{\mu}{\epsilon}} \dfrac{\lambda_g}{\lambda} \quad \text{for } H\text{-modes,} \\ = \zeta \dfrac{\kappa}{k} = \sqrt{\dfrac{\mu}{\epsilon}} \dfrac{\lambda}{\lambda_g} \quad \text{for } E\text{-modes.} \end{array} \right\} \quad (11a)$$

Since the above transmission-line description is applicable to every mode, the sub- and superscripts distinguishing the mode type and number will be omitted in this section. The parameters k, k_c, κ, and Z are termed the free-space wave number, the cutoff wave number, the guide wave number, and the characteristic impedance of the mode in question. Instead of the parameters k, k_c, and κ the corresponding wavelengths λ, λ_c, and λ_g are frequently employed. These are related by

$$\left. \begin{array}{l} k = \dfrac{2\pi}{\lambda}, \qquad k_c = \dfrac{2\pi}{\lambda_c}, \qquad \kappa = \dfrac{2\pi}{\lambda_g}, \\[2mm] \lambda_g = \dfrac{\lambda}{\sqrt{1 - \left(\dfrac{\lambda}{\lambda_c}\right)^2}}. \end{array} \right\} \qquad (11b)$$

The explicit dependence of the mode cutoff wave number k_c and mode functions **e** and **h** on the cross-sectional geometry of several uniform guides will be given in Chap. 2. Together with the knowledge of the wavelength λ of field excitation, these quantities suffice to determine completely the transmission-line behavior of an individual mode.

Since the voltage V and current I are chosen as rms quantities, and since the vector functions **e** and **h** are normalized over the cross section in accordance with Eq. (5), the average total mode power flow along the direction of propagation is Re (VI^*). Although the voltage V and current I suffice to characterize the behavior of a mode, it is evident that such a characterization is not unique. Occasionally it is desirable to redefine the relations [Eqs. (10)] between the fields and the voltage and current in order to correspond more closely to customary low-frequency definitions, or to simplify the equivalent circuit description of waveguide discontinuities. These redefinitions introduce changes of the form

$$V = \frac{\bar{V}}{N^{1/2}}, \qquad I = N^{1/2}\bar{I}, \qquad (12a)$$

where the scale factor $N^{1/2}$ is so chosen as to retain the form of the power expression as Re $(\bar{V}\bar{I}^*)$. On substitution of the transformations (12a) into Eqs. (11) it is apparent that the transmission-line equations retain the same form in the new voltage $\bar{V}$ and current $\bar{I}$ provided a new characteristic impedance

$$\bar{Z} = ZN = \frac{1}{\bar{Y}} \qquad (12b)$$

is introduced. Transformation relations of this kind are generally important only in the case of the dominant mode and even then only when absolute impedance comparisons are necessary. Most transmission-line properties depend on relative impedances; the latter are unaffected by transformations of the above type.

Equations (11) may be schematically represented by the transmission-line diagram of Fig. 1·2 wherein the choice of positive directions for V and I is indicated. To determine explicit solutions of Eqs. (11) it is convenient to eliminate either I or V and thus obtain the one-dimensional wave equations

$$\frac{d^2V}{dz^2} + \kappa^2 V = 0 \tag{13a}$$

or

$$\frac{d^2I}{dz^2} + \kappa^2 I = 0. \tag{13b}$$

Equations (13) define waves of two types: either propagating or attenuating with the distance z depending on whether the constant κ^2 is either positive or negative. Although both types of waves can be treated by the same formalism, the following applies particularly to the propagating type.

Impedance Descriptions.—The solutions to Eqs. (13) can be written as a superposition of the trigonometrical functions

$$\cos \kappa z, \quad \sin \kappa z. \tag{14}$$

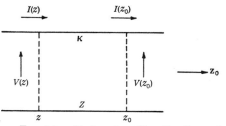

Fig. 1·2.—Choice of positive directions of voltage and current in a uniform transmission line.

By means of these so-called standing waves, the solutions to Eq. (11) can be expressed in terms of the voltage *or* current at two different points z_0 and z_1 as

$$V(z) = \frac{V(z_0) \sin \kappa(z_1 - z) + V(z_1) \sin \kappa(z - z_0)}{\sin \kappa(z_1 - z_0)}, \tag{15a}$$

$$I(z) = \frac{I(z_0) \sin \kappa(z_1 - z) + I(z_1) \sin \kappa(z - z_0)}{\sin \kappa(z_1 - z_0)}, \tag{15b}$$

or in terms of the voltage *and* current at the same point z_0 as

$$V(z) = V(z_0) \cos \kappa(z - z_0) - jZI(z_0) \sin \kappa(z - z_0), \tag{16a}$$
$$I(z) = I(z_0) \cos \kappa(z - z_0) - jYV(z_0) \sin \kappa(z - z_0). \tag{16b}$$

Equations (16) represent the voltage and current everywhere in terms of the voltage and current at a single point z_0. Since in many applications the absolute magnitudes of V and I are unimportant, it is desirable to introduce at any point z the ratio

$$\frac{1}{Y} \frac{I(z)}{V(z)} = Y'(z) = \frac{1}{Z'(z)} \tag{17}$$

called the relative, or normalized, admittance at z looking in the direction of increasing z. In terms of this quantity Eqs. (16) can be reexpressed, by division of Eqs. (16a) and (16b), in the form

$$Y'(z) = \frac{j + Y'(z_0) \cot \kappa(z_0 - z)}{\cot \kappa(z_0 - z) + jY'(z_0)}, \tag{18}$$

which is the fundamental transmission-line equation relating the relative admittance at any point z to that at any other point z_0.

Many graphical schemes have been proposed to facilitate computations with Eq. (18). One of the more convenient representations, the so-called circle diagram, or Smith chart, is shown in Fig. 1·3. For real κ this diagram represents Eq. (18) as a constant radius rotation of the complex quantity $Y'(z_0)$ into the complex quantity $Y'(z)$, the angle of rotation being $2\kappa(z_0 - z)$ radians. Since graphical uses of this diagram have been treated in sufficient detail elsewhere in this series,[1] we shall consider only a few special but important analytical forms of Eq. (18). For $Y'(z_0) = \infty$,

$$Y'(z) = -j \cot \kappa(z_0 - z); \tag{19a}$$

for $Y'(z_0) = 0$,

$$Y'(z) = +j \tan \kappa(z_0 - z); \tag{19b}$$

for $Y'(z_0) = 1$,

$$Y'(z) = 1. \tag{19c}$$

These are, respectively, the relative input admittances at z corresponding to a short circuit, an open circuit, and a "match" at the point z_0.

The fundamental admittance relation [Eq. (18)] can be rewritten as an impedance relation

$$Z'(z) = \frac{j + Z'(z_0) \cot \kappa(z_0 - z)}{\cot \kappa(z_0 - z) + jZ'(z_0)}. \tag{20}$$

The similarity in form of Eqs. (18) and (20) is indicative of the existence of a duality principle for the transmission-line equations (11). Duality in the case of Eqs. (11) implies that if V, I, Z are replaced respectively by I, V, Y, the equations remain invariant in form. As a consequence relative admittance relations deduced from Eqs. (11) have exactly the same form as relative impedance relations.

It is occasionally desirable to represent the admittance relation (18) by means of an equivalent circuit. The circuit equations for such a representation are obtained by rewriting Eqs. (16) in the form

$$\left. \begin{aligned} I(z) &= -jY \cot \kappa(z_0 - z)[V(z)] - jY \csc \kappa(z_0 - z)[-V(z_0)], \\ I(z_0) &= -jY \csc \kappa(z_0 - z)[V(z)] - jY \cot \kappa(z_0 - z)[-V(z_0)]. \end{aligned} \right\} \tag{21a}$$

---

[1] *Cf.* G. L. Ragan, *Microwave Transmission Circuits*, Vol. 9, Radiation Laboratory Series.

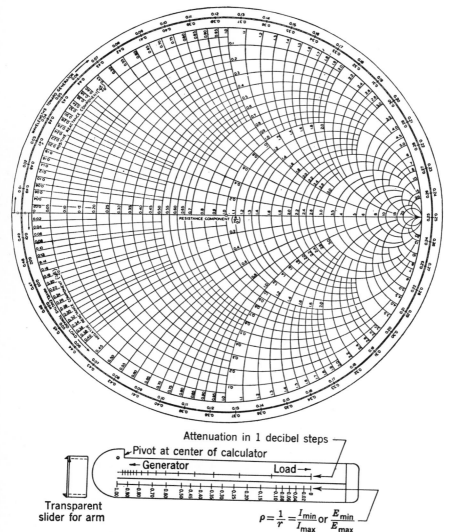

FIG. 1·3.—Circle diagram for uniform transmission lines.

The equivalent circuit is schematically represented by the π network shown in Fig. 1·4 which indicates both the positive choice of voltage and current directions as well as the admittance values of the circuit elements for a length $l = z_0 - z$ of transmission line.

By the duality replacements indicated above, Eqs. (21a) may be written in impedance form as

$$V(z) = -jZ \cot \kappa(z_0 - z)[I(z)] - jZ \csc \kappa(z_0 - z)[-I(z_0)], \\ V(z_0) = -jZ \csc \kappa(z_0 - z)[I(z)] - jZ \cot \kappa(z_0 - z)[-I(z_0)]. \quad (21b)$$

Hence an alternative equivalent circuit for a length l of transmission line may be represented by the T network shown in Fig. 1·4b wherein are indicated the impedance values of the circuit elements. The relation between the impedances at the points z and z_0 follows from the above circuit representations by the well-known combinatorial rules for impedances.

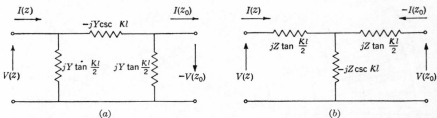

Fig. 1·4.—(a) π-Circuit for a length l of uniform transmission line; (b) T-circuit for a length l of uniform transmission line.

An alternative form of Eq. (18) useful for conceptual as well as computational purposes is obtained on the substitution

$$Y'(z) = -j \cot \theta(z). \tag{22a}$$

The resulting equation for $\theta(z)$ in terms of $\theta(z_0)$ is, omitting an additive multiple of 2π,

$$\theta(z) = \theta(z_0) + \kappa(z_0 - z). \tag{22b}$$

The quantity $\theta(z)$ represents the electrical "length" of a short-circuit line equivalent to the relative admittance $Y'(z)$. The fundamental transmission-line relation (18), expressed in the simple form of Eq. (22b), states that the length equivalent to the input admittance at z is the algebraic sum of the length equivalent to the output admittance at z_0 plus the electrical length of the transmission line between z and z_0. It should be noted that the electrical length corresponding to an arbitrary admittance is in general complex.

In addition to the relation between the relative admittances at the two points z and z_0 the relation between the frequency derivatives of the relative admittances is of importance. The latter may be obtained by differentiation of Eqs. (21) either as

$$\frac{\kappa \dfrac{dY'(z)}{d\kappa}}{1 + [jY'(z)]^2} = j\kappa(z_0 - z) + \frac{\kappa \dfrac{dY'(z_0)}{d\kappa}}{1 + [jY'(z_0)]^2}, \tag{23a}$$

or, since from Eq. (11b)

$$\frac{d\kappa}{\kappa} = \left(\frac{k}{\kappa}\right)^2 \frac{dk}{k} = \left(\frac{k}{\kappa}\right)^2 \frac{d\omega}{\omega} = -\left(\frac{\lambda_g}{\lambda}\right)^2 \frac{d\lambda}{\lambda},$$

as

$$\frac{dY'(z)}{1 + [jY'(z)]^2} = j\left(\frac{k}{\kappa}\right)^2 \kappa(z_0 - z)\frac{d\omega}{\omega} + \frac{dY'(z_0)}{1 + [jY'(z_0)]^2}. \qquad (23b)$$

It should be emphasized that Eq. (23b) determines the frequency derivative of the *relative* admittance. If the characteristic admittance Y varies with frequency, it is necessary to distinguish between the frequency derivatives of the relative admittance $Y'(z)$ and the absolute admittance $Y(z)$ by means of the relation

$$\omega \frac{dY(z)}{d\omega} = Y\left[\omega \frac{dY'(z)}{d\omega}\right] + Y'(z)\left(\omega \frac{dY}{d\omega}\right). \qquad (24)$$

Equations (22) to (24) are of importance in the computation of frequency sensitivity and Q of a waveguide structure.

1·4. Uniform Transmission Lines. *Scattering Descriptions.*—The scattering, just as the impedance, description of a propagating mode is based on Eqs. (10) to (11), wherein the mode fields are represented in terms of a voltage and a current. For the scattering description, however, solutions to the wave equations (13) are expressed as a superposition of exponential functions

$$e^{-j\kappa z} \quad \text{and} \quad e^{+j\kappa z}, \qquad (25)$$

which represent waves traveling in the direction of increasing and decreasing z. The resulting traveling-wave solutions can be represented as

$$V(z) = V_{\text{inc}} e^{-j\kappa(z-z_0)} + V_{\text{refl}} e^{+j\kappa(z-z_0)}, \qquad (26a)$$
$$ZI(z) = V_{\text{inc}} e^{-j\kappa(z-z_0)} - V_{\text{refl}} e^{+j\kappa(z-z_0)}, \qquad (26b)$$

where V_{inc} and V_{refl} are the complex amplitudes at $z = z_0$ of "incident" and "reflected" voltage waves, respectively.

Equations (26) constitute the complete description of the mode fields everywhere in terms of the incident and reflected amplitudes at a single point. Since many of the physical properties of the mode fields depend only on a ratio of incident and reflected wave amplitudes, it is desirable to introduce at any point z the ratio

$$\Gamma(z) = \frac{V_{\text{refl}}}{V_{\text{inc}}} e^{j2\kappa(z-z_0)} \qquad (27)$$

called the voltage reflection coefficient. The current reflection coefficient defined as the negative of the voltage reflection coefficient is also employed in this connection. However, in the following the reflection coefficient Γ is to be understood as the voltage coefficient.

In terms of Eqs. (26) and (27) the expression (7) for the total average power flow at any point z on a nondissipative uniform transmission line becomes

$$P = \text{Re}\,(VI^*) = \frac{|V_{\text{inc}}|^2}{Z} - \frac{|V_{\text{refl}}|^2}{Z} = \frac{|V_{\text{inc}}|^2}{Z}[1 - |\Gamma|^2], \tag{28}$$

which may immediately be interpreted as the difference between the incident and the reflected power flowing down the guide. Equation (28) makes evident the significance of $|\Gamma|^2$ as the power reflection coefficient, which, in turn, implies that $|\Gamma| < 1$.

The relation between the reflection coefficients at z and z_0 is simply

$$\Gamma(z) = \Gamma(z_0) e^{j2\kappa(z-z_0)}. \tag{29}$$

A graphical representation of Eq. (29) is afforded by the circle diagram shown in Fig. 1·3 from which both the amplitude and phase of the reflection coefficient may be obtained. The greater simplicity of the fundamental reflection-coefficient relation (29) as compared with the admittance relation (18) implies the advantage of the former for computations on transmission lines without discontinuities. The presence of discontinuities on the line leads to complications in description that usually are more simply taken into account on an admittance rather than a reflection-coefficient basis. In any case both methods are equivalent and, as seen by Eqs. (26) and (27), the connection between them follows from the relations

$$Y'(z) = \frac{1 - \Gamma(z)}{1 + \Gamma(z)} \quad \text{or} \quad \Gamma(z) = \frac{1 - Y'(z)}{1 + Y'(z)}. \tag{30}$$

It is frequently useful to employ a circuit representation of the connection between the scattering and impedance descriptions at any point z_0 of a transmission line. This representation is based on the fact, evident from Eqs. (26), that

$$V(z_0) = 2V_{\text{inc}} - ZI(z_0), \tag{31a}$$

or

$$I(z_0) = 2I_{\text{inc}} - YV(z_0), \tag{31b}$$

where

$$I_{\text{inc}} = YV_{\text{inc}}.$$

These relations are schematically represented by the circuits shown in Fig. 1·5a and b. Figure 1·5a indicates that the excitation at z_0 may be thought of as arising from a generator of constant voltage $2V_{\text{inc}}$ and internal impedance Z. The alternative representation in Fig. 1·5b shows the excitation as a generator of constant current $2I_{\text{inc}}$ and internal admittance Y.

A transmission-line description that is particularly desirable from the measurement point of view is based on the standing-wave pattern set up by the voltage or current distribution along the line. From Eqs. (26a) and (27) the amplitude of the voltage pattern at any point z is given by

$$|V(z)| = |V_{inc}| \sqrt{1 + |\Gamma|^2 + 2|\Gamma| \cos \Phi(z)}, \qquad (32)$$

where

$$\Gamma(z) = |\Gamma|e^{j\Phi(z)},$$

defines the amplitude $|\Gamma|$ and phase Φ of the reflection coefficient. Most probe types of standing-wave detectors read directly proportional to the voltage amplitude or its square. The ratio of the maximum to the mini-

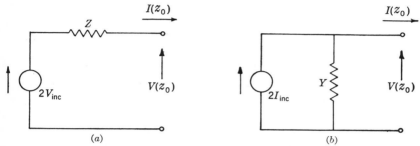

FIG. 1·5.—(a) Representation of an incident wave at z_0 as a constant-voltage generator; (b) representation of an incident wave at z_0 as a constant-current generator.

mum voltage amplitude is defined as the standing-wave ratio r and is given by Eq. (31) as

$$r = \frac{1 + |\Gamma|}{1 - |\Gamma|}, \qquad (33a)$$

and similarly the location of the minimum z_{min} is characterized by

$$\Phi(z_{min}) = \pi. \qquad (33b)$$

At any point z the relation between the reflection coefficient and the standing-wave parameters can then be expressed as

$$\Gamma(z) = -\frac{r-1}{r+1} e^{j2\kappa(z-z_{min})} = -\frac{r-1}{r+1} e^{j2\kappa d}. \qquad (34)$$

For the calculation of frequency sensitivity it is desirable to supplement the relation between the reflection coefficients at two points on a transmission line by the corresponding relation for the frequency derivatives. The latter is obtained by taking the derivative with respect to κ of the logarithm of both sides of Eq. (29). This yields

$$\frac{d\Gamma(z)}{\Gamma(z)} = \frac{d\Gamma(z_0)}{\Gamma(z_0)} + j2\kappa(z - z_0) \frac{d\kappa}{\kappa}. \qquad (35)$$

For the case of nondissipative transmission lines (κ real) it is useful to separate Eq. (35) into its real and imaginary parts as

$$\frac{\dfrac{d|\Gamma(z)|}{|\Gamma(z)|}}{\dfrac{d\omega}{\omega}} = \frac{\dfrac{d|\Gamma(z_0)|}{|\Gamma(z_0)|}}{\dfrac{d\omega}{\omega}}, \tag{36a}$$

$$\frac{d\Phi(z)}{\dfrac{d\omega}{\omega}} = \frac{d\Phi(z_0)}{\dfrac{d\omega}{\omega}} + 2\kappa(z - z_0)\left(\frac{k}{\kappa}\right)^2, \tag{36b}$$

since from Eq. (11b)

$$\frac{d\kappa}{\kappa} = \left(\frac{k}{\kappa}\right)^2 \frac{dk}{k} = \left(\frac{k}{\kappa}\right)^2 \frac{d\omega}{\omega}.$$

It is seen that on a relative change of frequency $d\omega/\omega$, the relative change $d|\Gamma|/|\Gamma|$ in amplitude of the reflection coefficient is identical at any two points z and z_0 on the transmission line. The absolute change $d\Phi$ in phase of the reflection coefficient at z differs from that at z_0 by an amount proportional to the change in electrical length of the intervening line. Equations (35) and (36) are equivalent to the corresponding Eqs. (23) for the admittance frequency sensitivity. The former are more suited for the investigation of broad-banding questions on long transmission lines, while the latter are more suited to the computation of Q's of short lengths of transmission lines or cavities.

1·5. Interrelations among Uniform Transmission-line Descriptions.—The interrelations among the impedance, relative admittance, reflection coefficient, and standing-wave characterizations of the voltage and current behavior on a uniform transmission line may be summarized as

$$\Gamma = |\Gamma|e^{j\Phi} = -\frac{r-1}{r+1}e^{j2\kappa d} = \frac{1-Y'}{1+Y'} = \frac{Z'-1}{Z'+1}, \tag{37a}$$

$$Y' = \frac{1}{Z'} = \frac{1-\Gamma}{1+\Gamma} = \frac{-j + r\cot\kappa d}{\cot\kappa d - jr}. \tag{37b}$$

On separation into real and imaginary parts these relations may be written in the form

$$|\Gamma| = \frac{r-1}{r+1} = \sqrt{\frac{(1-G')^2 + B'^2}{(1+G')^2 + B'^2}} = \sqrt{\frac{(R'-1)^2 + X'^2}{(R'+1)^2 + X'^2}}, \tag{38a}$$

$$\Phi = 2\kappa d + \pi = \tan^{-1}\left(\frac{2B'}{B'^2 + G'^2 - 1}\right)$$

$$= \tan^{-1}\left(\frac{2X'}{X'^2 + R'^2 - 1}\right), \tag{38b}$$

$$r = \frac{1+|\Gamma|}{1-|\Gamma|} = \frac{\sqrt{(1+G')^2 + B'^2} + \sqrt{(1-G')^2 + B'^2}}{\sqrt{(1+G')^2 + B'^2} - \sqrt{(1-G')^2 + B'^2}}$$

$$= \frac{\sqrt{(R'+1)^2 + X'^2} + \sqrt{(R'-1)^2 + X'^2}}{\sqrt{(R'+1)^2 + X'^2} - \sqrt{(R'-1)^2 + X'^2}}, \quad (38c)$$

$$G' = \frac{R'}{R'^2 + X'^2} = \frac{r}{r^2 \sin^2 \kappa d + \cos^2 \kappa d} = \frac{1 - |\Gamma|^2}{1 + 2|\Gamma|\cos\Phi + |\Gamma|^2}, \quad (38d)$$

$$B' = \frac{-X'}{R'^2 + X'^2} = \frac{(r^2 - 1)\cot \kappa d}{r^2 + \cot^2 \kappa d} = \frac{-2|\Gamma|\sin\Phi}{1 + 2|\Gamma|\cos\Phi + |\Gamma|^2}, \quad (38e)$$

$$R' = \frac{G'}{G'^2 + B'^2} = \frac{r}{r^2 \cos^2 \kappa d + \sin^2 \kappa d} = \frac{1 - |\Gamma|^2}{1 - 2|\Gamma|\cos\Phi + |\Gamma|^2}, \quad (38f)$$

$$X' = \frac{-B'}{G'^2 + B'^2} = \frac{(1 - r^2)\cot \kappa d}{r^2 \cot^2 \kappa d + 1} = \frac{2|\Gamma|\sin\Phi}{1 - 2|\Gamma|\cos\Phi + |\Gamma|^2}, \quad (38g)$$

where $Y' = G' + jB'$ = relative admittance at z.
$Z' = R' + jX'$ = relative impedance at z.
$\Gamma = |\Gamma|e^{j\Phi}$ = reflection coefficient at z.
r = voltage standing-wave ratio.
$d = z - z_{\min}$ = distance to standing-wave minimum.
$P_t = 1 - |\Gamma|^2$ = relative transmitted power.
$P_r = |\Gamma|^2$ = relative reflected power.

As previously stated Fig. 1·3 provides a graphical representation of most of the above relations. In addition the graph of the dependence of P_t, P_r, and $|\Gamma|$ on r, shown in Fig. 1·6, is often of use.

1·6. Uniform Transmission Lines with Complex Parameters.
a. Waveguides with Dissipation.—The presence of dissipation in either the dielectric medium or metallic walls of a waveguide modifies slightly the transmission-line description [Eq. (11)] of a propagating mode. This modification takes the form of a complex rather than an imaginary propagation constant γ and leads to transmission-line equations that may be written as

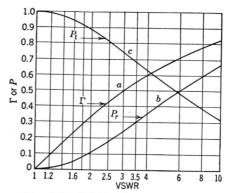

Fig. 1·6.—Relation between VSWR and (a) reflection coefficient Γ, (b) relative power reflected P_r, (c) relative power transmitted P_t.

$$\left.\begin{aligned} \frac{dV}{dz} &= -\gamma ZI, \\ \frac{dI}{dz} &= -\gamma YV. \end{aligned}\right\} \quad (39)$$

The complex propagation constant γ may be expressed as

$$\gamma = \alpha + j\beta = \sqrt{k_c^2 - k^2}, \tag{39a}$$

where the attenuation constant α, the inverse of which is the distance along z for the field to decay by $1/e$, and the wave number $\beta = 2\pi/\lambda_g$ are determined by the type of dissipation, the mode in question, and the geometry of the waveguide. The quantity 8.686α, the decibels of attenuation per unit length, or its inverse $1/8.686\alpha$, the loss length per decibel of attenuation, is frequently employed as a measure of attenuation instead of α. The characteristic impedance $Z = 1/Y$ is likewise complex and, for the same voltage-current definitions (10) as employed in the nondissipative case, is given by

$$Z = \frac{1}{Y} \begin{cases} = \dfrac{jw\mu}{\gamma} & \text{for } H\text{-modes} \\ = \dfrac{\gamma}{jw\epsilon} & \text{for } E\text{-modes,} \end{cases} \tag{39b}$$

where μ and ϵ, the permeability and dielectric constant of the medium filling the waveguide, may in general be complex.

Electric-type dissipation in the dielectric medium of a waveguide may be taken into account by introduction of a complex relative dielectric constant

$$\frac{\epsilon}{\epsilon_0} = \epsilon' - j\epsilon'', \tag{40}$$

where ϵ' is the relative dielectric constant and ϵ'' the loss factor. For a medium having a relative permeability of unity, the propagation constant is

$$\gamma = \sqrt{\left(\frac{2\pi}{\lambda_c}\right)^2 - \left(\frac{2\pi}{\lambda_0}\right)^2 \frac{\epsilon}{\epsilon_0}}. \tag{41}$$

In a waveguide having a cutoff wavelength $\lambda_c > \lambda_0$ the attenuation constant α is, therefore,

$$\alpha = \frac{\pi \lambda_g \epsilon''}{\lambda_0^2} = \frac{2\pi}{\lambda_{g0}} \sqrt{\frac{-1 + (1 + x^2)^{1/2}}{2}} = \frac{2\pi}{\lambda_{g0}} \sinh\left(\frac{\sinh^{-1} x}{2}\right), \tag{42a}$$

$$\alpha \cong \frac{\pi \lambda_{g0} \epsilon''}{\lambda_0^2}\left(1 - \frac{x^2}{8} \cdots \right), \qquad x \ll 1, \tag{42b}$$

$$\alpha \cong \frac{1}{\delta}\left(1 - \frac{1}{2x} \cdots \right), \qquad x \gg 1, \tag{42c}$$

TABLE 1·1.—PROPERTIES OF DIELECTRIC MATERIALS*

Substance	$f = 100$ cps $\lambda = 3 \times 10^8$ cm		$f = 3 \times 10^9$ cps $\lambda = 10$ cm		$f = 10^{10}$ cps $\lambda = 3$ cm		Uses
	ϵ'	ϵ''	ϵ'	ϵ''	ϵ'	ϵ''	
1. Ceramic and other inorganic materials:							
AlSiMag 243	6.30	0.0013	5.75	0.0002	5.40	0.0002	1 and 6
Steatite Ceramic F-66	6.25	0.0015	6.25	0.00055	....	....	1
TI-Pure 0-600	99.0	0.001	....	....	....	....	1
Tam Ticon T-J, T-L, T-M	96.0	0.0008	96.0	0.00034	....	....	1
Mixture of ceramics and polymers:							
Titanium dioxide (41.9%) / Polydichlorostyrene (58.1%)	3.50	0.0031	5.30	0.00060	5.30	0.00085	1
Titanium dioxide (65.3%) / Polydichlorostyrene (34.7%)	10.2	0.0016	10.2	0.00067	10.2	0.00132	1
Titanium dioxide (81.4%) / Polydichlorostyrene (18.6%)	23.6	0.0060	23.0	0.0013	23.0	0.00157	1
Fused quartz	3.85	0.0009	3.80	0.0001	3.80	0.0001	1
Ruby mica	5.4	0.0025	5.4	0.0003	....	....	7
Mycalex 1364	7.09	0.0059	6.91	0.00360	....	....	1
Mycalex K10	9.5	0.0170	11.3	0.004	11.3	0.004	1
Turx 52	7.04	0.0078	6.70	0.0052	6.69	0.0066	1
Turx 160	7.05	0.0063	6.83	0.00380	6.85	0.0049	1
AlSiMag 393	4.95	0.0038	4.95	0.00097	4.95	0.00097	1
2. Glasses and mixtures with glasses:							
Corning glass 707	4.00	0.0006	4.00	0.0019	3.99	0.0021	6
Corning glass 790	3.90	0.0006	3.84	0.00068	3.82	0.00094	6
Corning glass (C. Lab. No. 7141M)	4.15	0.0020	4.00	0.0010	4.00	0.0016	6
Corning glass 8871	8.45	0.0018	8.34	0.0026	8.05	0.0049	7
Polyglas P +	3.45	0.0014	3.35	0.00078	3.32	0.00084	1
Polyglas D + (Monsanto)	3.25	0.0005	3.22	0.00120	3.22	0.0013	1
Polyglas M	5.58	0.0140	4.86	0.0339	5.22	0.0660	1
Polyglas S	3.60	0.0011	3.55	0.0040	3.53	0.0046	1
3. Liquids:							
Water conductivity	....	....	77.00	0.150	....	....	3
Fractol A	66	2.17	2.15	0.00072	....	....	4
Cable oil 5314	2.28	0.001	2.23	0.0018	....	....	4
Transil oil 10C	2.24	0.001	2.18	0.0028	....	....	4
Dow Corning 200; 3.87 cp	2.57	0.0005	2.48	0.0048	....	....	4
Dow Corning 200; 300 cp	2.75	0.0005	2.69	0.010	....	....	4
Dow Corning 200; 7,600 cs	2.75	0.0005	2.71	0.0103	....	....	4
Dow Corning 500; 0.65 cs	2.20	0.0005	2.20	0.00145	....	....	4
Ignition sealing compound 4	2.80	0.0004	2.77	0.010	....	....	5
4. Polymers:							
Bakelite BM 120	4.87	0.030	3.70	0.0438	3.68	0.0390	3
Cibanite E	3.70	0.0038	3.47	0.0053	3.47	0.0075	1 and 2
Dielectene 100	3.62	0.0033	3.44	0.0039	....	....	1 and 2
Plexiglas	3.40	0.061	2.60	0.0057	2.59	0.0067	1
Polystyrene XMS10023	2.59	0.002	2.55	0.0005	....	....	1 and 2
Loalin (molding powder)	2.50	0.001	2.49	0.00022	....	....	1 and 2
Styron C-176	2.56	0.0008	2.55	0.00026	2.54	0.0003	1 and 2
Lustron D-276	2.53	0.0004	2.51	0.00041	....	....	1 and 2
Polystyrene D-334	2.56	0.0006	2.54	0.00024	....	....	1 and 2
Styramic	2.88	0.0025	2.65	0.00022	2.62	0.00023	1 and 2

TABLE 1·1.—PROPERTIES OF DIELECTRIC MATERIALS.*—(Continued)

Substance	$f = 100$ cps $\lambda = 3 \times 10^8$ cm		$f = 3 \times 10^9$ cps $\lambda = 10$ cm		$f = 10^{10}$ cps $\lambda = 3$ cm		Uses
	ϵ'	ϵ''	ϵ'	ϵ''	ϵ'	ϵ''	
Styraloy 22	2.40	0.0009	2.40	0.0032	2.40	0.0024	1 and 2
GE Resin #1421	2.56	0.001	2.53	0.0005	2.52	0.00056	1 and 2
Dow Exp. Plastic Q-200.5	2.55	0.0009	2.52	0.00044	....	....	1 and 2
Dow Exp. Plastic Q-385.5	2.51	0.0005	2.50	0.00063	2.49	0.0008	1 and 2
Dow Exp. Plastic Q-409	2.60	0.0010	2.60	0.00087	2.60	0.0012	1 and 2
Poly 2, 5-dichlorostyrene D-1385	2.63	0.0005	2.62	0.00023	2.60	0.00023	1 and 2
Thalid X-526-S	3.55	0.0144	2.93	0.0163	2.93	0.0159	1 and 2
Polyethylene	2.26	0.0006	2.26	0.00040	....	....	1 and 8
Polyethylene M702-R	2.25	0.0005	2.21	0.00019	....	....	1 and 8
Polyethylene KLW A-3305	2.25	0.0005	2.25	0.00022	....	....	1 and 8
"Teflon" Poly F-1114	2.1	0.0005	2.1	0.00015	2.08	0.00037	1, 2, and 8
5. Waxes:							
Acrawax C	2.60	0.0157	2.48	0.0015	2.45	0.0019	5
Paraffin wax (135° amp)	2.25	0.0013	2.22	0.0001	2.22	0.00020	5
Parowax	....	......	2.25	0.0002	2.25	0.00025	5
Cerese wax AA	2.34	0.0006	2.29	0.00088	2.26	0.0007	5

Uses:
1. For use as waveguide windows or coax beads, cable fittings.
2. For use as dielectric transformers or matching sections.
3. For use as attenuators or loading materials.
4. For liquid-filled lines.
5. For moistureproofing radar components.
6. For use in vacuum tubes.
7. For capacitor dielectrics.
8. Cable materials.
* Abstracted from Von Hipple et al., "Tables of Dielectric Materials," NDRC 14-237.

and the wave number β is

$$\beta = \frac{2\pi}{\lambda_{g0}} \sqrt{\frac{1 + (1 + x^2)^{1/2}}{2}} = \frac{2\pi}{\lambda_{g0}} \cosh\left(\frac{\sinh^{-1} x}{2}\right), \quad (43a)$$

$$\beta \cong \frac{2\pi}{\lambda_{g0}} \left(1 + \frac{x^2}{8} \cdots \right), \quad x \ll 1, \quad (43b)$$

$$\beta \cong \frac{1}{\delta} \left(1 + \frac{1}{2x} \cdots \right), \quad x \gg 1, \quad (43c)$$

where

$$\lambda_{g0} = \frac{\lambda_0}{\sqrt{\epsilon' - \left(\frac{\lambda_0}{\lambda_c}\right)^2}}, \quad x = \epsilon'' \frac{\lambda_{g0}^2}{\lambda_0^2},$$

and

$$\frac{1}{\delta} = \frac{\pi}{\lambda_0} \sqrt{2\epsilon''} = \frac{\pi}{\lambda_{g0}} \sqrt{2x} = \sqrt{\frac{\omega \mu_0 \sigma}{2}}.$$

The approximations (42b) and (43b) are valid for $\epsilon''/\epsilon' \ll 1$ and λ_0 not

TABLE 1·2.—ELECTRICAL CONDUCTIVITIES OF METALS*

$$\delta = 9.19 \times 10^{-6} \sqrt{\left(\frac{10^7}{\sigma}\right)} \lambda_0 \text{ meters,} \qquad \lambda_0 \text{ in meters.}$$

$$\mathcal{R} = 10.88 \times 10^{-3} \sqrt{\left(\frac{10^7}{\sigma}\right)} \frac{1}{\lambda_0} \text{ ohms,} \qquad \lambda_0 \text{ in meters.}$$

Material	Eff. cond. σ at $\lambda = 1.25$ cm in 10^7 mhos/m	DC cond. σ in 10^7 mhos/m
Aluminum:†		
Pure, commercial (machined surface)	1.97	3.25 (measured)
17S Alloy† (machined surface)	1.19	1.95 (measured)
24S Alloy (machined surface)	1.54	1.66 (measured)
Brass:		
Yellow (80-20) drawn waveguide	1.45	1.57 (measured)
Red (85-15) drawn waveguide	2.22	
Yellow round drawn tubing	1.36	1.56 (Eshbach)
Yellow (80-20) (machined surface)	1.17	1.57 (Eshbach)
Free machining brass (machine surface)	1.11	1.48 (measured)
Cadmium plate	1.04–0.89	1.33 *Hdbk. of Phys. and Chem.*
Chromium plate, dull	1.49–0.99	3.84 *Hdbk. of Phys. and Chem.*
Copper:		
Drawn OFC waveguide	4.00	5.48 (measured)
Drawn round tubing	4.10	4.50 (measured)
Machined surface†	4.65	5.50 (measured)
Copper plate	2.28–1.81	5.92 ⎫ *Hdbk. of Phys. and Chem.*
Electroformed waveguide†	3.15	5.92 ⎭
Gold plate	1.87	4.10 *Hdbk. of Phys. and Chem.*
Mercury	0.104	0.104 *Hdbk. of Phys. and Chem.*
Monel (machined surface)†	0.155	0.156 (measured)
Silver:		
Coin silver drawn waveguide	3.33	4.79 (measured)
Coin silver lined waveguide	1.87 ⎫	4.79 (assumed)
Coin silver (machined surface)†	2.66 ⎭	
Fine silver (machined surface)†	2.92 ⎫	6.14 *Hdbk. of Phys. and Chem.*
Silver plate	3.98–2.05 ⎭	
Solder, soft†	0.600	0.70 (measured)

* Abstracted from E. Maxwell, "Conductivity of Metallic Surfaces," *J. Applied Phys.*, July, 1947.
† Only one sample was tested.

too close to the cutoff wavelength λ_c. The approximations (42c) and (43c) apply to a metal, i.e., a strongly conducting dielectric with $\epsilon''/\epsilon' \gg 1$, and are expressed in terms of the skin depth δ rather than ϵ''. In each case the leading term provides a good approximation for most of the dielectrics and metals encountered in practice.

Measurements of the loss factor ϵ'' and conductivity σ at various wavelengths are displayed in Tables 1·1 and 1·2 for a number of dielectrics and metals. The conductivity properties of a nonmagnetic metal are

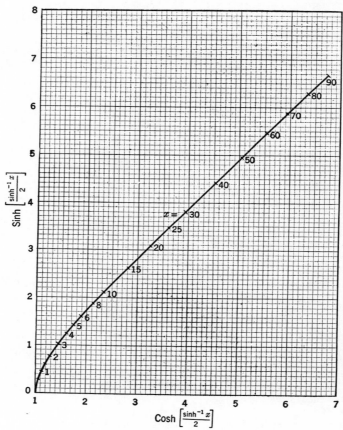

FIG. 1·7.—Phase and attenuation functions vs. x.

frequently described by its characteristic resistance $\mathfrak{R}$, which is related to its skin depth δ and conductivity σ by

$$\mathfrak{R} = \pi \sqrt{\frac{\mu_0}{\epsilon_0}} \frac{\delta}{\lambda_0} = 10.88 \times 10^{-3} \sqrt{\left(\frac{10^7}{\sigma}\right) \frac{1}{\lambda_0}} \quad \text{ohms,}$$

σ being measured in mhos per meter, δ and λ_0 in meters.

To facilitate computations of α and β a graph of the functions $\cosh\left(\dfrac{\sinh^{-1} x}{2}\right)$ and $\sinh\left(\dfrac{\sinh^{-1} x}{2}\right)$ is plotted vs. x in Fig. 1.7.

A complex relative permeability

$$\frac{\mu}{\mu_0} = \mu' - j\mu'' \tag{44}$$

may be introduced to account for dissipation of a magnetic type. For a medium of unit relative dielectric constant, the attenuation constant and wave number may be obtained from Eqs. (42) and (43) by the replacement of ϵ' and ϵ'' by μ' and μ'', respectively. The skin depth δ in this case is

$$\frac{\lambda_0}{2\pi}\sqrt{\frac{2}{\mu''}} = \sqrt{\frac{2}{\omega\epsilon_0\sigma_m}}, \tag{45}$$

where the conductivity σ_m accounts for the magnetic type of dissipation. Extensive tables of the loss factor μ'' or alternatively the conductivity σ_m are not as yet available.

The presence of dissipation of both the electric and magnetic type may be taken into account by introduction of both a complex relative dielectric constant and a complex permeability, as given by Eqs. (40) and (44). The attenuation constant and wave number may again be obtained from Eqs. (42) and (43) if ϵ' and ϵ'' therein are replaced by $\epsilon'\mu' - \epsilon''\mu''$ and $\epsilon''\mu' + \epsilon'\mu''$, respectively. In this case the skin depth δ is given by

$$\frac{1}{\delta^2} = \frac{\mu'}{\delta_e^2} + \frac{\epsilon'}{\delta_m^2}, \tag{46}$$

where

$$\delta_e = \sqrt{\frac{2}{\omega\mu_0\sigma}}, \qquad \delta_m = \sqrt{\frac{2}{\omega\epsilon_0\sigma_m}}.$$

When the medium is an ionized gas, it may be desirable to introduce a complex conductivity

$$\sigma = \sigma' - j\sigma'' \tag{47}$$

to describe both the dielectric and dissipative properties of the medium. For a medium of unit relative dielectric constant and permeability the attenuation constant and wave number can be obtained from Eqs. (42) and (43) on the replacement of ϵ' and ϵ'' therein by $1 - \sigma''/\omega\epsilon_0$ and $\sigma'/\omega\epsilon_0$, respectively.

The characteristic admittance of a propagating mode in a dissipative guide follows from the knowledge of the complex propagation constant. For example, in a dissipative dielectric medium the characteristic admittance for H-modes is given by Eqs. (39b), (42), and (43) as

$$Y = \sqrt{\frac{\epsilon_0}{\mu_0}}\frac{\lambda_0}{\lambda_{g0}}\left[\cosh\left(\frac{\sinh^{-1} x}{2}\right) - j\sinh\left(\frac{\sinh^{-1} x}{2}\right)\right], \tag{48a}$$

$$Y \cong \sqrt{\frac{\epsilon_0}{\mu_0}} \frac{\lambda_0}{\lambda_{g0}} \left[1 - j \left(\frac{\lambda_{g0}}{\lambda_0}\right)^2 \frac{\epsilon''}{2} \right], \qquad \frac{\epsilon''}{\epsilon'} \ll 1, \qquad (48b)$$

$$Y \cong \sqrt{\frac{\epsilon_0}{\mu_0}} \frac{\lambda_0}{2\pi\delta} (1 - j), \qquad \frac{\epsilon''}{\epsilon'} \gg 1. \qquad (48c)$$

The approximation (48b) is applicable to the case of small dissipation $\epsilon''/\epsilon' \ll 1$ and λ_0 not too close to λ_c, whereas the approximation (48c) applies to the metallic case.

The effect of dissipation in the metallic walls of a uniform waveguide is described by a complex propagation constant which may be obtained by explicit evaluation of the complex cutoff wavenumber for the waveguide. An alternative method particularly desirable for first-order computation is based on the formula for the attenuation constant

$$\alpha = -\frac{1}{2P} \frac{dP}{dz}, \qquad (49)$$

where P is the total power flow at z [cf. Eq. (7)] and therefore $-dP$ is the power dissipated in a section of waveguide of length dz. Equation (49) refers to a mode traveling in the positive z direction. From Eq. (49) it follows that the attenuation constant $\alpha = \alpha_m$ due to losses in the metallic guide walls is

$$\alpha_m = \frac{1}{2} \frac{\text{Re } (Z_m) \int |\mathbf{H}_{\text{tan}}|^2 \, ds}{\text{Re } (Z) \int\int |\mathbf{H}_t|^2 \, dS}, \qquad (50)$$

where Z_m the characteristic impedance of the metallic walls [cf. Eq. (48c)] is approximately the same for both E- and H-modes, and Z is the characteristic impedance of the propagating mode under consideration. In first-order computations $\mathbf{H}_{\text{tan}}$ and $\mathbf{H}_t$ are set equal to the nondissipative values of the magnetic field tangential to the guide periphery and transverse to the guide cross section, respectively. The line integral with respect to ds extends over the guide periphery, and the surface integral with respect to dS extends over the guide cross section.

The tangential and transverse components of the magnetic field of an E-mode can be expressed in terms of the mode function Φ defined in Eqs. (3) of Sec. 1·2. Hence by Eqs. (6), (9a), (48c), and (50) the attenuation constant of a typical E-mode in an arbitrary uniform guide with dissipative metallic walls is to a first order (omitting modal indices)

$$\alpha_m = \frac{1}{2} \frac{\mathcal{R}}{\zeta} \frac{k}{\kappa} \left[\frac{\int \left(\frac{\partial \Phi}{\partial \nu}\right)^2 ds}{\int\int \Phi^2 \, dS} \right] \qquad (50a)$$

where $\mathcal{R} = k\zeta\delta/2$ is the characteristic resistance of the metallic walls as tabulated in Table 1·2, and the derivative with respect to ν is along the outward normal at the guide periphery. The magnetic field components of a traveling H-mode can be expressed in terms of the mode function Ψ defined in Eqs. (4) of Sec. 1·2. Thus by Eqs. (6), (9b), (48c), and (50) the attenuation constant of a typical H-mode in an arbitrary uniform guide with dissipative metallic walls is to a first order (omitting modal indices)

$$\alpha_m = \frac{1}{2} \frac{\Re}{\zeta} \frac{\kappa}{k} \left[\frac{\int \left(\frac{\partial \Psi}{\partial s} \right)^2 ds}{k_c^2 \iint \Psi^2 \, dS} + \frac{k_c^2 \int \Psi^2 \, ds}{\kappa^2 \iint \Psi^2 \, dS} \right], \qquad (50b)$$

where the derivative with respect to s is along the tangent to the guide periphery. A useful alternative to Eq. (50a) for the attenuation constant of an E-mode is

$$\alpha_m = \frac{1}{2} \frac{\Re}{\zeta} \frac{k}{\kappa} \left[-\frac{1}{k_c^2} \frac{\delta k_c^2}{\delta \nu} \right], \qquad (50c)$$

where $\delta k_c^2 / \delta \nu$ represents the variation of the square of the mode cutoff wave number k_c with respect to an infinitesimal outward displacement of the guide periphery along the normal at each point. Equation (50c) permits the evaluation of the E-mode attenuation constant by simple differentiation of k_c^2 with respect to the cross-sectional dimensions of the guide. Although there is no simple dependence on k_c^2, the corresponding expression, alternative to Eq. (50b), for the attenuation constant of an H-mode may be written as

$$\alpha_m = \frac{1}{2} \frac{\Re}{\zeta} \frac{k}{\kappa} \left[-\frac{1}{k_c^2} \frac{\delta k_c^2}{\delta \nu} \left(\frac{k_c^2}{k^2} + \frac{f}{1-f} \right) \right], \qquad (50d)$$

where the factor

$$f = \frac{1}{k_c^2} \frac{\int \left(\frac{\partial \Psi}{\partial s} \right)^2 ds}{\iint \Psi^2 \, dS}$$

must be obtained by integration.

Explicit values for α_m are dependent upon the cross-sectional shape of the waveguide and the mode in question; several first-order values are indicated in Chap. 2 for different guide shapes. The corresponding first-order values for the wave number β are the same as in the nondissipative case. The attenuation constant due to the presence of dissipation in both the dielectric and metallic walls of a waveguide is to a first order the sum of the individual attenuation constants for each case.

With the knowledge of the complex propagation constant γ and the complex characteristic impedance Z to be associated with losses in either the dielectric medium or metallic walls, a transmission-line description of a propagating mode in a dissipative guide can be developed in close analogy with the nondissipative description of Secs. 1·3 and 1·4. In fact, the two descriptions are formally the same if the κ of Secs. 1·3 and 1·4 is replaced by $-j\gamma$. This implies that an impedance description for the dissipative case is based on the standing waves

$$\cosh \gamma z \quad \text{and} \quad \sinh \gamma z$$

and leads to a relation between the relative admittances at the points z and z_0 of the form

$$Y'(z) = \frac{1 + Y'(z_0) \coth \gamma(z_0 - z)}{\coth \gamma(z_0 - z) + Y'(z_0)} \tag{51}$$

rather than the previous form employed in Eq. (18). The circle diagram of Fig. 1·3 can again be employed to facilitate admittance computations; however, Eq. (51) can no longer be interpreted as a constant amplitude rotation of $Y'(z_0)$ into $Y'(z)$.

A special case of Eq. (51) with practical interest relates to a short-circuited dissipative line [$Y'(z_0) = \infty$]; in which case

$$Y'(z) = \frac{\coth \alpha l \csc^2 \beta l - j \cot \beta l \operatorname{csch}^2 \alpha l}{\cot^2 \beta l + \coth^2 \alpha l}, \tag{52a}$$

$$Y'(z) \cong \alpha l \csc^2 \beta l - j \cot \beta l, \qquad \alpha l \ll 1, \ \beta l \neq n\pi, \tag{52b}$$

where

$$\gamma = \alpha + j\beta, \qquad l = z_0 - z.$$

Relative values of input conductance and susceptance are indicated in these equations and are to be distinguished from the absolute values, since the characteristic admittance is complex. The approximation (52b) applies when $\alpha l \ll 1$. For dissipation such that $\alpha l > 3$, Eq. (51) states in general that $Y'(z) \cong 1$ independently of the value of $Y'(z_0)$.

Although Eq. (51) provides a straightforward means for admittance computations in dissipative transmission lines, such computations are tedious because of the complex nature of the propagation constant. In many practical problems dissipative effects are slight and hence have a small, albeit important, effect on admittance calculations. For such problems a perturbation method of calculation is indicated. In this method one performs an admittance calculation by first assuming the propagation constant to be purely imaginary, i.e., $\gamma = j\beta$ as for the case of no dissipation; one then accounts for the presence of dissipation by adding the admittance correction due to a perturbation α in γ. Thus in the case illustrated in Eq. (52b) one notes that the input admittance of a short-circuited length of slightly dissipative line is the sum of the unperturbed admittance $Y_0 = \coth j\beta l$ and the correction $(dY_0/d\gamma)\,\alpha$ due to the perturbation α in γ.

Equivalent-circuit representations of Eq. (51) can be obtained from those in Sec. 1·3 (*cf.* Fig. 1·4) by the replacement of κ therein by $-j\gamma$. Another useful representation of the equivalent network between the input and output points of a dissipative line of length l consists of a tandem connection of a nondissipative line of electrical length βl and a beyond cutoff line of electrical length $-j\alpha l$, the characteristic impedances of both lines being the same as that of the dissipative line.

The scattering description of a propagating mode in a dissipative guide is based on wave functions of the type

$$e^{-\gamma z} \quad \text{and} \quad e^{+\gamma z}.$$

These functions represent waves traveling in the direction of increasing and decreasing z, respectively, and attenuating as $e^{-\alpha|z|}$. A mode description can therefore be expressed in terms of an incident and reflected wave whose voltage amplitudes V_{inc} and V_{refl} are defined as in Eqs. (26) with κ replaced by $-j\gamma$. A reflection coefficient,

$$\Gamma(z) = \frac{V_{\text{refl}}}{V_{\text{inc}}} e^{2\gamma(z-z_0)}, \tag{53}$$

may likewise be defined such that at any two points z and z_0

$$\Gamma(z) = \Gamma(z_0) e^{2\gamma(z-z_0)}. \tag{54}$$

However, the total power flow at z is now given by

$$P = \text{Re}\ (VI^*) = P_{\text{inc}} \left(1 - |\Gamma|^2 - 2\Gamma_i \frac{Y_i}{Y_r}\right), \tag{55}$$

where

$$P_{\text{inc}} = Y_r |V_{\text{inc}}|^2 e^{-2\alpha(z-z_0)}.$$

The subscripts r and i denote the real and imaginary parts of a quantity, and Y is the complex characteristic admittance for the mode in question. From Eq. (55) it is evident that for dissipative lines $|\Gamma|^2$ can no longer be regarded as the power-reflection coefficient. Moreover, $|\Gamma|$ is not restricted to values equal or less than unity. The meaning of Γ as a reflection coefficient can be retained if the voltage and current on the dissipative line are defined so as to make the characteristic admittance real; in this event Eq. (155) reduces to the nondissipative result given in Eq. (28).

b. Waveguides beyond Cutoff.—The voltage and current amplitudes of a higher, or nonpropagating, mode in a waveguide are described by the transmission-line equations (39). In the absence of dissipation the propagation constant is real and equal to

$$\gamma = \frac{2\pi}{\lambda_c} \sqrt{1 - \left(\frac{\lambda_c}{\lambda}\right)^2}, \quad \lambda > \lambda_c. \tag{56}$$

The nondissipative decay of the mode fields in decibels per unit length (same unit as for λ_c) is therefore

$$\frac{54.57}{\lambda_c} \sqrt{1 - \left(\frac{\lambda_c}{\lambda}\right)^2}, \qquad \lambda > \lambda_c. \tag{57}$$

At low frequencies the rate of decay is independent of λ, the wavelength of field excitation, and dependent only on the geometry of the guide cross section. Values of the cutoff wavelength λ_c are given in Chap. 2 for several waveguide modes and geometries.

The characteristic impedance of a beyond-cutoff mode (i.e. $\lambda > \lambda_c$) may be obtained from Eqs. (39b) and (56) as

$$\left. \begin{array}{l} Z = \dfrac{1}{Y} = j\, \dfrac{\zeta}{\sqrt{\left(\dfrac{\lambda}{\lambda_c}\right)^2 - 1}}, \qquad \text{for } H\text{-modes,} \\[2em] Z = \dfrac{1}{Y} = -j\zeta \sqrt{\left(\dfrac{\lambda}{\lambda_c}\right)^2 - 1}, \qquad \text{for } E\text{-modes,} \end{array} \right\} \tag{58}$$

and is inductive for H-modes, capacitive for E-modes.

The knowledge of the propagation constant and characteristic impedance of a beyond-cutoff mode permits the application of the transmission-line analysis developed in Secs. 1·3 and 1·4, provided κ therein is replaced by $-j\gamma$ (γ real). The impedance description is given by Eq. (51), and the scattering description by Eq. (54). Several modifications resulting from the fact that γ is real and Z is imaginary have already been discussed in Sec. 1·6a.

The presence of dissipation within the dielectric medium or the walls of a beyond-cutoff waveguide introduces an imaginary part into the propagation constant γ. If dissipation is present only in the medium and is characterized by a complex dielectric constant, as in Eq. (40), we have for the propagation constant $\gamma = \alpha + j\beta$

$$\alpha = \frac{2\pi}{\lambda_{g0}} \cosh\left(\frac{\sinh^{-1} x}{2}\right), \tag{59a}$$

$$\alpha \cong \frac{2\pi}{\lambda_{g0}} \left(1 + \frac{x^2}{8} \cdots \right), \qquad \frac{\epsilon''}{\epsilon'} \ll 1, \tag{59b}$$

and

$$\beta = \frac{2\pi}{\lambda_{g0}} \sinh\left(\frac{\sinh^{-1} x}{2}\right), \tag{60a}$$

$$\beta \cong \frac{\pi \lambda_{g0}}{\lambda_0^2} \epsilon'' \left(1 - \frac{x^2}{8} \cdots \right), \qquad \frac{\epsilon''}{\epsilon'} \ll 1, \tag{60b}$$

where

$$\lambda_{g0} = \frac{\lambda_c}{\sqrt{1 - \epsilon'\left(\dfrac{\lambda_c}{\lambda_0}\right)^2}}, \qquad x = \frac{\lambda_{g0}^2}{\lambda_0^2}\epsilon''.$$

The approximations (59b) and (60b) apply to the case of small dissipation with $\epsilon''/\epsilon' \ll 1$ and λ not too close to λ_c. Equations (59) and (60) for a beyond-cutoff mode and Eqs. (42) and (43) for a propagating mode differ mainly in the replacement of the attenuation constant of the one case by the wave number of the other case and conversely. This correspondence

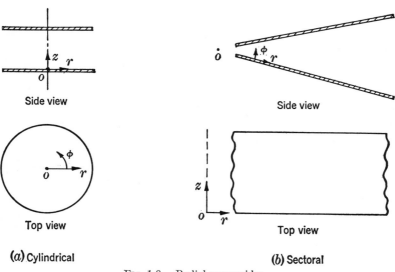

FIG. 1·8.—Radial waveguides.

between the two cases is general and applies as well to the other types of dissipation mentioned in Sec. 1·6a.

1·7. Field Representation in Nonuniform Radial Waveguides.—Nonuniform regions are characterized by the fact that cross sections transverse to the transmission direction are similar to but not identical with one another. A radial waveguide is a nonuniform cylindrical region described by an $r\phi z$ coordinate system; the transmission direction is along the radius r, and the cross sections transverse thereto are the ϕz cylindrical surfaces for which r is constant. Typical examples of radial waveguides are the cylindrical and cylindrical sector regions shown in Figs. 1·8a and b.

In the $r\phi z$ polar coordinate system appropriate to the radial waveguides of Figs. 1·8, the field equations for the electric and magnetic field components transverse to the radial direction r may be written as

$$\frac{\partial E_z}{\partial r} = -jk\zeta\left[-H_\phi + \frac{1}{k^2}\left(\frac{1}{r}\frac{\partial^2 H_z}{\partial \phi\, \partial z} - \frac{\partial^2 H_\phi}{\partial z^2}\right)\right],$$
$$\frac{1}{r}\frac{\partial}{\partial r}(rE_\phi) = -jk\zeta\left[H_z + \frac{1}{k^2}\left(\frac{1}{r^2}\frac{\partial^2 H_z}{\partial \phi^2} - \frac{1}{r}\frac{\partial^2 H_\phi}{\partial \phi\, \partial z}\right)\right],$$
(61a)

and

$$\frac{\partial H_z}{\partial r} = -jk\eta\left[E_\phi + \frac{1}{k^2}\left(\frac{\partial^2 E_\phi}{\partial z^2} - \frac{1}{r}\frac{\partial^2 E_z}{\partial \phi\, \partial z}\right)\right],$$
$$\frac{1}{r}\frac{\partial}{\partial r}(rH_\phi) = -jk\eta\left[-E_z + \frac{1}{k^2}\left(\frac{1}{r}\frac{\partial^2 E_\phi}{\partial \phi\, \partial z} - \frac{1}{r^2}\frac{\partial^2 E_z}{\partial \phi^2}\right)\right].$$
(61b)

The radial components follow from the transverse components as

$$j w \epsilon E_r = \frac{1}{r}\frac{\partial H_z}{\partial \phi} - \frac{\partial H_\phi}{\partial z},$$
$$-j w \mu H_r = \frac{1}{r}\frac{\partial E_z}{\partial \phi} - \frac{\partial E_\phi}{\partial z}.$$
(62)

A component form of the field equations is employed because the left-hand members of Eqs. (61) cannot be written in invariant vector form. The inability to obtain a transverse vector formulation, as in Eqs. (1), implies, in general, the nonexistence of a field representation in terms of transverse *vector* modes. The transverse field representation in a radial waveguide must consequently be effected on a scalar basis.

For the case where the magnetic field has no z-component, the transverse field may be represented as a superposition of a set of E-type modes. The transverse functional behavior of an E-type mode (*cf.* Sec. 2·7) is of the form

$$\begin{matrix}\cos\\ \sin\end{matrix} m\phi \begin{matrix}\cos\\ \sin\end{matrix} \frac{n\pi}{b} z,$$
(63)

where the mode indices m and n are determined by the angular aperture and height of the cylindrical ϕz cross section of the radial guide. The amplitudes of the transverse electric and magnetic fields of an E-type mode are characterized by a mode voltage V'_i and a mode current I'_i.

For the case of no z component of electric field, the fields can be represented in terms of a set of H-type modes whose transverse form, as shown in Sec. 2·7, is likewise characterized by functions of the form (63). The voltage and current amplitudes of the transverse electric and magnetic field intensity of an H-type mode are designated as V''_i and I''_i.

For the case of a general field both mode types are required, and these are not independent of one another. Incidentally, it is to be emphasized that the above classification into mode types is not based on the trans-

NONUNIFORM RADIAL WAVEGUIDES

mission direction. Relative to the r direction all modes are generally hybrid in that they possess both an E_r and an H_r component (cf. Sec. 2·7).

On substitution of the known transverse functional form of the modes into Eqs. (61) there are obtained the transmission-line equations

$$\begin{aligned} \frac{dV}{dr} &= -j\kappa ZI, \\ \frac{dI}{dr} &= -j\kappa YV, \end{aligned} \quad (64a)$$

for the determination of each of the mode amplitudes V and I. Because of the identity in form of the equations for all modes, the distinguishing sub- and superscripts have been omitted. The characteristic impedance Z and mode constant κ are given by

$$\begin{aligned} Z &= \frac{1}{Y} = \frac{\zeta}{r} \frac{\kappa_n^2}{\kappa k} N' \quad \text{for the } E\text{-type modes,} \\ Z &= \frac{1}{Y} = \zeta r \frac{\kappa k}{\kappa_n^2} N'' \quad \text{for the } H\text{-type modes,} \end{aligned} \quad (64b)$$

$$\kappa = \sqrt{k^2 - \left(\frac{n\pi}{b}\right)^2 - \left(\frac{m}{r}\right)^2}, \quad \kappa_n = \sqrt{k^2 - \left(\frac{n\pi}{b}\right)^2},$$

where N' and N'' are constants dependent on the cross-sectional dimensions of the radial waveguide and the definitions of V and I (cf. Sec. 2·7).

Because of the indicated variability with r of the propagation constant and characteristic impedance, Eqs. (64a) are called radial transmission-line equations. Correspondingly the mode amplitudes V and I are defined as the rms mode voltage and current; they furnish the basis for the reformulation of the field description in impedance terms. The variability with r of the line parameters implies a corresponding variability in the spatial periodicity of the fields along the transmission direction. The concept of wavelength on a radial line thus loses its customary significance.

Impedance Description of Dominant E-type Mode.—In practice, the frequency and excitation of the radial waveguide illustrated in Fig. 1·8a are often such that, almost everywhere, only the dominant E-type mode with $m = 0$ and $n = 0$ is present. The field configuration of this transverse electromagnetic mode is angularly symmetric with **E** parallel to the z-axis and **H** in the form of circles about the z-axis. The transverse mode fields are represented as

$$\begin{aligned} \mathbf{E}_t(r,\phi,z) &= -\frac{V(r)}{b}\mathbf{z}_0, \\ \mathbf{H}_t(r,\phi,z) &= \frac{I(r)}{2\pi r}\boldsymbol{\phi}_0, \end{aligned} \quad (65)$$

where $\mathbf{z}_0$ and $\boldsymbol{\phi}_0$ are unit vectors in the positive z and ϕ directions. The mode voltage V and current I obey Eqs. (64a) with $\kappa = k$ and $Z = \zeta b/2\pi r$ (cf. Sec. 2·7). On elimination of I from Eqs. (64a) the wave equation for V becomes

$$\frac{1}{r}\frac{d}{dr}\left(r\frac{dV}{dr}\right) + k^2 V = 0. \tag{66}$$

The two independent, standing-wave, solutions of this equation are the Bessel functions

$$J_0(kr) \quad \text{and} \quad N_0(kr),$$

wherein it is to be emphasized that $\lambda = 2\pi/k$ does not in general imply the existence of a fixed wavelength along the direction of propagation.

The impedance description of the E-type radial line is based on the above standing-wave solutions; the voltages and currents at the points r and r_0 follow from Eqs. (64a, b) as

$$\left.\begin{array}{l} V(r) = V(r_0)\,\mathrm{Cs}(x,y) - jZ_0 I(r_0)\,\mathrm{sn}(x,y), \\ ZI(r) = Z_0 I(r_0)\,\mathrm{cs}(x,y) - jV(r_0)\,\mathrm{Sn}(x,y), \end{array}\right\} \tag{67}$$

where

$$\mathrm{Cs}(x,y) = \frac{J_1(y)N_0(x) - N_1(y)J_0(x)}{2/\pi y},$$

$$\mathrm{cs}(x,y) = \frac{N_0(y)J_1(x) - J_0(y)N_1(x)}{2/\pi y},$$

$$\mathrm{Sn}(x,y) = \frac{J_1(y)N_1(x) - N_1(y)J_1(x)}{2/\pi y},$$

$$\mathrm{sn}(x,y) = \frac{J_0(y)N_0(x) - N_0(y)J_0(x)}{2/\pi y},$$

$$x = kr, \quad y = kr_0,$$

and $Z = \zeta b/2\pi r$ and $Z_0 = \zeta b/2\pi r_0$ are the characteristic impedances at r and r_0, respectively. These voltage-current relations may be schematically represented by the radial transmission-line diagram of Fig. 1·9, which also shows the positive directions of V and I.

Equations (67) may be converted to a more convenient form by introduction of the relative, or normalized, admittances

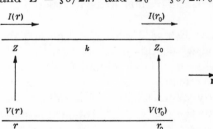

FIG. 1·9.—Choice of positive directions of voltage and current in a radial transmission line.

$$Y'(r) = \frac{ZI(r)}{V(r)} = \frac{Y(r)}{Y} \quad \text{and} \quad Y'(r_0) = \frac{Z_0 I(r_0)}{V(r_0)} = \frac{Y(r_0)}{Y_0} \tag{68}$$

at the radii r and r_0; these admittances are positive in the direction of increasing radius. By division of Eq. (67) one obtains the fundamental radial transmission-line relation for the lowest E-type mode as

$$Y'(r) = \frac{j + Y'(r_0)\,\zeta(x,y)\,\text{ct}(x,y)}{\text{Ct}(x,y) + jY'(r_0)\zeta(x,y)}, \qquad (69)$$

where

$$\left.\begin{aligned}
\text{ct}(x,y) &= \frac{J_1(x)N_0(y) - N_1(x)J_0(y)}{J_0(x)N_0(y) - N_0(x)J_0(y)} = \frac{1}{\text{tn}(x,y)} = \frac{\text{cs}(x,y)}{-\text{sn}(x,y)}, \\
\text{Ct}(x,y) &= \frac{J_1(y)N_0(x) - N_1(y)J_0(x)}{J_1(x)N_1(y) - N_1(x)J_1(y)} = \frac{1}{\text{Tn}(x,y)} = \frac{\text{Cs}(x,y)}{-\text{Sn}(x,y)}, \\
\zeta(x,y) &= \frac{J_0(x)N_0(y) - N_0(x)J_0(y)}{J_1(x)N_1(y) - N_1(x)J_1(y)} = \zeta(y,x) = \frac{\text{sn}(x,y)}{\text{Sn}(x,y)},
\end{aligned}\right\} \qquad (70)$$

and

$$x = kr, \qquad y = kr_0.$$

The ct and Ct functions are called the small and large radial cotangent functions; their inverses tn and Tn are the small and large radial tangent functions. The radial functions are asymmetric. The nature of the asymmetry is evident in the relation

$$\text{ct}(x,y)\,\zeta(x,y) = -\text{Ct}(y,x), \qquad (71)$$

which may be employed to obtain alternative forms of Eq. (69).

The radial functions are plotted vs. $y - x$ with y/x as a parameter in the graphs of Figs. 1·10 to 1·12. The curves of Figs. 1·10a and 1·11a apply when y is less than x, whereas those of Figs. 1·10b and 1·11b are for y greater than x. The symmetry of the functions $\zeta(x,y)$ permits the use of the single graph of Fig. 1·12, for both ranges of y/x. In addition to the graphs numerical values of the radial functions are given in Tables 1·3 for several values of y/x. These tables are incomplete, as many of the data from which the curves were plotted are not in a form convenient for tabulation.

The parametric values $y/x \approx 1$, but $y - x$ finite, correspond to the case of large radii. In this range $\text{ct}(x,y) \cong \text{Ct}(x,y) \cong \cot(y - x)$, and $\zeta(x,y) \cong 1$. Thus at large radii the radial and uniform transmission-line equations (69) and (18) are asymptotically identical. The transmission equations (69) permit the determination of the relative admittance at the input of a line of electrical length $y - x$ from a knowledge of the relative admittance at the output. A few examples will serve to illustrate both the use of Eq. (69) and the physical significance of the radial cotangent functions.

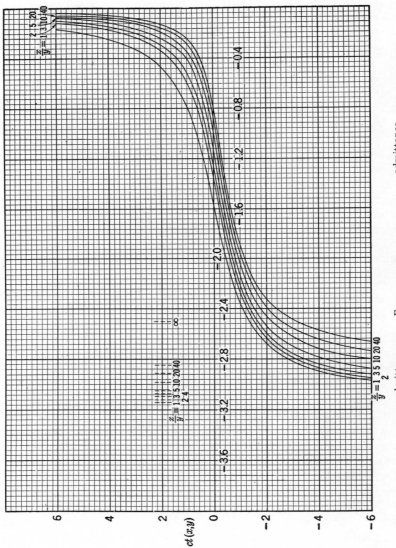

Fig. 1·10a.—Relative input admittance of an E-type radial line with infinite admittance termination ($y < x$). impedance of an H-type radial line with infinite impedance

SEC. 1·7] NONUNIFORM RADIAL WAVEGUIDES 35

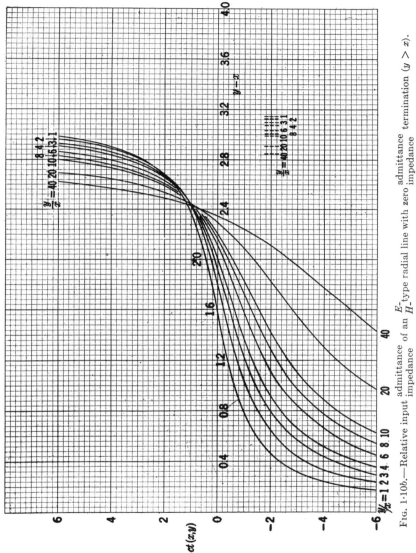

FIG. 1·10b.—Relative input admittance of an E-type radial line with zero admittance termination ($y > x$). H-type radial line with zero impedance

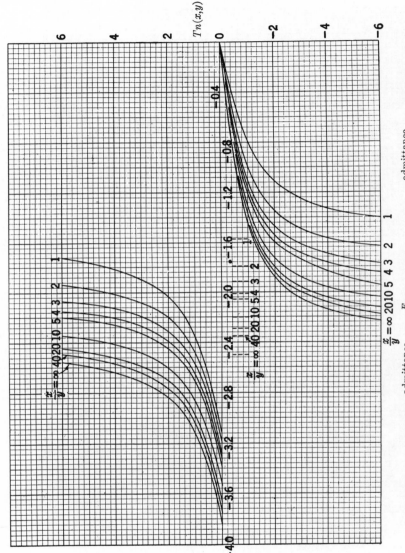

Fig. 1·11a.—Relative input admittance of an E-type radial line with zero admittance termination ($y < x$) impedance of an H-type radial line with zero impedance termination.

SEC. 1·7] NONUNIFORM RADIAL WAVEGUIDES 37

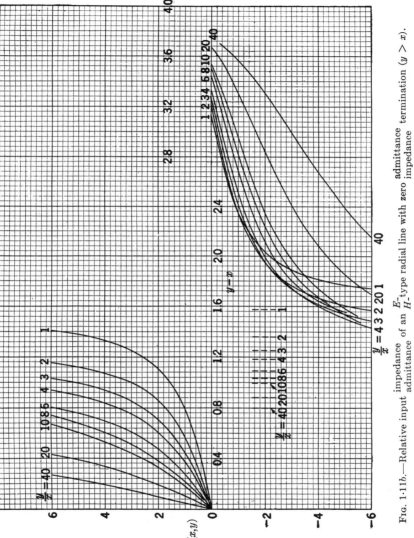

FIG. 1·11b.—Relative input $\genfrac{}{}{0pt}{}{\text{impedance}}{\text{admittance}}$ of an $\genfrac{}{}{0pt}{}{E\text{-}}{H\text{-}}$type radial line with zero $\genfrac{}{}{0pt}{}{\text{admittance}}{\text{impedance}}$ termination ($y > x$).

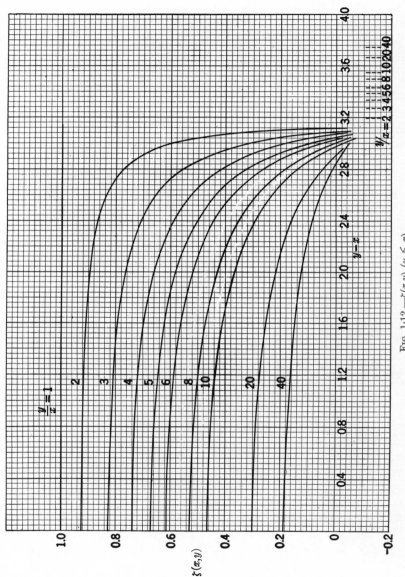

Fig. 1·12.—$\zeta(x,y)$ $(y \lessgtr x)$.

TABLE 1·3a.—VALUES OF THE RADIAL FUNCTIONS

$$\zeta(x,y) = \frac{J_0(x)N_0(y) - N_0(x)J_0(y)}{J_1(x)N_1(y) - N_1(x)J_1(y)}$$

$y - x$	y/x 2	3	4	5	6	8	10	20
0								
0.1	0.9242	0.8239	0.7393	0.6701	0.6142	0.5279	0.4653	0.3003
0.2	0.9240	0.8236	0.7388	0.6700	0.6137	0.5275	0.4644	0.2997
0.3	0.9238	0.8231	0.7382	0.6692	0.6128	0.5266	0.4635	
0.4	0.9234	0.8224	0.7370	0.6682	0.6117	0.5253	0.4624	0.2979
0.5	0.9230	0.8215	0.7361	0.6670	0.6102	0.5235	0.4607	
0.6	0.9225	0.8203	0.7345	0.6650	0.6083	0.5217	0.4587	0.2947
0.7	0.9218	0.8190	0.7326	0.6630	0.6060	0.5193	0.4563	
0.8	0.9210	0.8173	0.7304	0.6604	0.6034	0.5165	0.4535	0.2902
0.9	0.9201	0.8154	0.7280	0.6575	0.6003	0.5135	0.4502	
1.0	0.9190	0.8132	0.7250	0.6542	0.5969	0.5095	0.4464	0.2841
1.1	0.9178	0.8107	0.7217	0.6504	0.5926	0.5052	0.4422	
1.2	0.9164	0.8079	0.7179	0.6461	0.5880	0.5002	0.4373	0.2763
1.3	0.9148	0.8046	0.7136	0.6410	0.5827	0.4948	0.4318	
1.4	0.9130	0.8008	0.7087	0.6356	0.5768	0.4886	0.4256	0.2663
1.5	0.9109	0.7966	0.7032	0.6296	0.5700	0.4815	0.4187	
1.6	0.9085	0.7917	0.6968	0.6220	0.5624	0.4736	0.4109	0.2540
1.7	0.9057	0.7861	0.6895	0.6138	0.5538	0.4648	0.4020	
1.8	0.9025	0.7797	0.6812	0.6046	0.5439	0.4546	0.3920	0.2386
1.9	0.8987	0.7722	0.6716	0.5938	0.5327	0.4431	0.3808	0.2292
2.0	0.8942	0.7635	0.6606	0.5815	0.5197	0.4299	0.3676	0.2188
2.1	0.8890	0.7532	0.6476	0.5671	0.5048	0.4148	0.3532	
2.2	0.8826	0.7410	0.6322	0.5503	0.4873	0.3972	0.3363	0.1937
2.3	0.8748	0.7261	0.6138	0.5303	0.4667	0.3768	0.3165	
2.4	0.8649	0.7079	0.5914	0.5062	0.4421	0.3523	0.2934	0.1607
2.5	0.8523	0.8848	0.5637	0.4767	0.4122	0.3237	0.2660	
2.6	0.8354	0.6549	0.5284	0.4397	0.3751	0.2882	0.2328	0.1156
2.7	0.8116	0.6145	0.4821	0.3921	0.3280	0.2440	0.1920	
2.8	0.7760	0.5570	0.4186	0.3284	0.2662	0.1874	0.1394	0.05050
2.9	0.7163	0.4688	0.3261	0.2388	0.1814	0.1121	0.07212	
3.0	0.5960	0.3156	0.1789	0.1036	0.0578	0.0058	−0.01829	−0.05201
3.1	0.2271	−0.0159	−0.09289	−0.1248	−0.1395	−0.1501	−0.1498	
3.2	21.419	−1.2767	−0.7641	−0.5940	−0.5057	−0.4071	−0.3561	−0.2391
3.3	1.5745	6.2225	−5.1963	−2.1200	−1.4237	−0.9298	−0.7277	
3.4	1.2953	2.0733	4.1385	23.0896	−8.0395	−1.2394	−1.6024	−0.7001
3.5	1.1815	1.5505	2.1136	2.9965	4.8423	248.085	−6.1984	
3.6	1.1239	1.3451	1.6193	1.9537	2.3711	3.6484	6.4870	−3.8555
3.7	1.0890	1.2350	1.3953	1.5645	1.7431	2.1355	2.6020	
3.8	1.0654	1.1659	1.2668	1.3638	1.4548	1.6242	1.7831	2.4796
3.9	1.0484	1.1183	1.1829	1.2394	1.2882	1.3671	1.4251	
4.0	1.0354	1.0832	1.1234	1.1549	1.1788	1.2102	1.2230	1.1863
4.1	1.0251	1.0560	1.0786	1.0930	1.1007	1.1020	1.0917	
4.2	1.0166	1.0342	1.0434	1.0452	1.0415	1.0239	0.9985	0.8549
4.3	1.0094	1.0160	1.0145	1.0066	0.9946	0.9629	0.9281	
4.4	1.0031	1.0000	0.9900	0.9745	0.9558	0.9143	0.8739	0.6961
4.5	0.9976	0.9867	0.9688	0.9468	0.9228	0.8732	0.8258	
4.6	0.9925	0.9743	0.9498	0.9223	0.8938	0.8379	0.7861	0.5971
4.7	0.9878	0.9629	0.9324	0.9001	0.8676	0.8091	0.7509	
$3\pi/2$	0.9869	0.9614	0.9302	0.8975	0.8654	0.8008	0.7458	0.5540

TABLE 1·3b

$$\mathrm{et}(x,y) = \frac{J_1(x)N_0(y) - N_1(x)J_0(y)}{J_0(x)N_0(y) - N_0(x)J_0(y)}$$

$y - x$	y/x 2	3	4	5	6	8	10	20
0.1	14.3929	18.1697	21.6068	24.8284	27.8666	33.5884	39.0408	
0.2	7.1454	9.0320	10.7497	12.3512	13.8747	16.7351	19.4556	31.6025
0.3	4.7064	5.9623	7.1037	8.1707	9.1843	11.0954	12.8982	
0.4	3.4694	4.4092	5.2645	6.0610	6.8188	8.2481	9.5942	15.6326
0.5	2.7126	3.4623	4.1419	4.7769	5.3827	6.5234	7.5932	
0.6	2.1955	2.8155	3.2890	3.9102	4.4108	5.3553	6.2439	10.2299
0.7	1.8149	2.3462	2.8277	3.2763	3.7034	4.5071	5.2636	
0.8	1.5190	1.9814	2.4003	2.7904	3.1607	3.8540	4.5159	7.4615
0.9	1.2790	1.6874	2.0566	2.4006	2.7270	3.3403	3.9197	
1.0	1.0775	1.4423	1.7717	2.0781	2.3686	2.9157	3.4302	5.7380
1.1	0.9031	1.2318	1.5280	1.8033	2.0641	2.5559	3.0163	
1.2	0.7483	1.0464	1.3143	1.5632	1.7988	2.2421	2.6583	4.5250
1.3	0.6075	0.8793	1.1229	1.3490	1.5625	1.9643	2.3410	
1.4	0.4767	0.7254	0.9475	1.1531	1.3473	1.7120	2.0544	3.5895
1.5	0 3525	0.5805	0.7832	0.9707	1.1474	1.4787	1.7897	
1.6	0.2321	0.4413	0.6264	0.7970	0.9576	1.2581	1.5405	2.8039
1.7	0.1130	0.3048	0.4733	0.6282	0.7736	1.0455	1.3003	
1 8	−0.007287	0.1679	0.3206	0.4603	0.5912	0.8360	1.0637	2.0860
1 9	−0.1315	0.0275	0.1647	0.2894	0.4059	0.6228	0.8249	
2.0	−0.2628	−0.1199	0.00157	0.1109	0.2126	0.4009	0.5755	1.3537
2 1	−0.4051	−0.2789	−0.1741	−0.0809	0.004948	0.1626	0.3082	
2.2	−0 5636	−0.4555	−0.3689	−0.2938	−0.2256	−0.1025	0.01050	0.5030
2 3	−0 7458	−0.6584	−0.5931	−0.5390	−0.4916	−0.4097	−0.3359	
2 4	−0 9627	−0.9005	−0.8616	−0.8340	−0.8130	−0.7868	−0.7608	−0.6994
2 5	−1.2328	−1.2036	−1.2001	−1.2086	−1.2239	−1.2647	−1.3171	
2 6	−1.5875	−1.6061	−1.6550	−1.7178	−1.7890	−1.9441	−2.1135	−3.0338
2 7	−2.0885	−2.1851	−2.3221	−2.4798	−2.6498	−3.0162	−3.4100	
2 8	−2.8729	−3.1208	−3.4378	−3.7963	−4.1879	−5.0572	−6.0831	−12.8084
2 9	−4.3235	−4.9593	−5.7809	−6.7639	−7.9139	−10.8406	−15.0410	
3.0	−8.0675	−10.4990	−14.3840	−20.7680	−32.5992	−272.075	76.1741	20.4216
3 1	−44.0149	340.55	41.0035	24.3027	18.5507	13.9023	12.1482	
3 2	13.3283	10.4135	8.9151	8.0858	7.5960	7.1330	6.9363	7.3531
3.3	5.8266	5.3617	5.0912	4.9478	4.8821	4.8825	4.9673	
3.4	3.7146	3.6158	3.5801	3.5893	3.6272	3.8418	3.9108	4.8215
3.5	2.7030	2.7160	2.7563	2.8574	2.8902	3.0581	3.2393	
3.6	2.0993	2.1553	2.2282	2.3092	2.3962	2.5785	2.7655	3.6737
3.7	1.6912	1.7693	1.8541	1.9435	2.0354	2.2209	2.4066	
3.8	1.3915	1.4794	1.5701	1.6625	1.7555	1.9390	2.1204	2.9627
3.9	1.1580	1.2503	1.3430	1.4357	1.5276	1.7085	1.8829	
4.0	0.9674	1.0613	1.1540	1.2455	1.3354	1.5120	1.6789	2.4502
4.1	0.8061	0.8999	0.9914	1.0808	1.1682	1.3375	1.4992	
4.2	0.6651	0.7579	0.8475	0.9344	1.0190	1.1816	1.3370	2.0398
4.3	0.5386	0.6297	0.7169	0.8011	0.8826	1.0386	1.1874	
4.4	0.4222	0.5112	0.5958	0.6770	0.7553	0.9046	1.0467	1.6838
4.5	0.3128	0.3994	0.4304	0.5592	0.6342	0.7765	0.9121	
4.6	0.2078	0.2917	0.3703	0.4451	0.5168	0.6530	0.7808	1.3521
4.7	0.1049	0.1859	0.2613	0.3327	0.4008	0.5297	0.6505	
$3\pi/2$	0.09215	0.1708	0.2479	0.3192	0.3864	0.5154	0.6340	1.1676

TABLE 1·3b.—(Continued)

y − x \ x/y	2	3	4	5	6	8	10	20
−3.0	8.231	10.59		19.6				
−2.8	3.107	3.448		4.124				
−2.6	1.849	2.012		2.272				
−2.4	1.253	1.3633		1.521				
−2.2	0.8833	0.9752		1.093				
−2.0	0.6172	0.7017		0.8025				
−1.8	0.4032	0.4864		0.5803				
−1.6	0.2148	0.3012		0.3942				
−1.4	0.03534	0.1286		0.2254				
−1.2	−0.1498	−0.0456		0.06014				
−1.0	−0.3581	−0.2370		−0.1164				
−0.8	−0.6190	−0.4712		−0.3254				
−0.6	−0.9944	−0.8000		−0.6114				
−0.4	−1.667	−1.378		−1.099				
−0.2	−3.538	−2.967		−2.416				

TABLE 1.3c

$$\mathrm{Tn}(x,y) = \frac{J_1(x)N_1(y) - N_1(x)J_1(y)}{J_1(y)N_0(x) - N_1(y)J_0(x)}$$

y − x \ x/y	2	3	5	∞	y − x \ y/x	2	3	5	6
−3.0	0.2109	0.3020	0.4644	1.3037	0	0	0	0	0
−2.8	0.4486	0.5753	0.8023	2.2145	0.2	0.3058	0.4094	0.6180	0.7228
−2.6	0.7544	0.9510	1.323	4.8636	0.4	0.6497	0.8824	1.361	1.605
−2.4	1.201	1.569	2.369	208.08	0.6	1.0903	1.524	2.460	2.958
−2.2	2.012	2.961	6.189	−5.0362	0.8	1.754	2.600	4.654	5.873
−2.0	4.256	10.922	−15.5	−2.5757	1.0	3.038	5.189	13.133	20.93
−1.8	149.9	−7.634	−3.593	−1.7102	1.2	7.288	27.16	−25.74	−18.42
−1.6	−4.74	−2.863	−2.017	−1.2514	1.4	−31.00	−9.709	−6.979	−6.782
−1.4	−2.297	−1.721	−1.365	−0.9559	1.6	−5.123	−4.192	−4.078	−4.424
−1.2	−1.458	−1.183	−0.9901	−0.7425	1.8	−2.749	−2.636	−2.850	−3.007
−1.0	−1.0097	−0.8524	−0.7354	−0.5751	2.0	−1.811	−1.865	−2.142	−2.297
−0.8	−0.7148	−0.6175	−0.5432	−0.4357	2.2	−1.282	−1.384	−1.662	−1.806
−0.6	−0.4972	−0.4324	−0.3841	−0.3143	2.4	−0.9229	−1.036	−1.299	−1.430
−0.4	−0.3274	−0.2757	−0.2469	−0.2041	2.6	−0.6473	−0.7593	−1.001	−1.119
−0.2	−0.1514	−0.1344		−0.1005	2.8	−0.4150	−0.5206	−0.7390	−0.8446
					3.0	−0.2036	−0.2997	−0.4937	−0.5863

The relative input admittance of an E-type radial line of electrical length $y - x$ with a short circuit $[Y'(r_0) = \infty]$ at its output end is

$$Y'(r) = -j\,\mathrm{ct}(x,y). \tag{72}$$

For an open circuit $[Y'(r_0) = 0]$ at the output the relative input admittance is

$$Y'(r) = j\,\frac{1}{\mathrm{Ct}(x,y)} = j\mathrm{Tn}\,(x,y). \tag{73}$$

For an infinitely long E-type line the relative input admittance at any point r, looking in the direction of increasing radius, is found from Eq. (69) by placing $Y'(r_0) = 1$ and $r_0 = \infty$ to be

$$Y'(r) = -j\frac{H_1^{(2)}(kr)}{H_0^{(2)}(kr)}, \tag{74}$$

or

$$Y'(r) \cong 1 \qquad \text{for } kr \gg 1,$$
$$Z'(r) \cong \frac{\pi kr}{2} + jkr \ln \frac{2}{\gamma kr} \qquad \text{for } kr \ll 1,$$

where $\gamma = 1.781$. The input admittance of an infinite radial line is not in general equal to the characteristic admittance! For E-type lines it is seen to be complex with a negative imaginary, i.e., inductive, part. The relative input admittance looking in the direction of increasing radius is, for all noninfinite $Y'(0)$,

$$Y'(r) = -j\frac{J_1(kr)}{J_0(kr)}, \tag{75}$$

or

$$Y'(r) \cong -j \tan\left(kr - \frac{\pi}{4}\right) \qquad \text{for } kr \gg 1,$$
$$Y'(r) \cong -j\frac{kr}{2} \qquad \text{for } kr \ll 1.$$

The relative input admittance is negative imaginary and hence capacitative; it is to be remembered that the output is at a smaller radius than the input and hence the input admittance is counted negatively.

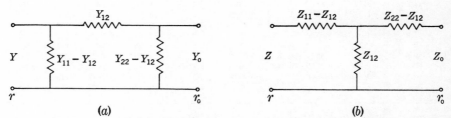

Fig. 1·13.—(a) π-Circuit representation of an E-type radial line of electrical length $y - x$; (b) T-circuit representation of an H-type radial line of electrical length $y - x$.

An alternative method of determining the relation between the input and output admittances of a radial line is afforded by the equivalent circuit representation for a length $y - x$ of line. From Eq. (67) one finds that the parameters of the π circuit representation (cf. Fig. 1·13a) of the E-type radial line are

SEC. 1·7]   NONUNIFORM RADIAL WAVEGUIDES   43

$$Y_{11} - Y_{12} = -jY\left[\operatorname{ct}(x,y) + \sqrt{\frac{y}{x}}\operatorname{cst}(x,y)\right],$$
$$Y_{22} - Y_{12} = -jY_0\left[-\operatorname{ct}(y,x) + \sqrt{\frac{x}{y}}\operatorname{cst}(x,y)\right], \quad (76)$$
$$Y_{12} = -j\sqrt{YY_0}\,\operatorname{cst}(x,y),$$

where

$$\operatorname{cst}^2(x,y) = \frac{1 + \operatorname{ct}(x,y)\,\operatorname{Ct}(x,y)}{\zeta(x,y)}.$$

The cst function may be termed the radial cosecant function since in the limit of large x and y it becomes identical with the ordinary cosecant function.

A useful formula for perturbation and frequency-sensitivity calculations, particularly in resonant radial structures, can be obtained from the differential form of Eq. (69). The differential change $dY'(r)$ in relative input admittance arising from either a change $dY'(r_0)$ of relative output admittance or a relative change dk/k in frequency, or both, is given by

$$\left(\frac{dY'(r)}{1 + [jY'(r)]^2} + \left\{jx + \frac{Y'(r)}{1 + [jY'(r)]^2}\right\}\frac{dk}{k}\right)\alpha(r)$$
$$= \left(\frac{dY'(r_0)}{1 + [jY'(r_0)]^2} + \left\{jy + \frac{Y'(r_0)}{1 + [jY'(r_0)]^2}\right\}\frac{dk}{k}\right)\alpha(r_0), \quad (77)$$

where

$$\alpha(r) = \frac{2}{\pi kr}\frac{1 + [jY'(r)]^2}{[J_1(kr) - jJ_0(kr)Y'(r)]^2}.$$

For large kr and kr_0 the ratio $\alpha(r_0)/\alpha(r)$ approaches unity, and hence in this range Eq. (77) and the corresponding uniform-line equation (23) become asymptotically identical. With a short circuit at r_0 the right-hand member of Eq. (77) simplifies to

$$\left[dZ'(r_0) + jy\frac{dk}{k}\right]\frac{2}{\pi y J_0^2(y)}. \quad (78)$$

Scattering Description of Dominant E-type Mode.—The scattering description of the dominant E-type mode in the radial guide of Fig. 1·8a is based on the traveling-wave solutions

$$H_0^{(2)}(kr) \quad \text{and} \quad H_0^{(1)}(kr)$$

of the radial-wave equation (66). The Hankel function solutions $H_0^{(2)}$ and $H_0^{(1)}$ represent waves traveling in the direction of increasing and decreasing radius and are the analogues of the exponential functions encountered in uniform lines. In terms of these functions the solutions of the radial transmission-line equations (64) for the dominant mode voltage and current can be written as

$$V(r) = V_{\text{inc}}H_0^{(2)}(kr) + V_{\text{refl}}H_0^{(1)}(kr),$$
$$jZI(r) = V_{\text{inc}}H_1^{(2)}(kr) + V_{\text{refl}}H_1^{(1)}(kr), \quad (79)$$

where V_{inc} and V_{refl} are the complex amplitudes of the incident and reflected waves traveling in the direction of increasing and decreasing r, respectively. The analogy with the uniform line Eqs. (26) becomes more evident by use of the relations

$$H_m^{(1)}(x) = (-j)^m h_m e^{+j\eta_m}, \qquad H_m^{(2)}(x) = (j)^m h_m e^{-j\eta_m}, \tag{80}$$

where

$$h_m = \sqrt{J_m^2(x) + N_m^2(x)}, \qquad \eta_m = \frac{m\pi}{2} + \tan^{-1}\frac{N_m(x)}{J_m(x)}.$$

The amplitude h_m and phase η_m of the Hankel functions are plotted and tabulated as functions of x for $m = 0$ and 1 in Fig. 1·14 and in Table 1·4.

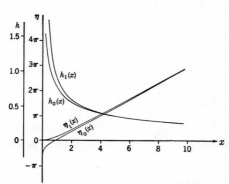

Fig. 1·14.—Amplitude and phase of Hankel functions of order zero and one.

A convenient measure of the voltage and current at any point r is obtained on introduction of a reflection coefficient defined as the ratio of the amplitudes of the reflected and incident waves. There exist two types of reflection coefficients: a voltage reflection coefficient

$$\Gamma_V(r) = \frac{V_{\text{refl}}}{V_{\text{inc}}}\frac{H_0^{(1)}(kr)}{H_0^{(2)}(kr)} = \frac{V_{\text{refl}}}{V_{\text{inc}}} e^{j2\eta_0(kr)} \tag{81a}$$

and a current reflection coefficient

$$\Gamma_I(r) = \frac{V_{\text{refl}}}{V_{\text{inc}}}\frac{H_1^{(1)}(kr)}{H_1^{(2)}(kr)} = -\frac{V_{\text{refl}}}{V_{\text{inc}}} e^{j2\eta_1(kr)}. \tag{81b}$$

In contradistinction to the case of a uniform line the voltage and current reflection coefficients are not negatives of each other.

The fundamental radial transmission-line equations for the reflection coefficients at any two points r and r_0 of an E-type line follow from Eqs. (81) either as

$$\Gamma_V(r) = \Gamma_V(r_0) e^{j2[\eta_0(kr) - \eta_0(kr_0)]} \tag{82a}$$

or as

$$\Gamma_I(r) = \Gamma_I(r_0) e^{j2[\eta_1(kr) - \eta_1(kr_0)]}. \tag{82b}$$

By division of Eqs. (79) the relation between the admittance and reflection coefficient at any point r is seen to be

$$Y'_+(r) = \frac{1 + \Gamma_I}{1 + \Gamma_V} = \frac{1 - \dfrac{Y_-}{Y_+}\Gamma_V}{1 + \Gamma_V} \tag{83a}$$

NONUNIFORM RADIAL WAVEGUIDES

TABLE 1·4

x	$h_0(x)$	$h_1(x)$	$\eta_0(x)$	$\eta_1(x)$	$J_0(x)$	$J_0(ix)$	$J_1(x)$	$-iJ_1(ix)$
0			$-90°$	$0°$	1.0000	1.0000	0.0000	0.0000
0.1	1.830	6.459	-57.0	0.5	0.9975	1.0025	0.0499	0.0501
0.2	1.466	3.325	-47.5	1.7	0.9900	1.0100	0.0995	0.1005
0.3	1.268	2.298	-39.5	3.7	0.9776	1.0226	0.1483	0.1517
0.4	1.136	1.792	-32.3	6.3	0.9604	1.0404	0.1960	0.2040
0.5	1.038	1.491	-25.4	9.4	0.9385	1.0635	0.2423	0.2579
0.6	0.9628	1.283	-18.7	12.8	0.9120	1.0420	0.2867	0.3137
0.7	0.9016	1.151	-12.2	16.6	0.8812	1.1263	0.3290	0.3719
0.8	0.8507	1.045	-5.9	20.7	0.8463	1.1665	0.3688	0.4329
0.9	0.8075	0.9629	-0.4	24.9	0.8075	1.2130	0.4060	0.4971
1.0	0.7703	0.8966	6.6	29.4	0.7652	1.2661	0.4401	0.5652
1.1	0.7377	0.8421	12.7	34.0	0.7196	1.3262	0.4709	0.6375
1.2	0.7088	0.7963	18.8	38.7	0.6711	1.3937	0.4983	0.7147
1.3	0.6831	0.7572	24.8	43.6	0.6201	1.4693	0.5220	0.7973
1.4	0.6599	0.7234	30.8	48.5	0.5669	1.5534	0.5420	0.8861
1.5	0.6389	0.6938	36.8	53.5	0.5118	1.6467	0.5579	0.9817
1.6	0.6198	0.6675	42.7	58.6	0.4554	1.7500	0.5699	1.0848
1.7	0.6023	0.6441	48.6	63.8	0.3980	1.864	0.5778	1.1963
1.8	0.5861	0.6230	54.6	69.0	0.3400	1.990	0.5815	1.3172
1.9	0.5712	0.6040	60.4	74.2	0.2818	2.128	0.5812	1.4482
2.0	0.5573	0.5866	66.3	79.5	0.2239	2.280	0.5767	1.5906
2.2	0.5323	0.5560	78.0	89.1	0.1104	2.629	0.5560	1.914
2.4	0.5104	0.5298	89.7	101.0	0.0025	3.049	0.5202	2.298
2.6	0.4910	0.5071	101.4	111.8	-0.0968	3.553	0.4708	2.755
2.8	0.4736	0.4872	113.0	122.8	-0.1850	4.157	0.4097	3.301
3.0	0.4579	0.4694	124.6	133.8	-0.2601	4.881	0.3391	3.953
3.5	0.4245	0.4326	153.6	161.5	-0.3801	7.378	0.1374	6.206
4.0	0.3975	0.4034	182.4	189.4	-0.3972	11.302	-0.0660	9.759
5.0	0.3560	0.3594	240.1	245.7	-0.1776	27.24	-0.3276	24.34
6.0	0.3252	0.3274	297.6	302.3	0.1507	67.23	-0.2767	61.34
7.0	0.3012	0.3027	355.1	359.1	0.3001	168.6	-0.0047	156.04
8.0	0.2818	0.2829	412.5	416.0	0.1717	427.6	0.2346	399.9
9.0	0.2658	0.2666	469.9	473.0	-0.0903	1093.6	0.2453	1030.9
10.00	0.2522	0.2528	527.2	53).1	-0.2459		0.0435	

or conversely

$$\Gamma_V(r) = \frac{Z'_+(r) - 1}{Z'_-(r) + 1}, \qquad \Gamma_I(r) = -\frac{1 - Y'_+(r)}{1 + Y'_-(r)}, \qquad (83b)$$

where

$$Y'_+(r) = \frac{1}{Z'_+(r)} = \frac{I(r)}{Y_+V(r)}, \qquad Y'_-(r) = \frac{1}{Z'_-(r)} = \frac{I(r)}{Y_-V(r)},$$

and

$$Y_+ = -jY \frac{H_1^{(2)}(kr)}{H_0^{(2)}(kr)} = Y \frac{h_1}{h_0} e^{-i(\eta_1 - \eta_0)},$$

$$Y_- = jY \frac{H_1^{(1)}(kr)}{H_0^{(1)}(kr)} = Y \frac{h_1}{h_0} e^{+i(\eta_1 - \eta_0)}.$$

With the aid of Eqs. (82) and (83) radial-line calculations may be performed with either the voltage or current reflection coefficient in a manner similar to uniform line calculations. In fact for large kr it is apparent that the two types of calculations are identical, since $\Gamma_V = -\Gamma_I$ and $Y_+ = Y_- = Y$.

The computation of frequency sensitivity on a long radial line is facilitated by a knowledge of the differential forms of Eqs. (82). From Eqs. (82a) and (80) it follows that

$$\frac{d\Gamma_V(r)}{\Gamma_V(r)} = \frac{d\Gamma_V(r_0)}{\Gamma_V(r_0)} + j\frac{4}{\pi}\left[\frac{1}{h_0^2(kr)} - \frac{1}{h_0^2(kr_0)}\right]\frac{dk}{k} \quad (84)$$

gives the relation between the changes $d\Gamma_V(r)$ and $d\Gamma_V(r_0)$ in the input and output reflection coefficients due to the relative frequency shift dk/k. As for the case of the corresponding uniform line relation (35), Eq. (84) can be decomposed into an amplitude and phase part from which it is apparent that the relative change in reflection coefficient is the same at all points of a nondissipative line.

The description of the dominant E-type mode in the waveguide shown in Fig. 1·8b is somewhat different from that just described. Because of the vanishing tangential electric field at the guide walls the dominant mode is no longer angularly symmetric and hence $m \neq 0$. The transmission-line description is based on the mth-order Bessel and Hankel functions but otherwise is formally identical with that just described. Since no table or plots of the mth-order radial cotangent, etc., functions are available,[1] no details of this dominant E-type description will be presented. The transmission-line description of the higher angular modes in radial lines likewise depends on mth-order Bessel functions.

Description of Dominant H-type Mode.—Frequently in the regions shown in Fig. 1·8 the frequency and excitation are such that, almost everywhere, only the lowest H-type mode is present. The field configuration of this mode is angularly symmetric about the z-axis, and hence $m = 0$. For regions of infinite height in the z direction the magnetic field is parallel to the z-axis, the electric field lines are circles or circular arcs about the z-axis, and $n = 0$. For the particular case of the waveguide shown in Fig. 1·8a the field configurations of the dominant E- and H-type modes are, therefore, dual to each other. For regions of finite height there is an additional radial component in the magnetic field of the dominant H-type mode, and $n \neq 0$. Duality between the fields of the two dominant-mode types no longer exists. However, whatever the height, the transmission-line description of the dominant H-type

---

[1] *Cf.* H. S. Bennett, "Transmission Line Characteristics of the Sectoral Horn," *Proc. I.R.E.*, **37**, 738 (1949).

mode in the waveguides of Fig. 1·8 is dual to that of the dominant E-type mode in Fig. 1·8a. Duality in this case implies that in Eqs. (64a) the V, ZI, and k of the dominant E-type description are replaced by the I, YV, and κ of the dominant H-type description. The dependence of the characteristic admittance Y and mode constant κ of an H-type mode on the cross-sectional dimensions of a radial waveguide are given in Sec. 2·7.

As a consequence of duality all relative admittance relations derived above for the dominant E-type mode in the guide of Fig. 1·8a are identical with the relative impedance relations for a dominant H-type mode, provided k is everywhere replaced by κ. For example, the fundamental input-output impedance relation for a length $y - x$ of the dominant H-type radial line is obtained from Eq. (69) by duality as

$$Z'(r) = \frac{j + Z'(r_0)\zeta(x,y)\,\mathrm{ct}(x,y)}{\mathrm{Ct}(x,y) + jZ'(r_0)\zeta(x,y)}, \tag{85}$$

where $x = \kappa r$ and $y = \kappa r_0$. In addition the relative admittance parameters (76) of the π circuit representation (cf. Fig. 1·13a) for a length $y - x$ of a dominant E-type radial line become the relative impedance parameters for the T circuit representation (cf. Fig. 1·13b) of a length $y - x$ of the dominant H-type line. It is also evident that the scattering description of the dominant H-type mode follows from that of the dominant E-type mode by the aforementioned duality replacements. In employing duality one should remember that an infinite admittance, or short circuit, becomes on the duality replacement an infinite impedance, or open circuit, and conversely.

1·8. Field Representation in Nonuniform Spherical Waveguides.— Another type of nonuniform region that permits a field representation in terms of an infinite set of known transmission modes is the spherical waveguide depicted in Fig. 1·15a or b. The transmission direction is along the radius r, and the $\theta\phi$ cross sections transverse thereto are either spherical surfaces as in Fig. 1·15a or spherical sectors bounded by cones of aperture $2\theta_1$, and $2\theta_2$ as in Fig. 1·15b. The Maxwell equations for the electric and magnetic fields transverse to the radial direction, which is characterized by the unit vector $\mathbf{r}_0$, may be written in invariant vector notation as

$$\left. \begin{array}{l} \dfrac{1}{r}\dfrac{\partial}{\partial r}(r\mathbf{E}_t) = -jk\zeta\left(\boldsymbol{\varepsilon} + \dfrac{i\boldsymbol{\nabla}\boldsymbol{\nabla}_t}{k^2}\right)\cdot(\mathbf{H}_t \times \mathbf{r}_0), \\[6pt] \dfrac{1}{r}\dfrac{\partial}{\partial r}(r\mathbf{H}_t) = -jk\eta\left(\boldsymbol{\varepsilon} + \dfrac{i\boldsymbol{\nabla}\boldsymbol{\nabla}_t}{k^2}\right)\cdot(\mathbf{r}_0 \times \mathbf{E}_t), \end{array} \right\} \tag{86a}$$

the longitudinal components being expressed in terms of the transverse components as

$$\left. \begin{array}{l} jk\eta E_r = \boldsymbol{\nabla}_t \cdot (\mathbf{H}_t \times \mathbf{r}_0), \\ jk\zeta H_r = \boldsymbol{\nabla}_t \cdot (\mathbf{r}_0 \times \mathbf{E}_t). \end{array} \right\} \tag{86b}$$

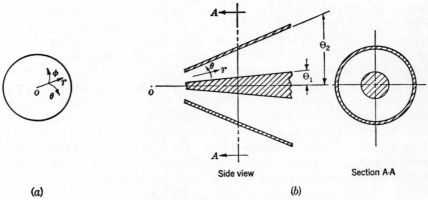

Fig. 1·15.—Spherical waveguides: (a) Spherical; (b) Conical.

The notation is the same as that employed in Eqs. (1) and (2) save that $_t\nabla$ and ∇_t, the gradients transverse to $\mathbf{r}_0$, are defined by

$$_t\nabla = \nabla - \mathbf{r}_0 \frac{\partial}{\partial r},$$

$$\nabla_t = \nabla - \frac{1}{r^2}\frac{\partial}{\partial r} r^2 \mathbf{r}_0.$$

Alternatively, in the spherical coordinate system, r, θ, ϕ appropriate to these geometries, the field equations for the transverse components may be written as

$$\left.\begin{aligned}
\frac{1}{r}\frac{\partial}{\partial r}(rE_\theta) &= -jk\zeta\left\{H_\phi + \frac{1}{k^2 r^2}\left[\frac{\partial}{\partial\theta}\frac{1}{\sin\theta}\frac{\partial}{\partial\theta}(\sin\theta\, H_\phi)\right.\right.\\
&\qquad\qquad\qquad\qquad\left.\left. - \frac{\partial}{\partial\theta}\frac{1}{\sin\theta}\frac{\partial}{\partial\phi}H_\theta\right]\right\}, \\
\frac{1}{r}\frac{\partial}{\partial r}(rE_\phi) &= -jk\zeta\left\{-H_\theta + \frac{1}{k^2 r^2}\left[\frac{1}{\sin^2\theta}\frac{\partial^2}{\partial\phi\,\partial\theta}(\sin\theta\, H_\phi)\right.\right.\\
&\qquad\qquad\qquad\qquad\left.\left. - \frac{1}{\sin^2\theta}\frac{\partial^2}{\partial\phi^2}H_\theta\right]\right\}, \\
\frac{1}{r}\frac{\partial}{\partial r}(rH_\theta) &= -jk\eta\left\{-E_\phi + \frac{1}{k^2 r^2}\left[\frac{\partial}{\partial\theta}\frac{1}{\sin\theta}\frac{\partial}{\partial\phi}E_\theta\right.\right.\\
&\qquad\qquad\qquad\qquad\left.\left. - \frac{\partial}{\partial\theta}\frac{1}{\sin\theta}\frac{\partial}{\partial\theta}(\sin\theta\, E_\phi)\right]\right\}, \\
\frac{1}{r}\frac{\partial}{\partial r}(rH_\phi) &= -jk\eta\left\{+E_\theta + \frac{1}{k^2 r^2}\left[\frac{1}{\sin^2\theta}\frac{\partial^2}{\partial\phi^2}E_\theta\right.\right.\\
&\qquad\qquad\qquad\qquad\left.\left. - \frac{1}{\sin^2\theta}\frac{\partial^2}{\partial\phi\,\partial\theta}(\sin\theta\, E_\phi)\right]\right\},
\end{aligned}\right\} \quad (87)$$

and for the longitudinal components as

$$\left.\begin{aligned} jk\eta E_r &= \frac{1}{r\sin\theta}\left[\frac{\partial}{\partial\theta}(\sin\theta\, H_\phi) - \frac{\partial}{\partial\phi}H_\theta\right], \\ jk\zeta H_r &= \frac{1}{r\sin\theta}\left[\frac{\partial}{\partial\phi}E_\theta - \frac{\partial}{\partial\theta}(\sin\theta\, E_\phi)\right]. \end{aligned}\right\} \quad (88)$$

The dependence of the fields on the transverse coordinates may be integrated out because the knowledge of the boundary conditions on the curve or curves s (if any) bounding the transverse cross sections implies the knowledge of the form of the transverse mode fields. As in the case of a uniform region this may be done by introduction of an infinite set of orthogonal vector functions which are of two types: $\mathbf{e}'_i(\theta,\phi)$ and $\mathbf{e}''_i(\theta,\phi)$. The E-mode functions $\mathbf{e}'_i(\theta,\phi)$ are defined by

$$\left.\begin{aligned} \mathbf{e}'_i &= -r\,{}_t\nabla\Phi_i, \\ \mathbf{h}'_i &= \mathbf{r}_0 \times \mathbf{e}'_i, \end{aligned}\right\} \quad (89a)$$

where

$$\left.\begin{aligned} r^2\nabla_t \cdot {}_t\nabla\Phi_i + k'^2_{ci}\Phi_i &= 0, \\ \Phi_i &= 0 \text{ on } s \quad \text{if } k'_{ci} \neq 0, \\ \frac{\partial\Phi_i}{\partial s} &= 0 \text{ on } s \quad \text{if } k'_{ci} = 0, \end{aligned}\right\} \quad (89b)$$

The H-mode functions $\mathbf{e}''_i(\theta,\phi)$ are defined by

$$\left.\begin{aligned} \mathbf{e}''_i &= \mathbf{r} \times {}_t\nabla\Psi_i, \\ \mathbf{h}''_i &= \mathbf{r}_0 \times \mathbf{e}''_i, \end{aligned}\right\} \quad (90a)$$

where

$$r^2\nabla_t \cdot {}_t\nabla\Psi_i + k''^2_{ci}\Psi_i = 0,$$
$$\frac{\partial\Psi_i}{\partial\nu} = 0 \text{ on } s. \quad (90b)$$

The two-dimensional scalar operator $r^2\nabla_t \cdot {}_t\nabla$ is represented in spherical coordinates by

$$r^2\nabla_t \cdot {}_t\nabla = \frac{1}{\sin\theta}\frac{\partial}{\partial\theta}\sin\theta\frac{\partial}{\partial\theta} + \frac{1}{\sin^2\theta}\frac{\partial^2}{\partial\phi^2}.$$

The subscript i denotes the double index mn and is indicative of the two-dimensional nature of the mode functions. The vector $\mathbf{v}$ is the outward-directed normal to s in the plane of the cross section. For unbounded cross sections, as in Fig. 1·15a, the boundary conditions on Φ_i and Ψ_i are replaced by periodicity requirements.

The explicit dependence of the mode functions $\mathbf{e}'_i$ and $\mathbf{e}''_i$ on the cross-sectional coordinates θ, ϕ will be given in Sec. 2·8. At this point we shall state only that the **e** functions possess the same vector orthogonality properties as in Eq. (5) for the uniform case. In the spherical case the domain of integration is the entire spherical transverse cross section having the angular surface element $dS = \sin \theta \, d\theta \, d\phi$. The representation of the transverse field in terms of the above set of vector modes is given by

$$\left.\begin{aligned}\mathbf{E}_t(r,\theta,\phi) &= \sum_i V'_i(r) \frac{\mathbf{e}'_i}{r} + \sum_i V''_i(r) \frac{\mathbf{e}''_i}{r}, \\ \mathbf{H}_t(r,\theta,\phi) &= \sum_i I'_i(r) \frac{\mathbf{h}'_i}{r} + \sum_i I''_i(r) \frac{\mathbf{h}''_i}{r},\end{aligned}\right\} \quad (91)$$

and that of the longitudinal fields follows from Eqs. (86b) to (91) as

$$\left.\begin{aligned}jk\eta E_r &= \sum_i \frac{k'^2_{ci}}{r^2} I'_i \Phi_i, \\ jk\zeta H_r &= \sum_i \frac{k''^2_{ci}}{r^2} V''_i \Psi_i.\end{aligned}\right\} \quad (92)$$

The mode amplitudes V_i and I_i are obtained from the orthogonality properties of the mode functions as

$$\left.\begin{aligned}V_i &= \iint r\mathbf{E}_t \cdot \mathbf{e}_i \, dS, \\ I_i &= \iint r\mathbf{H}_t \cdot \mathbf{h}_i \, dS,\end{aligned}\right\} \quad (93)$$

where, since the amplitude relations apply to both mode types, the mode superscript is omitted.

The substitution of Eqs. (91) into (86a) and use of Eqs. (5) lead to the defining equations (omitting the mode sub- and superscripts)

$$\left.\begin{aligned}\frac{dV}{dr} &= -j\kappa Z I, \\ \frac{dI}{dr} &= -j\kappa Y V,\end{aligned}\right\} \quad (94)$$

which determine the variation with r of the as yet unknown amplitudes V and I. As before, these equations are of transmission-line form and constitute the basis for the designation of V and I as the mode voltage and current. The propagation wave number κ and characteristic impedance Z of the ith mode are given by

NONUNIFORM SPHERICAL WAVEGUIDES

$$\kappa = \sqrt{k^2 - \frac{k_c^2}{r^2}} \quad \text{where } k_c^2 = n(n+1).$$

$$\left.\begin{aligned} Z = \frac{1}{Y} = \zeta\frac{k}{\kappa} & \quad \text{for } H\text{-modes,} \\ Z = \frac{1}{Y} = \zeta\frac{\kappa}{k} & \quad \text{for } E\text{-modes,} \end{aligned}\right\} \tag{95}$$

where the numerical value of n is determined by the cross-sectional shape and the θ dependence of the mode in question.

Spherical Transmission Lines.—The frequency and excitation of a spherical waveguide of the type illustrated in Fig. 1·15 may be such that the fields therein are almost everywhere characterized by only a single mode. For such a mode the electric and magnetic fields transverse to the transmission direction r may be represented as

$$\mathbf{E}_t(r,\theta,\phi) = V(r)\frac{\mathbf{e}(\theta,\phi)}{r},$$

$$\mathbf{H}_t(r,\theta,\phi) = I(r)\frac{\mathbf{h}(\theta,\phi)}{r},$$

where the mode voltage V and current I obey the spherical transmission-line equations (94), and where $\mathbf{e}$ and $\mathbf{h}$ are known orthogonal vector functions characteristic of the cross-sectional form of the mode. The knowledge of the latter functions reduces the problem of field description to that of the determination of the behavior of V and I on a spherical transmission line. Spherical transmission lines are distinguished by the numerical value of n [*cf.* Eqs. (95)], this mode index being indicative of the θ variation of the mode fields. In the waveguide of Fig. 1·15a the dominant mode is a dipole field characterized by $n = 1$, whereas in the waveguide of Fig. 1·15b the field of the dominant mode is angularly symmetric and $n = 0$ (*cf.* Sec. 2·8). Although a transmission-line analysis of these dominant modes can be presented along the lines developed in the preceding sections, a detailed treatment will not be carried out because of the lack of appropriate numerical tables. The transmission-line behavior of typical spherical modes will, however, be sketched.

The wave equations that describe the mode behavior on a spherical transmission line may be obtained for the case of E-modes by elimination of V'_i from Eqs. (94) as

$$\frac{d^2 I'_i}{dr^2} + \left[k^2 - \frac{n(n+1)}{r^2}\right] I'_i = 0 \tag{96a}$$

and for the H-modes by elimination of I''_i as

$$\frac{d^2 V''_i}{dr^2} + \left[k^2 - \frac{n(n+1)}{r^2}\right] V''_i = 0. \tag{96b}$$

The corresponding equations for V_i' and I_i'' are not so simple, and hence these amplitudes are best obtained from I_i' and V_i'' with the aid of Eqs. (94). Solutions to Eqs. (96) may be expressed in terms of the standing waves

$$\hat{J}_n(kr) \quad \text{and} \quad \hat{N}_n(kr), \tag{97a}$$

where

$$\left.\begin{aligned}
\hat{J}_n(x) &= x^{n+1}\left(-\frac{1}{x}\frac{d}{dx}\right)^n\left(\frac{\sin x}{x}\right), \\
\hat{N}_n(x) &= -x^{n+1}\left(-\frac{1}{x}\frac{d}{dx}\right)^n\left(\frac{\cos x}{x}\right), \\
\hat{J}_n(x)\hat{N}_n'(x) &- \hat{N}_n(x)\hat{J}_n'(x) = 1.
\end{aligned}\right\} \tag{97b}$$

The functions $\hat{J}_n(x)$ and $\hat{N}_n(x)$ are closely related to the half-order Bessel functions; typical functions are

$$\left.\begin{aligned}
\hat{J}_0(x) &= \sin x, & \hat{N}_0(x) &= -\cos x, \\
\hat{J}_1(x) &= -\cos x + \frac{\sin x}{x}, & \hat{N}_1(x) &= -\sin x - \frac{\cos x}{x}.
\end{aligned}\right\} \tag{97c}$$

In terms of these solutions the current and voltage of an E-mode at any point r of a spherical line follow from Eqs. (94) as

$$\left.\begin{aligned}
I(r) &= I(r_0)(\hat{J}_n\hat{N}_{n0}' - \hat{N}_n\hat{J}_{n0}') - j\eta V(r_0)(\hat{J}_{n0}\hat{N}_n - \hat{N}_{n0}\hat{J}_n), \\
V(r) &= V(r_0)(\hat{J}_{n0}\hat{N}_n' - \hat{N}_{n0}\hat{J}_n') - j\zeta I(r_0)(\hat{J}_{n0}'\hat{N}_n' - \hat{N}_{n0}'\hat{J}_n'),
\end{aligned}\right\} \tag{98}$$

where $I(r_0)$ and $V(r_0)$ are the corresponding mode current and voltage at any other point r_0 and

$$\hat{J}_n = \hat{J}_n(kr), \qquad \hat{J}_{n0} = \hat{J}_n(kr_0), \qquad \hat{J}_n' = \frac{d}{dx}\hat{J}_n(x)\bigg]_{x=kr}, \quad \cdots.$$

As for the case of the uniform line the voltage-current relations may be schematically represented by a transmission-line diagram similar to that of Fig. 1·2 which indicates the choice of positive directions of V and I (if z is replaced by r). The relations given in Eqs. (98) between the mode voltage and current at two points on a spherical E-line may be rephrased in impedance terms. On introduction of the relative impedance at any point r,

$$Z'(r) = \frac{V(r)}{\zeta I(r)} = \frac{Z(r)}{\zeta},$$

which is counted positive in the direction of increasing r, and division of Eqs. (98), one has

$$Z'(r) = \frac{j + Z'(r_0)\,\mathrm{ct}_s(x,y)\zeta_s(x,y)}{\mathrm{Ct}_s(x,y) + jZ'(r_0)\zeta_s(x,y)}, \tag{99}$$

SEC. 1·8]   NONUNIFORM SPHERICAL WAVEGUIDES   53

where $x = kr$, $y = kr_0$, and

$$\left.\begin{aligned}
\operatorname{ct}_s(x,y) &= \frac{\hat{J}_n(y)\hat{N}'_n(x) - \hat{N}_n(y)\hat{J}'_n(x)}{\hat{J}_n(x)\hat{N}_n(y) - \hat{N}_n(x)\hat{J}_n(y)} = \frac{1}{\operatorname{tn}_s(x,y)}, \\
\operatorname{Ct}_s(x,y) &= \frac{\hat{J}_n(x)\hat{N}'_n(y) - \hat{N}_n(x)\hat{J}'_n(y)}{\hat{J}'_n(x)\hat{N}'_n(y) - \hat{N}'_n(x)\hat{J}'_n(y)} = \frac{1}{\operatorname{Tn}_s(x,y)}, \\
\zeta_s(x,y) &= \frac{\hat{J}_n(x)\hat{N}_n(y) - \hat{N}_n(x)\hat{J}_n(y)}{\hat{J}'_n(x)\hat{N}'_n(y) - \hat{N}'_n(x)\hat{J}'_n(y)} = \zeta_s(y,x), \\
\operatorname{ct}_s(x,y)\, \zeta_s(x,y) &= -\operatorname{Ct}_s(y,x).
\end{aligned}\right\} \quad (99a)$$

The functions ct_s and Ct_s are termed the small and large spherical cotangents, respectively; their inverses, tn_s and Tn_s, are correspondingly called small and large spherical tangents. As in the case of the radial functions of Sec. 1·7 this nomenclature is based on the asymptotic identity of the spherical and trigonometric functions at large kr. Plots of the spherical functions are not as yet available.[1]

Equations (99) apply to any E-mode. For conical regions of the type indicated in Fig. 1·15b the lowest mode is transverse electromagnetic and $n = 0$. For this mode

$$\operatorname{ct}_s(x,y) = \operatorname{Ct}_s(x,y) = \cot(y - x),$$
$$\zeta_s(x,y) = 1,$$

and therefore the transmission-line description reduces to that of a uniform line with a propagation wave number k and characteristic impedance ζ.

From Eqs. (94) and (95) it is evident that the transmission-line description of an H-mode follows from that of an E-mode on the duality replacements of I, V, ζ of the latter by V, I, η of the former. Consequently the relation between the relative admittances at two points on an H-mode spherical line is given by Eq. (99) with $Z'(r)$ replaced by $Y'(r)$. In the case of H-modes n is always unequal to zero.

The scattering description of the nth mode on a spherical transmission line is based on the spherical Hankel function solutions

$$\begin{aligned}
\hat{H}_n^{(1)}(kr) &= \hat{J}_n(kr) + j\hat{N}_n(kr), \\
\hat{H}_n^{(2)}(kr) &= \hat{J}_n(kr) - j\hat{N}_n(kr)
\end{aligned} \quad (100)$$

of the spherical wave equations (96). The former solution represents an ingoing and the latter an outgoing traveling wave. For the case of an E-mode the solution of Eq. (96a) for the mode current may be written

$$I(r) = I_{\text{inc}}\hat{H}_n^{(2)}(kr) + I_{\text{refl}}\hat{H}_n^{(1)}(kr), \quad (101a)$$

[1] Cf. P. R. Desikachar, "Impedance Relations in Spherical Transmission Lines," Master's thesis, Polytechnic Institute of Brooklyn, (1948).

and hence
$$-j\eta V(r) = I_{\text{inc}}\hat{H}_n^{(2)\prime}(kr) + I_{\text{refl}}\hat{H}_n^{(1)\prime}(kr), \tag{101b}$$

where the prime denotes the derivative with respect to the argument. The generator, or exciting source, is assumed to be at small r. To emphasize the analogy with exponential functions the amplitude $\hat{h}_n$ and phase $\hat{\eta}_n$ of the spherical Hankel functions are defined by

$$\begin{aligned}\hat{H}_n^{(1)}(x) &= (-j)^{n+1}\hat{h}_n e^{+j\hat{\eta}_n}, \\ \hat{H}_n^{(2)}(x) &= (+j)^{n+1}\hat{h}_n e^{-j\hat{\eta}_n}, \end{aligned} \tag{102}$$

where
$$\hat{h}_n = \sqrt{\hat{J}_n^2(x) + \hat{N}_n^2(x)}, \qquad \hat{\eta}_n = (n+1)\frac{\pi}{2} + \tan^{-1}\frac{N_n(x)}{\hat{J}_n(x)}.$$

A current reflection coefficient

$$\Gamma_I(r) = \frac{I_{\text{refl}}}{I_{\text{inc}}}\frac{\hat{H}_n^{(1)}(kr)}{\hat{H}_n^{(2)}(kr)} = (-1)^{n+1}\frac{I_{\text{refl}}}{I_{\text{inc}}}e^{j2\hat{\eta}_n(kr)} \tag{103}$$

can then be employed to characterize the field conditions at the point r. The relation between the current reflection coefficients at the point r and r_0 follows as

$$\Gamma_I(r) = \Gamma_I(r_0)e^{j2[\hat{\eta}_n(kr) - \hat{\eta}_n(kr_0)]}. \tag{104}$$

Hence, with the knowledge of tables[1] of the spherical amplitude and phase functions $\hat{h}_n$ and $\hat{\eta}_n$, a scattering description can be developed in exact analogy with that of the uniform and radial lines (*cf.* Secs. 1·4 and 1·7). In fact for an $n = 0$ E-mode in the conical guide of Fig. 1·15b

$$\hat{h}_0 = 1 \qquad \text{and} \qquad \hat{\eta}_0 = kr,$$

and therefore the two scattering descriptions are identical. A similar scattering description can, of course, be developed for the H-modes on a spherical transmission line.

[1] Tables of the amplitude and phase functions can be found in *ibid*. Also *cf.* Morse, Lowan, Feshbach, and Lax, "Scattering and Radiation from Spheres," OSRD report reprinted by U.S. Navy Dept., Office of Research and Inventions, Washington, D.C., 1946. The amplitude functions in this report are defined somewhat differently from those in Eqs. (102).

CHAPTER 2

TRANSMISSION-LINE MODES

2·1. Mode Characteristics.—As outlined in the preceding chapter the description of the electromagnetic fields within a waveguide can be reformulated in terms of the voltage and current amplitudes of a set of mode functions e_i indicative of the possible transverse field distributions in the waveguide. The resulting transmission-line description, though formally independent of the form of the mode functions, depends quantitatively on the characteristic impedance and propagation constant of the individual modes. In many cases these two fundamental mode characteristics are simply interrelated so that a knowledge of only the mode propagation constant is necessary. For the case of waveguides with walls of finite conductivity the mode propagation constant $\gamma = \alpha + j\beta$ is complex (cf. Sec. 1·6). The attenuation constant α and the wavenumber β depend upon the cross-sectional dimensions, the conductivity, and the excitation wavelength λ of the given waveguide. These mode characteristics must be known explicitly for quantitative transmission-line considerations. Their computation requires a knowledge of the mode field distribution or, equivalently, of the mode function.

In this chapter the explicit form of the electric and magnetic field distribution in the various modes or, equivalently, of the vector mode functions e_i will be presented for several uniform and nonuniform waveguides. The customary engineering assumption of $\exp(j\omega t)$ for the time dependence of the fields, with suppression of the time factor $\exp(j\omega t)$, is adhered to. Electric and magnetic field intensities are expressed as rms quantities. Concomitantly V_i and I_i, the voltage and current amplitudes of the normalized mode functions e_i, are rms quantities. As a consequence the total mode power flow in the transmission direction is given by Re $(V_i I_i^*)$. First-order values of the attenuation constant α and the cutoff wavelength λ_c will be stated for modes in several types of waveguide. In addition maximum electric-field intensities, power expressions, etc., will be indicated in several cases. Although the principal concern of this chapter is to provide quantitative data for transmission-line computations in waveguides of different cross sections, the presentation includes the requisite mode information for the theoretical computation of the equivalent circuit parameters of waveguide discontinuities (cf. Sec. 3·5).

Plots of the electric- and magnetic-field distributions are desirable as an aid to the visualization of the field distributions in the various modes. Mode patterns of two types are useful in this connection. The one type indicates the electric- and magnetic-field strengths on transverse and longitudinal planes within a waveguide; the other shows the magnetic-field intensity or, equivalently, the current density on the inner surface of the guide. The former of these patterns readily furnishes qualitative information as to the location of points of maximum field strength, power flow, etc. The latter yields information on current flow and is desirable in connection with questions of dissipation and of coupling by apertures in the guide walls.

Since a number of mode patterns will be presented in this chapter, it is desirable to say a few words as to the construction of the patterns. Each pattern depicts the instantaneous field distribution in a traveling waveguide mode. For a given mode it is desirable to indicate quantitatively the intensity of the electric- and magnetic-field distribution on a specified plane. Such information can be portrayed in the usual type of flux plots only if the field lines are divergenceless on the given plane, the intensity of the field being then indicated by the density of the field lines. Mode fields generally are not divergenceless in a given viewing plane since field lines generally leave and enter the plane. As a result many of the mode patterns are not true flux plots and hence do not indicate the field intensity everywhere. Nevertheless, wherever possible the density of the field lines has been drawn so as to represent the field intensity. For example, this convention has been adhered to in regions where no lines enter or leave the viewing plane. These regions are generally apparent from a comparison of the field distribution in the various sectional views.

The mode patterns are drawn so that the relative scale of different views is correct as is also the direction of the field lines. The following conventions have been adhered to:

1. Electric field lines are solid.
2. Magnetic field lines are short dashes.
3. Lines of electric current flow are long dashes.

Lines of zero intensity have generally been omitted from the mode patterns for the sake of clarity. The location of these omissions should be apparent and taken into account to preserve the flux plot.

2·2. Rectangular Waveguides. *a. E-modes.*—A uniform waveguide of rectangular cross section is described by the cartesian coordinate system xyz shown in Fig. 2·1. Transmission is along the z direction.

In a rectangular waveguide of inner dimensions a and b, the E-mode functions $\mathbf{e}'_i$ normalized over the cross section in the sense of Eqs. (1·5) are derivable from the scalar functions

$$\Phi_i = \frac{2}{\pi}\frac{1}{\sqrt{m^2\frac{b}{a}+n^2\frac{a}{b}}} \sin\frac{m\pi}{a}x \sin\frac{n\pi}{b}y, \qquad (1)$$

where $m, n = 1, 2, 3, \cdots$. By Eqs. (1·3), (1·10), and (1) the field distribution of the E_{mn}-mode is given as

$$\left.\begin{aligned}
E_x &= -\frac{2V'_i}{a}\frac{m}{\sqrt{m^2\frac{b}{a}+n^2\frac{a}{b}}} \cos\frac{m\pi}{a}x \sin\frac{n\pi}{b}y, \\
E_y &= -\frac{2V'_i}{b}\frac{n}{\sqrt{m^2\frac{b}{a}+n^2\frac{a}{b}}} \sin\frac{m\pi}{a}x \cos\frac{n\pi}{b}y, \\
E_z &= -j\frac{\zeta I'_i \lambda}{ab}\sqrt{m^2\frac{b}{a}+n^2\frac{a}{b}} \sin\frac{m\pi}{a}x \sin\frac{n\pi}{b}y,
\end{aligned}\right\} \quad (2a)$$

$$\left.\begin{aligned}
H_x &= \frac{2I'_i}{b}\frac{n}{\sqrt{m^2\frac{b}{a}+n^2\frac{a}{b}}} \sin\frac{m\pi}{a}x \cos\frac{n\pi}{b}y, \\
H_y &= -\frac{2I'_i}{a}\frac{m}{\sqrt{m^2\frac{b}{a}+n^2\frac{a}{b}}} \cos\frac{m\pi}{a}x \sin\frac{n\pi}{b}y, \\
H_z &= 0,
\end{aligned}\right\} \quad (2b)$$

where the field variation along the z direction is determined by the transmission-line behavior of the mode voltage $V'_i(z)$ and current $I'_i(z)$ [cf. Eq. (1·11)]. The components of the orthonormal vector $\mathbf{e}'_i$ are given directly by E_x and E_y on omission of the amplitude factor V'_i.

The cutoff wavelength for the E_{mn}-mode is

$$\lambda'_{ci} = \frac{2\sqrt{ab}}{\sqrt{m^2\frac{b}{a}+n^2\frac{a}{b}}}. \qquad (3)$$

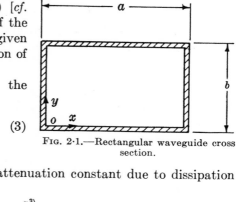

FIG. 2·1.—Rectangular waveguide cross section.

From Eq. (1·50) the E_{mn}-mode attenuation constant due to dissipation in the guide walls is

$$\alpha = \frac{2\Re}{\zeta a}\left(\frac{m^2+n^2\frac{a^3}{b^3}}{m^2+n^2\frac{a^2}{b^2}}\right)\frac{1}{\sqrt{1-\left(\frac{\lambda}{\lambda'_{ci}}\right)^2}}, \qquad (4)$$

where the characteristic resistance $\mathcal{R}$ is a measure of the conductivity properties of the metal walls (*cf.* Table 1·2). The attenuation constant is a minimum at the wavelength $\lambda = 0.577\lambda'_{ci}$.

The maximum value of the *transverse* rms electric-field intensity of the E_{mn}-mode is

$$|E_{\max}| = \frac{2|V'_i|}{\sqrt{ab}} \frac{1}{\dfrac{mb}{na} + \dfrac{na}{mb}} \tag{5a}$$

and occurs at values of x and y for which

$$\tan \frac{m\pi}{a} x = \pm \frac{mb}{na}, \qquad \tan \frac{n\pi}{b} y = \pm \frac{na}{mb}. \tag{5b}$$

Since an E-mode possesses a longitudinal component of electric field, the maximum of the *total* electric field is dependent on the impedance conditions in the guide.

For a matched nondissipative guide the total average power flow along the positive z direction is given by

$$P_i = \mathrm{Re}\,(V'_i I'^*_i) = \frac{\eta}{\sqrt{1 - \left(\dfrac{\lambda}{\lambda'_{ci}}\right)^2}} |V'_i|^2. \tag{6}$$

In Fig. 2·2 are portrayed the field distributions of the E_{11}, E_{21}, and E_{22}-modes in a rectangular guide of dimensions $a/b = 2.25$ and excitation such that $\lambda_g/a = 1.4$. The mode patterns on the left-hand side of the figure depict the electric and magnetic lines within transverse and longitudinal sections of the guide. The right-hand patterns show the magnetic field and current lines on the inner surfaces at the top and side of the guide.

b. H-modes.—The H-mode functions $\mathbf{e}''_i$, normalized over the cross-section, are derivable from scalar functions

$$\Psi_i = \frac{\sqrt{\epsilon_m \epsilon_n}}{\pi} \frac{1}{\sqrt{m^2 \dfrac{b}{a} + n^2 \dfrac{a}{b}}} \cos \frac{m\pi}{a} x \cos \frac{n\pi}{b} y, \tag{7}$$

where

$$m, n = 0, 1, 2, 3 \cdots, \qquad \text{mode } m = n = 0 \text{ excluded,}$$
$$\epsilon_m = 1 \quad \text{if } m = 0,$$
$$\epsilon_m = 2 \quad \text{if } m \neq 0.$$

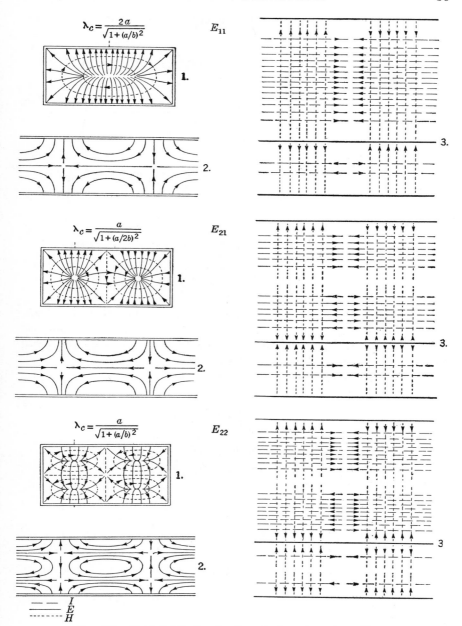

Fig. 2·2.—Field distribution for E-modes in rectangular guides.
1. Cross-sectional view
2. Longitudinal view
3. Surface view

The field components of the H_{mn}-mode follow from Eqs. (1·4), (1·10), and (7) as

$$E_x = V_i'' \frac{\sqrt{\epsilon_m \epsilon_n}}{b} \frac{n}{\sqrt{m^2 \frac{b}{a} + n^2 \frac{a}{b}}} \cos \frac{m\pi}{a} x \sin \frac{n\pi}{b} y,$$

$$E_y = -V_i'' \frac{\sqrt{\epsilon_m \epsilon_n}}{a} \frac{m}{\sqrt{m^2 \frac{b}{a} + n^2 \frac{a}{b}}} \sin \frac{m\pi}{a} x \cos \frac{n\pi}{b} y, \qquad (8a)$$

$$E_z = 0,$$

$$H_x = I_i'' \frac{\sqrt{\epsilon_m \epsilon_n}}{a} \frac{m}{\sqrt{m^2 \frac{b}{a} + n^2 \frac{a}{b}}} \sin \frac{m\pi}{a} x \cos \frac{n\pi}{b} y,$$

$$H_y = I_i'' \frac{\sqrt{\epsilon_m \epsilon_n}}{b} \frac{n}{\sqrt{m^2 \frac{b}{a} + n^2 \frac{a}{b}}} \cos \frac{m\pi}{a} x \sin \frac{n\pi}{b} y, \qquad (8b)$$

$$H_z = -j\eta V_i'' \frac{\lambda \sqrt{\epsilon_m \epsilon_n}}{2ab} \sqrt{m^2 \frac{b}{a} + n^2 \frac{a}{b}} \cos \frac{m\pi}{a} x \cos \frac{n\pi}{b} y.$$

The z dependence of the field components is determined by the transmission-line behavior of $V_i''(z)$ and $I_i''(z)$ [*cf.* Eq. (1·11)]. On omission of the amplitude factor V_i'' the components of the orthonormal vector function $\mathbf{e}_i''$ are given directly by E_x and E_y.

The cutoff wavelength of the H_{mn}-mode is

$$\lambda_{ci}'' = \frac{2\sqrt{ab}}{\sqrt{m^2 \frac{b}{a} + n^2 \frac{a}{b}}} \qquad (9)$$

and is exactly the same as for the E_{mn}-mode.

The attenuation constant for a propagating H_{mn}-mode due to dissipation in the guide walls is

$$\alpha = \frac{\mathfrak{R}}{\zeta b} \left[\frac{\epsilon_n m^2 \frac{b}{a} + \epsilon_m n^2}{m^2 \frac{b}{a} + n^2 \frac{a}{b}} \sqrt{1 - \left(\frac{\lambda}{\lambda_{ci}''}\right)^2} + \frac{\left(\epsilon_n + \epsilon_m \frac{b}{a}\right)\left(\frac{\lambda}{\lambda_{ci}''}\right)^2}{\sqrt{1 - \left(\frac{\lambda}{\lambda_{ci}''}\right)^2}} \right], \qquad (10)$$

where the characteristic resistance $\mathfrak{R}$ (*cf.* Table 1·2) is a measure of the conductivity of the metal walls.

The maximum electric-field intensity in a H_{mn}-mode is for $m \neq 0$, $n \neq 0$

$$|E_{\max}| = \frac{2|V_i''|}{\sqrt{ab}} \frac{1}{\frac{mb}{na} + \frac{na}{mb}} \qquad (11a)$$

and occurs at values of x and y for which

$$\tan \frac{m\pi}{a}x = \pm \frac{na}{mb}, \qquad \tan \frac{n\pi}{b}y = \pm \frac{mb}{na}.$$

For $m \neq 0$, $n = 0$, the maximum electric field is

$$|E_{\max}| = \sqrt{\frac{2}{ab}}\, |V_i''| \tag{11b}$$

and occurs at integral multiples of $x = a/2m$. For $m = 0$, $n \neq 0$ the maximum field has the same magnitude as in Eq. (11b) but occurs at integral multiples of $y = b/2n$.

In terms of the rms mode voltage V_i'' the total power carried by a traveling H_{mn}-wave in a nondissipative guide is given by

$$P_i = \eta \sqrt{1 - \left(\frac{\lambda}{\lambda_{ci}''}\right)^2}\, |V_i''|^2. \tag{12}$$

The H_{10}-mode is the dominant mode in rectangular guide and hence will be considered in some detail. Instead of the voltage V_{10}'' and current I_{10}'' employed in Eqs. (8) a voltage V and current I more closely related to low-frequency definitions can be introduced by the transformations

$$V_{10}'' = \sqrt{\frac{a}{2b}}\, V, \qquad I_{10}'' = \sqrt{\frac{2b}{a}}\, I.$$

In terms of V and I the nonvanishing field components of the H_{10}-mode are

$$\left.\begin{aligned} E_y &= -\frac{V}{b} \sin \frac{\pi}{a}x, \\ H_x &= \frac{2I}{a} \sin \frac{\pi}{a}x, \\ H_z &= -j\eta \frac{\lambda}{2a} \frac{V}{b} \cos \frac{\pi}{a}x. \end{aligned}\right\} \tag{13}$$

For guide walls of finite conductivity the attenuation constant of the H_{10}-mode is

$$\alpha = \frac{\mathcal{R}}{\zeta b} \left[\frac{1 + \frac{2b}{a}\left(\frac{\lambda}{2a}\right)^2}{\sqrt{1 - \left(\frac{\lambda}{2a}\right)^2}}\right] \tag{14a}$$

and the guide wavelength is

$$\lambda_g = \frac{\lambda}{\sqrt{1 - \left(\dfrac{\lambda}{2a}\right)^2}} \tag{14b}$$

to a first order. The maximum electric-field intensity occurs at the center of the guide and has the magnitude $|V|/b$. The rms voltage V in a traveling wave is related to the total power flow by

$$P = \frac{|V|^2}{\zeta \dfrac{\lambda_g}{\lambda} \dfrac{2b}{a}}$$

for a nondissipative guide.

Mode patterns for traveling H_{10}, H_{11}, H_{21}-modes in rectangular guides are displayed in Fig. 2·3. As in the case of Fig. 2·2, $a/b = 2.25$ and $\lambda_g/a = 1.4$. The mode patterns on the left indicate the field distribution on cross-sectional and longitudinal planes; those on the right depict the electric-current distribution on the inner surface of the guide.

c. Modes in a Parallel Plate Guide. The modes in a parallel plate guide of height b may be regarded as appropriate limiting forms of modes in a rectangular guide of height b as the width a of the latter becomes infinite. As noted in Secs. 2·2a and b, the modes in a rectangular guide of height b and width a form a discrete set. However, as the width of the rectangular guide becomes infinite the corresponding set of modes assumes both a discrete and continuous character; the mode index n characteristic of the mode variation along the finite y dimension is discrete, whereas the index m characteristic of the variation along the infinite x dimension becomes continuous. The complete representation of a general field in a parallel plate guide requires both the discrete and continuous modes. For simplicity, we shall consider only those discrete modes required for the representation of fields having no variation in the x direction. For the representation of more general fields, reference should be made to Sec. 2·6 wherein a typical representation in terms of continuous modes is presented.

The discrete E-modes in a parallel plate guide of height b are derivable from the scalar functions

$$\Phi_i = \sqrt{\frac{\epsilon_n}{b}} \frac{\sin \dfrac{n\pi y}{b}}{\dfrac{n\pi}{b}}, \qquad 0 < y < b$$

$$n = 0, 1, 2, 3, \cdots.$$
$$\epsilon_n = 1 \quad \text{if } n = 0,$$
$$\epsilon_n = 2 \quad \text{if } n \neq 0,$$

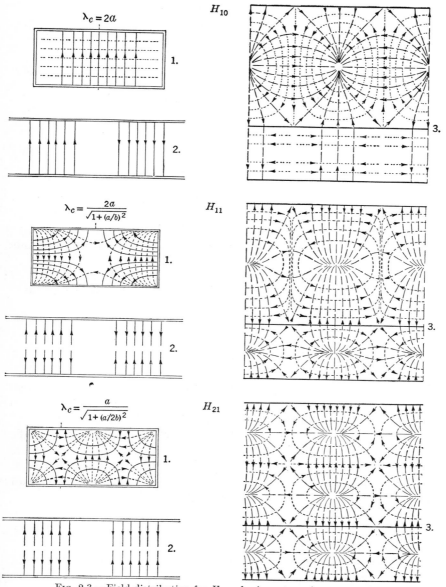

FIG. 2·3.—Field distribution for H-modes in rectangular waveguide.
1. Cross-sectional view
2. Longitudinal view
3. Surface view

Hence, by Eqs. (1·3) and (1·10) the field components of the E_{0n}-mode follow as

$$\left.\begin{aligned} E_x &= 0, \\ E_y &= -V'_i \sqrt{\frac{\epsilon_n}{b}} \cos \frac{n\pi y}{b}, \\ E_z &= -j\zeta I'_i \frac{n\lambda}{2b} \sqrt{\frac{\epsilon_n}{b}} \sin \frac{n\pi y}{b}, \end{aligned}\right\} \quad (15a)$$

$$\left.\begin{aligned} H_x &= I'_i \sqrt{\frac{\epsilon_n}{b}} \cos \frac{n\pi y}{b}, \\ H_y &= 0, \\ H_z &= 0, \end{aligned}\right\} \quad (15b)$$

where the z dependence of the mode fields is determined by the transmission-line behavior of the mode voltage V'_i and current I'_i given in Eqs. (1·11). The components of the orthonormal (in the y dimension only) vector $\mathbf{e}'_i$ are obtained from E_x and E_y on omission of the mode amplitude V'_i.

The cutoff wavelength of the E_{0n}-mode in parallel plate guide is $\lambda'_{ci} = 2b/n$. The attenuation constant of the E_{0n}-mode caused by finite conductivity of the parallel plates is

$$\alpha = \frac{\epsilon_n}{b} \frac{\mathcal{R}}{\zeta} \frac{1}{\sqrt{1 - \left(\frac{\lambda}{\lambda'_{ci}}\right)^2}}, \quad (15c)$$

where the dependence of the characteristic resistance $\mathcal{R}$ on the conductivity of the plates may be obtained from Table 1·2. It should be noted that these results are special cases of Eqs. (3) and (4) with $m = 0$.

The E_{00}-mode is the principal, or *TEM*, mode in a parallel plate guide. The principal mode is characterized by an infinite cutoff wavelength λ_c and hence by a guide wavelength λ_g identical with the space wavelength λ. Instead of the voltage V'_{00} and current I'_{00}, more customary definitions for the principal mode are obtained on use of the substitutions

$$V'_{00} = \frac{V}{\sqrt{b}} \quad \text{and} \quad I'_{00} = \sqrt{b}\, I.$$

In terms of the new mode amplitudes V and I the nonvanishing field components of the principal mode are

$$E_y = -\frac{V}{b},$$
$$H_x = I,$$

as derivable from a scalar function y/b by means of Eq. (1·3). It is evident that V represents the voltage between plates and I the current flowing per unit width of each plate in the conventional manner. The attenuation constant, descriptive of dissipation in the plates, for the principal mode reduces to

$$\alpha = \frac{1}{b}\frac{\Re}{\zeta}.$$

The discrete H-modes in a parallel plate guide of height b are derivable from the scalar mode functions

$$\Psi_i = \sqrt{\frac{2}{b}}\frac{\cos\frac{n\pi y}{b}}{\frac{n\pi}{b}}, \qquad 0 < y < b,$$
$$n = 1, 2, 3, \cdots.$$

Hence, by Eqs. (1·4) and (1·10) the field components of the H_{0n}-mode follow as

$$\left.\begin{array}{l} E_x = V_i''\sqrt{\dfrac{2}{b}}\sin\dfrac{n\pi y}{b}, \\ E_y = 0, \\ E_z = 0, \end{array}\right\} \quad (16a)$$

$$\left.\begin{array}{l} H_x = 0, \\ H_y = I_i''\sqrt{\dfrac{2}{b}}\sin\dfrac{n\pi y}{b}, \\ H_z = -j\eta V_i''\dfrac{n\lambda}{2b}\sqrt{\dfrac{2}{b}}\cos\dfrac{n\pi y}{b}, \end{array}\right\} \quad (16b)$$

where as above the z dependence of the mode fields is determined from the transmission-line behavior of the mode amplitudes V_i'' and I_i''. The components of the orthonormal vector $\mathbf{e}_i''$ are obtained from E_x and E_y on omission of the mode amplitude V_i''.

The cutoff wavelength λ_{ci}'' of the H_{0n}-mode is $2b/n$, the same as for the E_{0n}-mode. The attenuation constant of the H_{0n}-mode due to finite conductivity of the guide plates is

$$\alpha = \frac{2}{b}\frac{\Re}{\zeta}\frac{\left(\dfrac{\lambda}{\lambda_{ci}''}\right)^2}{\sqrt{1 - \left(\dfrac{\lambda}{\lambda_{ci}''}\right)^2}}, \quad (16c)$$

where again the characteristic resistance $\Re$ of the metallic plates may be obtained from Table 1·2. The decrease of the attenuation constant with increasing frequency is to be noted.

2·3. Circular Waveguides. *a. E-modes.*—A uniform waveguide of circular cross section is most conveniently described by a polar coordinate system $r\phi z$ as shown in Fig. 2·4.

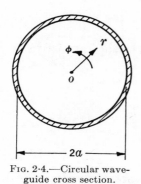

FIG. 2·4.—Circular waveguide cross section.

For a cross section of radius a the E-mode vector functions $\mathbf{e}_i'$ normalized over the cross section in accordance with Eqs. (1·5) are derivable from the scalar functions

$$\Phi_i = \sqrt{\frac{\epsilon_m}{\pi}} \frac{J_m\left(\frac{\chi_i r}{a}\right)}{\chi_i J_{m+1}(\chi_i)} \frac{\cos}{\sin} m\phi, \qquad (17)$$

where

$m = 0, 1, 2, 3, \cdots,$
$\epsilon_m = 1 \quad \text{if } m = 0,$
$\epsilon_m = 2 \quad \text{if } m \neq 0,$

and $\chi_i = \chi_{mn}$, the nth nonvanishing root of the mth-order Bessel function $J_m(\chi)$, is tabulated in Table 2·1 for several values of m and n.

TABLE 2·1.—ROOTS OF $J_m(\chi) = 0$

$$\chi_{mn} = \left(m + 2n - \frac{1}{2}\right)\frac{\pi}{2} - \frac{4m^2 - 1}{4\pi(m + 2n - \frac{1}{2})} - \frac{(4m^2 - 1)(28m^2 - 31)}{48\pi^3(m + 2n - \frac{1}{2})^3} \cdots$$

n \ m	0	1	2	3	4	5	6	7
1	2.405	3.832	5.136	6.380	7.588	8.771	9.936	11.086
2	5.520	7.016	8.417	9.761	11.065	12.339	13.589	14.821
3	8.654	10.173	11.620	13.015	14.372			
4	11.792	13.323	14.796					

On use of Eqs. (1·3), (1·10), and (17) the field components of the E_{mn}-mode become in polar coordinates

$$\left.\begin{aligned} E_r &= -V_i' \sqrt{\frac{\epsilon_m}{\pi}} \frac{J_m'\left(\frac{\chi_i r}{a}\right)}{a J_{m+1}(\chi_i)} \frac{\cos}{\sin} m\phi, \\ E_\phi &= \pm V_i' \sqrt{\frac{\epsilon_m}{\pi}} \frac{m}{\chi_i} \frac{J_m\left(\frac{\chi_i r}{a}\right)}{r J_{m+1}(\chi_i)} \frac{\sin}{\cos} m\phi, \\ E_z &= -j\zeta \frac{\lambda \chi_i}{2\pi a} I_i' \sqrt{\frac{\epsilon_m}{\pi}} \frac{J_m\left(\frac{\chi_i r}{a}\right)}{a J_{m+1}(\chi_i)} \frac{\cos}{\sin} m\phi, \end{aligned}\right\} \qquad (18a)$$

$$H_r = \mp I_i' \sqrt{\frac{\epsilon_m}{\pi}} \frac{m}{\chi_i} \frac{J_m\left(\frac{\chi_i r}{a}\right)}{rJ_{m+1}(\chi_i)} \begin{matrix}\sin\\ \cos\end{matrix} m\phi,$$

$$H_\phi = -I_i' \sqrt{\frac{\epsilon_m}{\pi}} \frac{J_m'\left(\frac{\chi_i r}{a}\right)}{aJ_{m+1}(\chi_i)} \begin{matrix}\cos\\ \sin\end{matrix} m\phi, \quad (18b)$$

$$H_z = 0.$$

The z variation of the mode fields is determined by the transmission-line behavior of V_i' and I_i' given in Eqs. (1·11). As is evident from Eqs. (18) the E_{mn}-mode ($m \neq 0$) is degenerate and consists of two modes with even or odd angular dependence. Though not explicitly shown each of these modes is, of course, characterized by a different mode voltage and current. The polar components of the vector $\mathbf{e}_i'$ are obtained from E_r and E_ϕ on omission of the amplitude V_i'.

The cutoff wavelength of the degenerate E_{mn}-mode is

$$\lambda_{ci}' = \frac{2\pi}{\chi_{mn}} a, \quad (19)$$

where the roots χ_{mn} are given in Table 2·1.

The E-mode attenuation constant due to finite conductivity of the guide walls is

$$\alpha = \frac{\mathcal{R}}{\zeta a} \frac{1}{\sqrt{1 - \left(\frac{\lambda}{\lambda_{ci}'}\right)^2}}, \quad (20)$$

where the frequency-dependent characteristic resistance $\mathcal{R}$ of the metal walls may be obtained from Table 1·2.

In terms of the rms voltage V_i' the total power carried by an E_{mn}-mode in a matched nondissipative guide is

$$P_i = \frac{\eta}{\sqrt{1 - \left(\frac{\lambda}{\lambda_{ci}'}\right)^2}} |V_i'|^2. \quad (21)$$

Mode patterns of the instantaneous field distribution in traveling waves of the E_{01}-, E_{11}-, and E_{21}-modes are shown in Fig. 2·5. The excitation frequency is such that $\lambda_g/a = 4.2$. The left-hand views of Fig. 2·5 depict the electric- and magnetic-field intensities on transverse and longitudinal planes in which the radial electric field is maximum. The right-hand view shows a development of the magnetic-field and electric-current distribution on half the guide circumference.

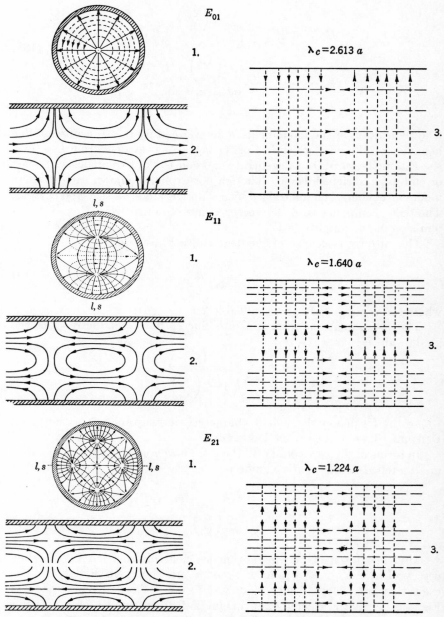

FIG. 2·5.—Field distribution for E-modes in circular waveguide.
1. Cross-sectional view
2. Longitudinal view through plane l-l
3. Surface view from s-s

SEC. 2·3] CIRCULAR WAVEGUIDES 69

b. *H-modes.*—The $\mathbf{e}''_i$ vector mode functions normalized over the circular cross section are derivable from scalar functions

$$\Psi_i = \sqrt{\frac{\epsilon_m}{\pi}} \frac{1}{\sqrt{\chi_i'^2 - m^2}} \frac{J_m\left(\frac{\chi_i' r}{a}\right)}{J_m(\chi_i')} \frac{\cos}{\sin} m\phi, \qquad (22)$$

where

$$m = 0, 1, 2, 3, \cdots$$

and $\chi_i' = \chi_{mn}'$ is the nth nonvanishing root of the derivative of the mth-order Bessel function. Several of the lower-order roots are given in Table 2·2.

TABLE 2·2.—ROOTS OF $J_m'(\chi') = 0$

$$\chi_{mn}' = \left(m + 2n - \frac{3}{2}\right)\frac{\pi}{2} - \frac{4m^2 + 3}{4\pi(m + 2n - \frac{3}{2})} - \frac{112m^4 + 328m^2 - 9}{48\pi^3(m + 2n - \frac{3}{2})^3} \cdots, \quad m > 0$$

n \ m	0	1	2	3	4	5	6	7
1	3.832	1.841	3.054	4.201	5.317	6.416	7.501	8.578
2	7.016	5.331	6.706	8.015	9.282	10.520	11.735	12.932
3	10.173	8.536	9.969	11.346	12.682	13.987		
4	13.324	11.706	13.170					

From Eqs. (1·4), (1·10), and (22) the field components of an H_{mn}-mode are found to be

$$\left.\begin{aligned}
E_r &= \pm V_i'' \sqrt{\frac{\epsilon_m}{\pi}} \frac{m}{\sqrt{\chi_i'^2 - m^2}} \frac{J_m\left(\frac{\chi_i' r}{a}\right)}{rJ_m(\chi_i')} \frac{\sin}{\cos} m\phi, \\
E_\phi &= V_i'' \sqrt{\frac{\epsilon_m}{\pi}} \frac{\chi_i'}{\sqrt{\chi_i'^2 - m^2}} \frac{J_m'\left(\frac{\chi_i' r}{a}\right)}{aJ_m(\chi_i')} \frac{\cos}{\sin} m\phi, \\
E_z &= 0,
\end{aligned}\right\} \quad (23a)$$

$$\left.\begin{aligned}
H_r &= -I_i'' \sqrt{\frac{\epsilon_m}{\pi}} \frac{\chi_i'}{\sqrt{\chi_i'^2 - m^2}} \frac{J_m'\left(\frac{\chi_i' r}{a}\right)}{aJ_m(\chi_i')} \frac{\cos}{\sin} m\phi, \\
H_\phi &= \pm I_i'' \sqrt{\frac{\epsilon_m}{\pi}} \frac{m}{\sqrt{\chi_i'^2 - m^2}} \frac{J_m\left(\frac{\chi_i' r}{a}\right)}{rJ_m(\chi_i')} \frac{\sin}{\cos} m\phi, \\
H_z &= -j\eta \frac{\lambda \chi_i'}{2\pi a} V_i'' \sqrt{\frac{\epsilon_m}{\pi}} \frac{\chi_i'}{\sqrt{\chi_i'^2 - m^2}} \frac{J_m\left(\frac{\chi_i' r}{a}\right)}{aJ_m(\chi_i')} \frac{\cos}{\sin} m\phi.
\end{aligned}\right\} \quad (23b)$$

As evident from Eqs. (23) the H_{mn}-mode ($m \neq 0$) is degenerate; i.e., there exist two H_{mn}-modes, one of odd and one of even angular dependence. The polar components of the normalized mode vectors $\mathbf{e}_i''$ are obtained from E_r and E_ϕ on omission of the z dependent voltage amplitudes V_i''.

The cutoff wavelength of the degenerate H_{mn}-mode is

$$\lambda_{ci}'' = \frac{2\pi}{\chi_i'} a, \tag{24}$$

the roots χ_i' being given in Table 2·2. Except for the degenerate case $m = 0$ the cutoff wavelengths differ from those of the E_{mn}-modes.

The H_{mn} attenuation constant due to dissipation in the guide walls is

$$\alpha = \frac{\Re}{\zeta a} \left[\frac{m^2}{\chi_i'^2 - m^2} + \left(\frac{\lambda}{\lambda_{ci}''}\right)^2 \right] \frac{1}{\sqrt{1 - \left(\frac{\lambda}{\lambda_{ci}''}\right)^2}}. \tag{25}$$

The frequency-dependent characteristic resistance $\Re$ of the metal walls may be obtained from Table 1·2.

The total average power carried by a traveling H_{mn}-mode in a matched nondissipative guide is expressed in terms of the mode voltage V_i'' as

$$P_i = \eta \sqrt{1 - \left(\frac{\lambda}{\lambda_{ci}''}\right)^2} |V_i''|^2. \tag{26}$$

The dominant mode in circular guide is the H_{11}. In a nondissipative guide the wavelength of propagation of the H_{11}-mode is

$$\lambda_g = \frac{\lambda}{\sqrt{1 - \left(\frac{\lambda}{3.41a}\right)^2}}. \tag{27}$$

The maximum rms electric-field intensity of the H_{11}-mode occurs at the axis of the guide and has a magnitude

$$E_{\max} = \frac{|V_{11}''|}{1.50a}. \tag{28}$$

In terms of $E_{\max}$ the maximum average power carried by the H_{11}-mode is

$$P_{\max} = 3.97 \times 10^{-3} \sqrt{1 - \left(\frac{\lambda}{3.41a}\right)^2} a^2 E_{\max}^2, \tag{29}$$

where all units are MKS.

Mode patterns of the instantaneous field distribution in traveling H_{01}, H_{11}, and H_{21}-modes are shown in Fig. 2·6 for $\lambda_g/a = 4.2$. The transverse and longitudinal views are in the plane of the maximum

CIRCULAR WAVEGUIDES

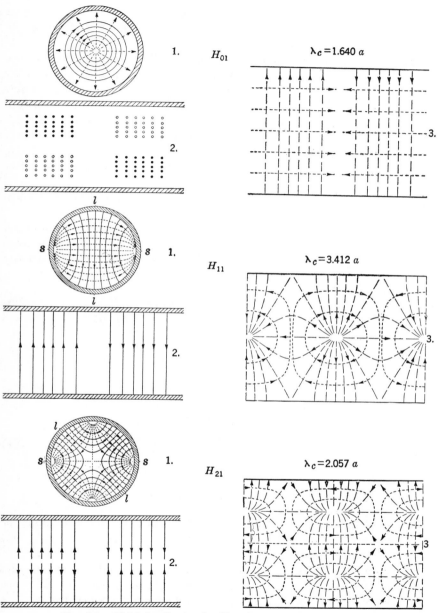

FIG. 2·6.—Field distribution for H-modes in circular waveguide.
1. Cross-sectional view
2. Longitudinal view through plane l-l
3. Surface view from s-s

radial electric field. The views on the right-hand side of Fig. 2·6 depict the magnetic-field and electric-current distribution on half the guide circumference.

2·4. Coaxial Waveguides. *a. E-modes.*—The description of the uniform coaxial guide depicted in Fig. 2·7 is closely related to that of the circular guide considered in the previous section. The coaxial guide is described by a polar coordinate $r\phi z$ system in which the outer and inner conductors are at radii a and b, respectively, and the transmission direction is along the z-axis.

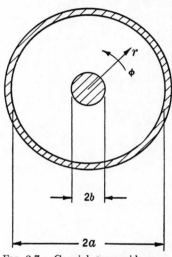

FIG. 2·7.—Coaxial waveguide cross section.

The vector functions $\mathbf{e}_i'$ normalized in accordance with Eqs. (1·5) and characteristic of the E-mode fields are derivable [*cf.* Eqs. (1·3)] as gradients of scalar functions. The scalar function appropriate to the lowest E-mode in a coaxial guide is

$$\Phi_{00} = \frac{\ln r}{\sqrt{2\pi \ln \frac{a}{b}}}. \tag{30}$$

Hence by Eqs. (1·10), the field components of this mode are

$$\left. \begin{array}{l} E_r = V_{00}' \dfrac{1}{\sqrt{2\pi \ln \dfrac{a}{b}}} \dfrac{1}{r}, \\[2ex] H_\phi = I_{00}' \dfrac{1}{\sqrt{2\pi \ln \dfrac{a}{b}}} \dfrac{1}{r}, \\[2ex] E_\phi = E_z = H_r = H_z = 0. \end{array} \right\} \tag{31}$$

This transverse electromagnetic or *TEM*-mode is the dominant, or principal, mode in coaxial guide. Its cutoff wavelength is infinite, and hence $\lambda_g = \lambda$ for the dominant mode.

A more customary definition for the voltage and current of the principal coaxial mode is obtained by multiplication and division, respectively, of the normalized voltage and current in Eq. (31) by

$$N^{\frac{1}{2}} = \sqrt{\frac{\ln \dfrac{a}{b}}{2\pi}}.$$

In terms of the new voltage V and current I the nonvanishing components of the dominant mode become

$$E_r = \frac{V}{r \ln \dfrac{a}{b}},$$

$$H_\phi = \frac{I}{2\pi r}. \tag{32}$$

The maximum electric-field intensity in the principal mode occurs at the surface of the inner conductor and is of magnitude

$$|E_{\max}| = \frac{|V|}{b \ln \dfrac{a}{b}}. \tag{33}$$

For constant outer radius and voltage the maximum electric-field intensity is a minimum when $a/b = 2.72$. On a matched coaxial line the rms voltage V is related to the total average power flow P by

$$P = \frac{2\pi}{\ln \dfrac{a}{b}} \eta |V|^2. \tag{34}$$

For constant outer radius and $|E_{\max}|$, the power flow is a maximum when $a/b = 1.65$. The attenuation constant of the dominant mode due to dissipation in the inner and outer conductors is

$$\alpha = \left(\frac{\Re_a}{a} + \frac{\Re_b}{b}\right) \frac{1}{2\zeta \ln \dfrac{a}{b}}, \tag{35}$$

where $\Re_a$ and $\Re_b$ (cf. Table 1·2) are the characteristic resistances of the metals of which the outer and inner conductors are constituted. For fixed outer radius and wavelength the attenuation constant is a minimum when $a/b = 3.6$ provided $\Re_a = \Re_b$.

The normalized $\mathbf{e}'_i$ vector functions characteristic of the higher E-modes are derivable from the scalar functions

$$\Phi_i = Z_m\left(\chi_i \frac{r}{b}\right) \frac{\cos}{\sin} m\phi, \tag{36}$$

where

$$Z_m\left(\chi_i \frac{r}{b}\right) = \frac{\sqrt{\pi \epsilon_m}}{2} \frac{J_m\left(\chi_i \dfrac{r}{b}\right) N_m(\chi_i) - N_m\left(\chi_i \dfrac{r}{b}\right) J_m(\chi_i)}{\left[\dfrac{J_m^2(\chi_i)}{J_m^2(c\chi_i)} - 1\right]^{\frac{1}{2}}},$$

$$m = 0, 1, 2, 3, \cdots,$$
$$\epsilon_m = 1 \quad \text{if } m = 0,$$
$$\epsilon_m = 2 \quad \text{if } m \neq 0.$$

The quantity $\chi_i = \chi_{mn}$ is the nth nonvanishing root of the Bessel-Neumann combination $Z_m(c\chi_i)$, where $c = a/b$. The quantities $(c - 1)\chi_i$ are tabulated in Table 2·3 as a function of the ratio c for several values of m and n.

TABLE 2·3.—ROOTS OF $J_m(c\chi)N_m(\chi) - N_m(c\chi)J_m(\chi) = 0$
Tabulated in the form $(c - 1)\chi_{mn}$, $n > 0$

c \ mn	01	11	21	31	02	12	22	32
1.0	3.142	3.142	3.142	3.142	6.283	6.283	6.283	6.283
1.1	3.141	3.143	3.147	3.154	6.283	6.284	6.286	6.289
1.2	3.140	3.146	3.161	3.187	6.282	6.285	6.293	6.306
1.3	3.139	3.150	3.182	3.236	6.282	6.287	6.304	6.331
1.4	3.137	3.155	3.208	3.294	6.281	6.290	6.317	6.362
1.5	3.135	3.161	3.237	3.36	6.280	6.293	6.332	6.397
1.6	3.133	3.168	3.27	3.43	6.279	6.296	6.349	6.437
1.8	3.128	3.182	3.36	3.6	6.276	6.304	6.387	6.523
2.0	3.123	3.197	3.4	3.7	6.273	6.312	6.43	6.62
2.5	3.110	3.235	.....	.....	6.266	6.335	.....	6.9
3.0	3.097	3.271	.....	.....	6.258	6.357		
3.5	3.085	3.305	.....	.....	6.250	6.381		
4.0	3.073	3.336	.....	.....	6.243	6.403		

c \ mn	03	13	23	33	04	14	24	34
1.0	9.425	9.425	9.425	9.425	12.566	12.566	12.566	12.566
1.1	9.425	9.425	9.427	9.429	12.566	12.567	12.568	12.569
1.2	9.424	9.426	9.431	9.440	12.566	12.567	12.571	12.578
1.3	9.424	9.427	9.438	9.457	12.566	12.568	12.577	12.590
1.4	9.423	9.429	9.447	9.478	12.565	12.570	12.583	12.606
1.5	9.423	9.431	9.458	9.502	12.565	12.571	12.591	12.624
1.6	9.422	9.434	9.469	9.528	12.564	12.573	12.600	12.644
1.8	9.420	9.439	9.495	9.587	12.563	12.577	12.619	12.689
2.0	9.418	9.444	9.523	9.652	12.561	12.581	12.640	12.738
2.5	9.413	9.460	.....	9.83	12.558	12.593	.....	12.874
3.0	9.408	9.476	.....	10.0	12.553	12.605	.....	13.02
3.5	9.402	9.493	.....	10.2	12.549	12.619	.....	13.2
4.0	9.396	9.509	.....	.....	12.545	12.631	.....	13.3

Cf. H. B. Dwight, "Tables of Roots for Natural Frequencies in Coaxial Cavities," *Jour. Math. Phys.*, **27** (No. 1), 84–89 (1948).

The cutoff wavelength of the E_{mn}-mode may be expressed in terms of the tabulated values $(c - 1)\chi_{mn}$ as

$$\lambda'_{ci} = \frac{2\pi}{(c - 1)\chi_{mn}}(a - b) \approx \frac{2(a - b)}{n}, \qquad n = 1, 2, 3, \cdots . \quad (37)$$

The field components of the E_{mn}-mode follow from Eqs. (1·3), (1·10), and (36) as

$$\left. \begin{array}{l} E_r = -V'_i \dfrac{\chi_i}{b} Z'_m\left(\chi_i \dfrac{r}{b}\right) \begin{array}{l}\cos\\ \sin\end{array} m\phi, \\[2mm] E_\phi = \pm V'_i \dfrac{m}{r} Z_m\left(\chi_i \dfrac{r}{b}\right) \begin{array}{l}\sin\\ \cos\end{array} m\phi, \\[2mm] E_z = -j\zeta \dfrac{\lambda}{\lambda'_{ci}} I'_i \dfrac{\chi_i}{b} Z_m\left(\chi_i \dfrac{r}{b}\right) \begin{array}{l}\cos\\ \sin\end{array} m\phi, \end{array} \right\} \quad (38a)$$

$$\left. \begin{array}{l} H_r = \mp I'_i \dfrac{m}{r} Z_m\left(\chi_i \dfrac{r}{b}\right) \begin{array}{l}\sin\\ \cos\end{array} m\phi, \\[2mm] H_\phi = -I'_i \dfrac{\chi_i}{b} Z'_m\left(\chi_i \dfrac{r}{b}\right) \begin{array}{l}\cos\\ \sin\end{array} m\phi, \\[2mm] H_z = 0. \end{array} \right\} \quad (38b)$$

From the form of Eqs. (38) it is apparent that the E_{mn}-mode ($m > 0$) is degenerate and may have either of two possible polarizations, each distinguished by a different voltage and current amplitude. The polar components of the $\mathbf{e}'_i$ vector are obtained from Eq. (38a) on omission of the z-dependent voltage amplitude V'_i.

The attenuation constant due to finite conductivity of the inner and outer conductor is

$$\alpha = \left[\frac{\dfrac{\mathcal{R}_a}{a} \dfrac{J_m^2(\chi_i)}{J_m^2(c\chi_i)} + \dfrac{\mathcal{R}_b}{b}}{\dfrac{J_m^2(\chi_i)}{J_m^2(c\chi_i)} - 1} \right] \frac{1}{\zeta \sqrt{1 - \left(\dfrac{\lambda}{\lambda'_{ci}}\right)^2}}, \quad (39)$$

where $\mathcal{R}_a$ and $\mathcal{R}_b$ (cf. Table 1·2) are the characteristic resistances of the inner and outer conductors, respectively.

In terms of the rms voltage V'_i the total power transported by a traveling E_{mn}-mode in a matched nondissipative guide is

$$P = \frac{\eta}{\sqrt{1 - \left(\dfrac{\lambda}{\lambda'_{ci}}\right)^2}} |V'_i|^2. \quad (40)$$

The instantaneous field distribution in traveling waves of the E_{01}, E_{11}, E_{21} type are shown in Fig. 2·8. The mode patterns are all drawn for the case $a/b = 3$ and $\lambda_g/a = 4.24$. The left-hand views portray the electric and magnetic field distributions in the transverse and longitudinal planes on which the radial electric field is a maximum. The right-hand patterns show developed views of the magnetic-field and electric-current distribution on half the circumference of the outer conductor.

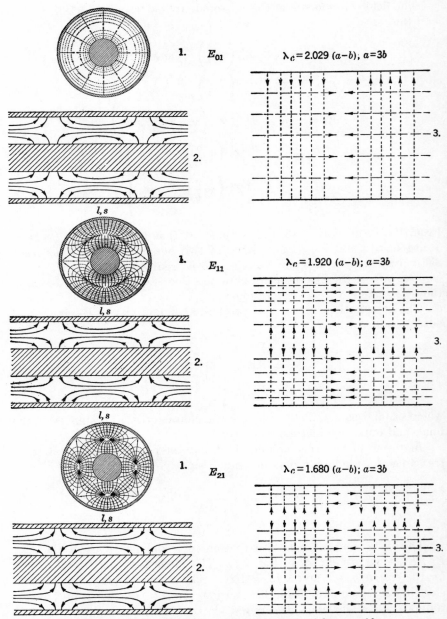

Fig. 2·8.—Field distribution for E-modes in coaxial waveguide.
1. Cross-sectional view
2. Longitudinal view through plane l-l
3. Surface view from s-s

b. *H-modes.*—The vector functions $\mathbf{e}_i''$ normalized in accordance with Eqs. (1·5) are derivable from scalar functions

$$\Psi_i = \mathcal{Z}_m\left(\chi_i' \frac{r}{b}\right) \begin{array}{c}\cos\\ \sin\end{array} m\phi, \tag{41}$$

where

$$\mathcal{Z}_m\left(\chi_i' \frac{r}{b}\right) = \frac{\sqrt{\pi \epsilon_m}}{2} \frac{J_m\left(\chi_i' \frac{r}{b}\right) N_m'(\chi_i') - N_m\left(\chi_i' \frac{r}{b}\right) J_m'(\chi_i')}{\left\{\left[\frac{J_m'(\chi_i')}{J_m'(c\chi_i')}\right]^2 \left[1 - \left(\frac{m}{c\chi_i'}\right)^2\right] - \left[1 - \left(\frac{m}{\chi_i'}\right)^2\right]\right\}^{1/2}},$$

$$m = 0, 1, 2, 3, \cdots,$$

and $\chi_i' = \chi_{mn}'$ is the nth root of the derivative of the Bessel-Neumann combination $\mathcal{Z}_m(c\chi_i')$ with $c = a/b$. For $n = 1$ the quantities $(c + 1)\chi_{m1}'$ are tabulated as a function of the ratio c in Table 2·4; for $n > 1$ the quantities $(c - 1)\chi_{mn}$ are tabulated in Table 2·5.

TABLE 2·4.—First Root of $J_m'(c\chi')N_m'(\chi') - N_m'(c\chi')J_m'(\chi') = 0$
Tabulated in the form $(c + 1)\chi_{m1}'$, $(m > 0)$

c \ $m1$	11	21	31
1.0	2.000	4.000	6.000
1.1	2.001	4.001	6.002
1.2	2.002	4.006	6.008
1.3	2.006	4.011	6.012
1.4	2.009	4.015	6.017
1.5	2.013	4.020	6.018
1.6	2.018	4.025	6.011
1.8	2.024	4.026	5.986
2.0	2.031	4.023	5.937
2.5	2.048	3.980	5.751
3.0	2.056	3.908	5.552
3.5	2.057	3.834	5.382
4.0	2.055	3.760	5.240

In terms of the tabulated values, the cutoff wavelength of an H_{m1}-mode can be expressed as

$$\left.\begin{array}{l}\lambda_{ci}'' = \dfrac{2\pi}{(c + 1)\chi_{m1}'} (a + b),\\[6pt] \lambda_{ci}'' \simeq \dfrac{\pi(a + b)}{m} \quad \text{for } m = 1, 2, 3, \cdots.\end{array}\right\} \tag{42a}$$

and for an H_{mn}-mode as

$$\left.\begin{array}{l}\lambda_{ci}'' = \dfrac{2\pi}{(c - 1)\chi_{mn}'} (a - b),\\[6pt] \lambda_{ci}'' \simeq \dfrac{2(a - b)}{(n - 1)} \quad \text{for } n = 2, 3, 4, \cdots.\end{array}\right\} \tag{42b}$$

It is evident from these equations that the H_{11}-mode is the dominant H-mode in coaxial guide. The cutoff wavelength of the H_{01}-mode is identical with that of the E_{11}-mode, i.e., $\chi'_{01} = \chi_{11}$ and can be obtained from Table 2·3.

TABLE 2·5.—Higher Roots of $J'_m(c\chi')N'_m(\chi') - N'_m(c\chi')J'_m(\chi') = 0$
Tabulated in the form $(c - 1)\chi'_{mn}$, $(n > 1)$

mn \ c	02*	12	22	32	03*	13	23	33
1.0	3.142	3.142	3.142	3.142	6.283	6.283	6.283	6.283
1.1	3.143	3.144	3.148	3.156	6.284	6.284	6.287	6.290
1.2	3.145	3.151	3.167	3.193	6.285	6.288	6.296	6.309
1.3	3.150	3.161	3.194	3.249	6.287	6.293	6.309	6.337
1.4	3.155	3.174	3.229	3.319	6.290	6.299	6.326	6.372
1.5	3.161	3.188	3.27	3.40	6.293	6.306	6.346	6.412
1.6	3.167	3.205	3.32	3.49	6.296	6.315	6.369	6.458
1.8	3.182	3.241	3.4	3.7	6.304	6.333	6.419	6.56
2.0	3.197	3.282	3.5	......	6.312	6.353	6.47	6.67
2.5	3.235	3.396	......	......	6.335	6.410	6.6	7.0
3.0	3.271	3.516	......	......	6.357	6.472	6.8	
3.5	3.305	3.636	......	......	6.381	6.538	7.0	
4.0	3.336	3.753	......	......	6.403	6.606		

mn \ c	04*	14	24	34	05*	15	25	35
1.0	9.425	9.425	9.425	9.425	12.566	12.566	12.566	12.566
1.1	9.425	9.426	9.427	9.429	12.567	12.567	12.568	12.570
1.2	9.426	9.428	9.433	9.442	12.567	12.569	12.573	12.579
1.3	9.427	9.431	9.442	9.461	12.568	12.571	12.579	12.593
1.4	9.429	9.435	9.454	9.484	12.570	12.574	12.588	12.611
1.5	9.431	9.440	9.467	9.511	12.571	12.578	12.598	12.631
1.6	9.434	9.446	9.482	9.541	12.573	12.582	12.609	12.654
1.8	9.439	9.458	9.515	9.609	12.577	12.591	12.634	12.704
2.0	9.444	9.471	9.552	9.684	12.581	12.601	12.661	12.761
2.5	9.460	9.509	9.665	9.990	12.593	12.629	12.739	12.92
3.0	9.476	9.550	9.77	10.1	12.605	12.660	12.82	13.09
3.5	9.493	9.593	9.89	......	12.619	12.692	12.91	13.3
4.0	9.509	9.638	10.0	......	12.631	12.725	13.0	13.5

* The first nonvanishing root χ'_{on} is designated as $n = 2$ rather than $n = 1$. The roots χ'_{on+1} and $\chi_{on}(n > 0)$ are identical.

The electric- and magnetic-field components of an H_{mn}-mode are given by Eqs. (1·4), (1·10), and (41) as

$$\left.\begin{aligned} E_r &= \pm V''_i \frac{m}{r} \mathcal{Z}_m\left(\chi'_i \frac{r}{b}\right) \begin{matrix} \sin \\ \cos \end{matrix} m\phi, \\ E_\phi &= V''_i \frac{\chi'_i}{b} \mathcal{Z}'_m\left(\chi'_i \frac{r}{b}\right) \begin{matrix} \cos \\ \sin \end{matrix} m\phi, \\ E_z &= 0, \end{aligned}\right\} \quad (43a)$$

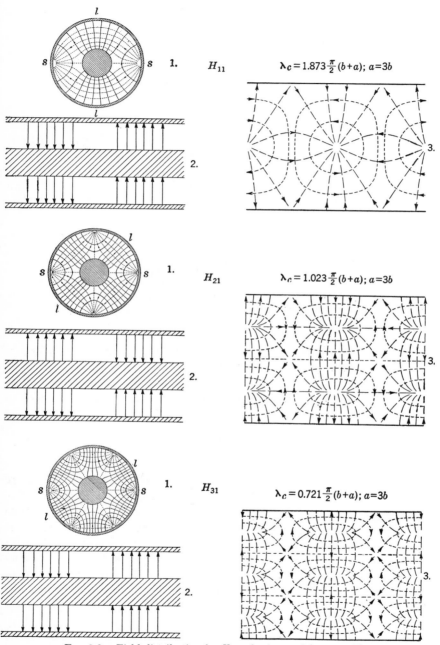

FIG. 2·9.—Field distribution for H-modes in coaxial waveguide.
1. Cross-sectional view
2. Longitudinal view through plane l-l
3. Surface view from s-s

$$H_r = -I_i'' \frac{\chi_i'}{b} Z_m' \left(\chi_i' \frac{r}{b}\right) \begin{matrix}\cos\\ \sin\end{matrix} m\phi,$$

$$H_\phi = \pm I_i'' \frac{m}{r} Z_m \left(\chi_i' \frac{r}{b}\right) \begin{matrix}\sin\\ \cos\end{matrix} m\phi, \qquad (43b)$$

$$H_z = -j\eta \frac{\lambda}{\lambda_{ci}''} V_i'' \frac{\chi_i'}{b} Z_m \left(\chi_i' \frac{r}{b}\right) \begin{matrix}\cos\\ \sin\end{matrix} m\phi.$$

As in the case of the E-modes, the H_{mn}-mode ($m > 0$) may possess either of two polarizations, each characterized by a different voltage and current. The polar components of the normalized $\mathbf{e}_i''$ vectors are obtained from E_r and E_ϕ of Eqs. (43a) on omission of the z-dependent amplitude V_i''.

For a traveling H_{mn}-mode the attenuation constant due to finite conductivity of the guide walls is given by

$$\alpha = \frac{\left[\dfrac{\mathcal{R}_a}{a} \dfrac{J_m'^2(\chi_i')}{J_m'^2(c\chi_i')} \left(\dfrac{b}{a}\right)^2 + \dfrac{\mathcal{R}_b}{b}\right] \dfrac{1}{\zeta} \dfrac{m^2}{\chi_i'^2} \sqrt{1 - \left(\dfrac{\lambda}{\lambda_{ci}''}\right)^2} + \left[\dfrac{\mathcal{R}_a}{a} \dfrac{J_m'^2(\chi_i')}{J_m'^2(c\chi_i')} + \dfrac{\mathcal{R}_b}{b}\right] \dfrac{1}{\zeta} \dfrac{(\lambda/\lambda_{ci}'')^2}{\sqrt{1 - (\lambda/\lambda_{ci}'')^2}}}{\dfrac{J_m'^2(\chi_i')}{J_m'^2(c\chi_i')}\left[1 - \left(\dfrac{m}{c\chi_i'}\right)^2\right] - \left[1 - \left(\dfrac{m}{\chi_i'}\right)^2\right]}, \qquad (44)$$

where $\mathcal{R}_a$ and $\mathcal{R}_b$ are the characteristic resistances of the metals of which the outer and inner conductors are composed (*cf.* Table 1·2).

The total average power carried by a traveling H_{mn}-mode in a matched nondissipative coaxial guide is expressed in terms of the rms voltage V_i'' as

$$P = \eta \sqrt{1 - \left(\frac{\lambda}{\lambda_{ci}''}\right)^2} |V_i''|^2. \qquad (45)$$

Mode patterns of the instantaneous field distribution in traveling waves of the H_{11}, H_{21}, and H_{31} coaxial modes are shown in Fig. 2·9. The patterns pertain to coaxial guides with $a/b = 3$ and $\lambda_g/a = 4.24$. The patterns on the left depict the electric and magnetic field distribution in transverse and longitudinal planes on which the transverse electric field is a maximum; those on the right portray the magnetic-field and electric-current distribution on half the circumference of the outer conductor.

2·5. Elliptical Waveguides.[1]—An elliptical waveguide is a uniform region in which the transverse cross section is of elliptical form. As

---

[1] *Cf.* L. J. Chu, "Electromagnetic Waves in Elliptic Hollow Pipes," *Jour. Applied Phys.*, **9**, September, 1938. Stratton, Morse, Chu, Hutner, *Elliptic Cylinder and Spheroidal Wavefunctions*, Wiley, 1941.

illustrated in Fig. 2·10 elliptic coordinates $\xi\eta$ (the coordinate η is not to be confused with the free-space admittance η employed elsewhere in this volume) describe the cross section and the coordinate z the transmission direction. The rectangular coordinates xy of the cross section are related to the coordinates of the confocal ellipse ξ and confocal hyperbola η by

$$x = q \cosh\xi \cos\eta, \\ y = q \sinh\xi \sin\eta, \quad (46)$$

where $2q$ is the focal distance. The boundary ellipse is defined by the coordinate $\xi = a$ with major axis, minor axis, and eccentricity given by $2q \cosh a$, $2q \sinh a$, and $e = 1/\cosh a$, respectively. The case of a circular boundary is described by $e = q = 0$ with q/e finite.

The mode functions $\mathbf{e}'_i$ characteristic of the E-modes are derivable from scalar functions of the form

$$_e\Phi_i = Re_m(\xi, {}_e\chi_i) Se_m(\eta, {}_e\chi_i), \quad (47a)$$
$$_o\Phi_i = Ro_m(\xi, {}_o\chi_i) So_m(\eta, {}_o\chi_i), \quad (47b)$$

where $m = 0, 1, 2, 3, \cdots$ and $_e\chi_i = {}_e\chi_{mn}$ is the nth nonvanishing root of the even mth-order radial Mathieu function $Re_m(a,\chi)$, whereas $_o\chi_i = {}_o\chi_{mn}$ is the nth nonvanishing root of the odd mth-order radial Mathieu function $Ro_m(a,\chi)$. The functions $Se_m(\eta,\chi)$ and $So_m(\eta,\chi)$ are even and odd angular Mathieu functions. In the limit of small $\chi = k_c q$ the Mathieu functions degenerate into circular functions as follows:

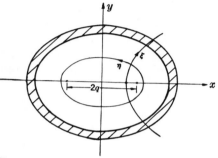

FIG. 2·10.—Elliptical waveguide cross section.

$$\lim_{\chi \to 0} Se_m(\eta,\chi) = \cos m\phi, \\ \lim_{\chi \to 0} So_m(\eta,\chi) = \sin m\phi, \\ \lim_{\chi \to 0} Re_m(\xi,\chi) = \lim_{\chi \to 0} Ro_m(\xi,\chi) = \sqrt{\frac{\pi}{2}} J_m(k_c r), \quad (48)$$

and, correspondingly, the confocal coordinates ξ and η become the polar coordinates r and ϕ.

The field components of the even $_eE_{mn}$-mode follow from Eqs. (1·3), (1·10), and (47a) as

$$E_\xi = -V'_i \frac{Re'_m(\xi, {}_e\chi_i) Se_m(\eta, {}_e\chi_i)}{q \sqrt{\cosh^2 \xi - \cos^2 \eta}}, \\ E_\eta = -V'_i \frac{Re_m(\xi, {}_e\chi_i) Se'_m(\eta, {}_e\chi_i)}{q \sqrt{\cosh^2 \xi - \cos^2 \eta}}, \\ E_z = -j\zeta \frac{\lambda}{_e\lambda'_{ci}} I'_i\, _ek'_{ci} Re_m(\xi, {}_e\chi_i) Se_m(\eta, {}_e\chi_i), \quad (49a)$$

$$H_\xi = +I'_i \frac{Re_m(\xi,{}_e\chi_i)Se'_m(\eta,{}_e\chi_i)}{q\sqrt{\cosh^2\xi - \cos^2\eta}},$$
$$H_\eta = -I'_i \frac{Re'_m(\xi,{}_e\chi_i)Se_m(\eta,{}_e\chi_i)}{q\sqrt{\cosh^2\xi - \cos^2\eta}},$$
$$H_z = 0.$$
(49b)

The prime denotes the derivative with respect to ξ or η, and ${}_ek'_{ci} = 2\pi/{}_e\lambda'_{ci}$ denotes the cutoff wavenumber of the even mn mode. The z dependence of the fields is determined by the transmission-line behavior of the mode amplitudes V'_i and I'_i. The odd ${}_0E_{mn}$-mode is represented in the same manner as in Eqs. (49) save for the replacement of the even Mathieu functions by the odd. The components of the mode functions $\mathbf{e}'_i$ are obtained from E_ξ and E_η of Eqs. (49a) on omission of the amplitude V'_i and insertion of a normalization factor.

The $\mathbf{e}''_i$-mode functions characteristic of the H-modes in elliptical guides are derivable from scalar functions of the form

$$_e\Psi_i = Re_m(\xi,{}_e\chi'_i)Se_m(\eta,{}_e\chi'_i), \quad (50a)$$
$$_0\Psi_i = Ro_m(\xi,{}_0\chi'_i)So_m(\eta,{}_0\chi'_i), \quad (50b)$$

where $m = 0, 1, 2, 3, \cdots$, and ${}_e\chi'_i = {}_e\chi'_{mn}$ and ${}_0\chi'_i = {}_0\chi'_{mn}$ are the nth nonvanishing roots of the derivatives of the radial Mathieu functions as defined by

$$Re'_m(a,{}_e\chi'_i) = 0,$$
$$Ro'_m(a,{}_0\chi'_i) = 0.$$
(51)

The field components of the even ${}_eH_{mn}$-mode are given by Eqs. (1·4), (1·10), and (50a) as

$$E_\xi = -V''_i \frac{Re_m(\xi,{}_e\chi'_i)Se'_m(\eta,{}_e\chi'_i)}{q\sqrt{\cosh^2\xi - \cos^2\eta}},$$
$$E_\eta = V''_i \frac{Re'_m(\xi,{}_e\chi'_i)Se_m(\eta,{}_e\chi'_i)}{q\sqrt{\cosh^2\xi - \cos^2\eta}},$$
$$E_z = 0,$$
(52)

$$H_\xi = -I''_i \frac{Re'_m(\xi,{}_e\chi'_i)Se_m(\eta,{}_e\chi'_i)}{q\sqrt{\cosh^2\xi - \cos^2\eta}},$$
$$H_\eta = -I''_i \frac{Re_m(\xi,{}_e\chi'_i)Se'_m(\eta,{}_e\chi'_i)}{q\sqrt{\cosh^2\xi - \cos^2\eta}},$$
$$H_z = -j\frac{\lambda}{{}_e\lambda''_{ci}}\frac{V''_i}{\zeta}{}_ek''_{ci}Re_m(\xi,{}_e\chi'_i)Se_m(\eta,{}_e\chi'_i).$$
(53)

The field components of the odd ${}_0H_{mn}$-mode are obtained from Eqs. (52) and (53) on replacement of the even Mathieu functions by the odd. The components of the $\mathbf{e}''_i$ mode functions are obtained from E_ξ and E_η on omission of the mode amplitude V''_i and addition of a normalization factor.

The cutoff wavelengths of the E- and H-modes may be expressed in terms of the roots χ_i and the semifocal distance q as

$$\lambda_{ci} = \frac{2\pi q}{\chi_i}, \tag{54}$$

on omission of the various mode designations. An alternative expression in terms of the eccentricity e of the boundary ellipse is obtained by use of the elliptic integral formula for the circumference

$$s = \frac{q}{e} \int_0^{2\pi} \sqrt{1 - e^2 \cos^2 \eta} \, d\eta = \frac{4q}{e} E(e) \tag{55}$$

of the boundary ellipse. The ratio λ_c/s is plotted vs. e in Fig. 2·11 for several of the even and odd E- and H-modes of largest cutoff wavelengths.

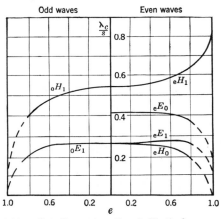

FIG. 2·11.—Cutoff wavelengths of elliptical waveguide.

It is evident from the figure that the $_eH_{11}$-mode is the dominant mode in an elliptical waveguide. The splitting of the degenerate modes ($m > 0$, $e = 0$) of a circular guide into even and odd modes is also evident from this figure.

Computation of power flow and attenuation in elliptical guides involves numerical integration of the Mathieu functions over the guide cross section. The reader is referred to Chu's paper quoted above for quantitative information.

Mode patterns of the transverse electric and magnetic field distribution of several of the lower modes in an elliptical guide of eccentricity $e = 0.75$ are shown in Fig. 2·12. The patterns are for the $_eH_{01}$, $_eH_{11}$, $_oH_{11}$, $_eE_{01}$, $_eE_{11}$, and $_oE_{11}$-modes.

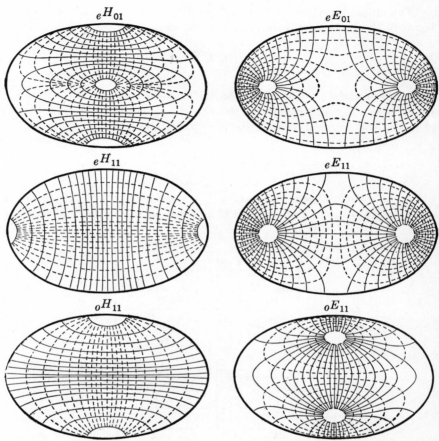

Fig. 2·12.—Field distribution of modes in elliptical waveguide. Cross-sectional view.

2·6. Space as a Uniform Waveguide. *a. Fields in Free Space.*—Free space may be regarded as a uniform waveguide having infinite cross-sectional dimensions. A transmission-line description of the fields within free space can be developed in a manner similar to that of the preceding sections. The cross-sectional directions will be described by the xy coordinates and the transmission direction by the z coordinate of a rectangular coordinate system as shown in Fig. 2·13. In this coordinate system the general electromagnetic field can be expressed as a superposition of an infinite set of E- and H-modes closely related to those employed for rectangular guides (*cf.* Sec. 2·2). In the absence of geometrical structures imposing periodicity requirements on the field, the required modes form a continuous set of plane waves, each wave being characterized by a wave number **k** indicative of the direction of propagation and wavelength.

The $\mathbf{e}'_i$ vector functions characteristic of the E-modes are derivable from scalar functions of the form

$$\Phi_i = \Psi_i = j \frac{e^{j(k_x x + k_y y)}}{\sqrt{k_x^2 + k_y^2}}, \quad (56)$$

where $-\infty < k_x < \infty$, $-\infty < k_y < \infty$, and i is an index indicative of the mode with wave numbers k_x in the x direction and k_y in the y direction. The question of normalization will be left open for the moment.

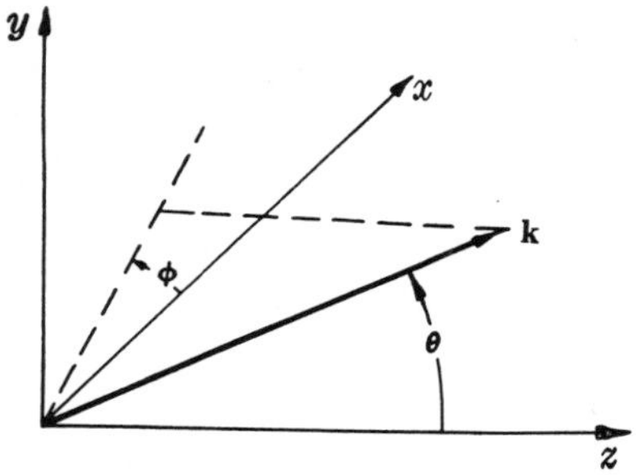

FIG. 2·13.—Coordinate systems for waves in space. Rectangular coordinates x,y,z. Polar and azimuthal angles θ, ϕ.

By Eqs. (1·3), (1·10), and (56) the field components of the E_i-mode can be written as

$$\left.\begin{aligned} E_x &= V'_i \frac{k_x}{\sqrt{k_x^2 + k_y^2}} e^{j(k_x x + k_y y)}, \\ E_y &= V'_i \frac{k_y}{\sqrt{k_x^2 + k_y^2}} e^{j(k_x x + k_y y)}, \\ E_z &= \frac{I'_i}{\omega \epsilon} \sqrt{k_x^2 + k_y^2}\, e^{j(k_x x + k_y y)}, \end{aligned}\right\} \quad (57a)$$

$$\left.\begin{aligned} H_x &= -I'_i \frac{k_y}{\sqrt{k_x^2 + k_y^2}} e^{j(k_x x + k_y y)}, \\ H_y &= I'_i \frac{k_x}{\sqrt{k_x^2 + k_y^2}} e^{j(k_x x + k_y y)}, \\ H_z &= 0. \end{aligned}\right\} \quad (57b)$$

The z dependence of this mode is determined by the transmission-line behavior of the mode voltage V'_i and current I'_i, which obey the uniform-line equations (1·11). On introduction of the new variables

$$\left.\begin{aligned} k_x &= k \sin \theta \cos \phi, & \tan \phi &= \frac{k_y}{k_x}, \\ k_y &= k \sin \theta \sin \phi, & \sin \theta &= \frac{\sqrt{k_x^2 + k_y^2}}{k}, \\ k &= \frac{2\pi}{\lambda}, \end{aligned}\right\} \quad (58)$$

it is evident from Fig. 2·13 that an E_i-mode with $\sqrt{k_x^2 + k_y^2} < k$, represents a propagating plane wave whose wave vector **k** is characterized by polar angle θ and azimuthal angle ϕ. The magnetic field of this mode is linearly polarized and lies wholly in the transverse xy plane at the angle $\phi + \pi/2$.

The $\mathbf{e}_i''$ vector function characteristic of an H_i-mode is derivable from a scalar function (56) of the same form as for the E_i-mode. The field components of the H_i-mode follow from Eqs. (1·4), (1·10), and (56) as

$$\left. \begin{array}{l} E_x = V_i'' \dfrac{k_y}{\sqrt{k_x^2 + k_y^2}}\, e^{j(k_x x + k_y y)}, \\[6pt] E_y = -V_i'' \dfrac{k_x}{\sqrt{k_x^2 + k_y^2}}\, e^{j(k_x x + k_y y)}, \\[6pt] E_z = 0, \end{array} \right\} \quad (59a)$$

$$\left. \begin{array}{l} H_x = I_i'' \dfrac{k_x}{\sqrt{k_x^2 + k_y^2}}\, e^{j(k_x x + k_y y)}, \\[6pt] H_y = I_i'' \dfrac{k_y}{\sqrt{k_x^2 + k_y^2}}\, e^{j(k_x x + k_y y)}, \\[6pt] H_z = \dfrac{V_i''}{\omega \mu} \sqrt{k_x^2 + k_y^2}\, e^{j(k_x x + k_y y)}, \end{array} \right\} \quad (59b)$$

where again the z dependence of the mode field is given by the transmission-line behavior of the mode voltage V_i'' and current I_i''. On substitution of Eqs. (58) it is seen that a traveling H_i-mode with $\sqrt{k_x^2 + k_y^2} < k$ is a plane wave whose wave vector $\mathbf{k}$ points in the direction specified by the polar angle θ and the azimuthal angle ϕ. The electric field of the H_i-mode lies entirely in the xy plane and is linearly polarized at the angle $\phi - \pi/2$.

The cross-sectional wave numbers of both the E_i- and H_i-modes are identical and equal to

$$k_{ci} = \sqrt{k_x^2 + k_y^2} = \frac{2\pi}{\lambda_{ci}} = \frac{2\pi}{\lambda} \sin \theta. \quad (60a)$$

By Eqs. (1·11b) it follows that

$$\lambda_g = \frac{\lambda}{\cos \theta}. \quad (60b)$$

From the dependence of the cross-sectional wave number on λ it is apparent that mode propagation ceases when $\sqrt{k_x^2 + k_y^2} > k$, i.e., when θ is imaginary.

The power flow per unit area in the z direction can be expressed in terms of the rms mode voltage V_i as

$$P_z = \frac{\sqrt{\dfrac{\epsilon}{\mu}}}{\sqrt{1 - \left(\dfrac{\lambda}{\lambda_{ci}}\right)^2}} |V_i'|^2 = \frac{\eta}{\cos \theta} |V_i'|^2 \quad (61a)$$

for the case of a propagating E_i-mode or as

$$P_z = \sqrt{\frac{\epsilon}{\mu}} \sqrt{1 - \left(\frac{\lambda}{\lambda_{ci}}\right)^2} |V_i''|^2 = \eta \cos\theta |V_i''|^2 \tag{61b}$$

for the case of a propagating H_i-mode.

The dominant E- and H-modes are obtained from Eqs. (57) and (59) by first placing $k_x = 0$ and then $k_y = 0$, or conversely. These modes are evidently *TEM* waves with the nonvanishing components

$$\left. \begin{array}{ll} E_y = V_0'(z), & E_x = V_0''(z), \\ H_x = -I_0'(z), & H_y = I_0''(z) \end{array} \right\} \tag{62}$$

and are seen to be polarized at right angles to one another. The wavelength λ_g of propagation in the z direction is equal to the free-space wavelength λ for these modes.

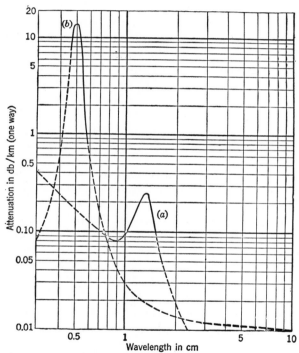

Fig. 2·14.—Atmospheric attenuation of plane waves. (a) Atmosphere composed of 10 gm. of water vapor per cubic meter. (b) Atmosphere composed of 20 per cent oxygen at a total pressure of 76 cm. Hg.

The attenuation constant of plane waves in free space is determined solely by the dielectric losses in the atmosphere of the space. Because of the importance of this type of attenuation at ultrahigh frequencies theoretical curves of the attenuation (8.686α), in decibels per kilometer, due to presence of oxygen and water vapor, are shown in Fig. 2·14 as a function of wavelength.

b. Field in the Vicinity of Gratings.—The analysis of the electromagnetic field in the vicinity of periodic structures, such as gratings in free space, is in many respects simpler than in the case where no periodicity exists. Instead of a continuous infinity of modes, as in Sec. 2·6a, a denumerably infinite set of E- and H-modes is present with only discrete values of k_x and k_y. When the excitation and geometrical structure is such that the field has a spatial periodicity of period a and b in the transverse x and y directions, respectively, the only permissible values of k_x and k_y are

$$k_x = \frac{2\pi m}{a}, \quad m = 0, \pm 1, \pm 2, \cdots, \\ k_y = \frac{2\pi n}{b}, \quad n = 0, \pm 1, \pm 2, \cdots. \quad (63)$$

A situation of this sort obtains, for example, when a uniform plane wave falls normally upon a planar grating having a structural periodicity of length a in the x direction and b in the y direction. The E_i- and H_i-modes in such a space are given by Eqs. (57) and (59) with k_x and k_y as in Eqs. (63). The cutoff wavelengths of the E_i- and H_i-modes are both equal to

$$\lambda_{ci} = \frac{2\pi}{\sqrt{k_x^2 + k_y^2}} = \frac{\sqrt{ab}}{\sqrt{m^2 \frac{b}{a} + n^2 \frac{a}{b}}}, \quad (64)$$

which is characteristic of the wavelength below which the plane waves are propagating and above which they are damped. On introduction of the substitution (58) it is apparent that these modes can be interpreted in terms of a discrete set of plane waves (diffracted orders) defined by the angles θ and ϕ. In many practical cases the excitation and dimensions are such that $\lambda > a > b$, and hence only a single mode, one of the dominant modes shown in Eqs. (62), can be propagated. As a consequence the field is almost everywhere described by the voltage and current of this one mode.

A similar mode analysis can be applied to describe the fields in the vicinity of a periodic structure when the excitation consists of a plane wave incident at the oblique angles θ' and ϕ'. If the spatial periodicity of the structure is again defined by the periods a and b in the x and y directions, the only permissible E- and H-modes, shown in Eqs. (57) and (59), are those for which

$$k_x = \frac{2\pi m}{a} - k'_x, \quad m = 0, \pm 1, \pm 2, \cdots, \\ k_y = \frac{2\pi n}{b} - k'_y, \quad n = 0, \pm 1, \pm 2, \cdots. \quad (65)$$

where $k'_x = k \sin \theta' \cos \phi'$ and $k'_y = k \sin \theta' \sin \phi'$ are the wave numbers of the incident excitation. The cross-sectional wave number of both the E_i- and the H_i-mode equals

$$k_{ci} = 2\pi \sqrt{\left(\frac{m}{a} - \frac{\sin \theta' \cos \phi'}{\lambda}\right)^2 + \left(\frac{n}{b} - \frac{\sin \theta' \sin \phi'}{\lambda}\right)^2} = \frac{2\pi}{\lambda_{ci}}, \quad (66)$$

and as in the previous case these wave numbers characterize a discrete set of plane waves (or diffracted orders).

Frequently the excitation and dimensions are such that only the dominant mode ($m = n = 0$) with $k_x = k'_x$ and $k_y = k'_y$ is propagating. For example, if $\phi' = 0$ and $a > b$, this situation obtains, as can be seen from Eq. (66), when

$$\lambda > a(1 + \sin \theta) = \lambda_{ci},$$

i.e., when the next higher diffraction order does not propagate. Under these conditions the dominant mode voltage and current describe the field almost everywhere.

The components of the $\mathbf{e}'_i$ vectors are obtained from the E_x and E_y components of the mode fields of Eqs. (57a) and (59a) on omission of the mode amplitudes. The normalization of the $\mathbf{e}_i$ vector functions has not been explicitly stated since it depends on whether the mode index i is continuous or discrete. For a continuous index i, $\mathbf{e}_i$ must be divided by 2π to obtain a vector function normalized over the infinite cross section to a delta function of the form $\delta(k_x - k'_x) \delta(k_y - k'_y)$.[1] For discrete index i, $\mathbf{e}_i$ should be divided by $\sqrt{ab}$ to obtain, in accordance with Eqs. (1·5), a vector function normalized to unity over a cross section of dimensions a by b.

2·7. Radial Waveguides. *a. Cylindrical Cross Sections.*—An example of a nonuniform region in which the transverse cross sections are complete cylindrical surfaces of height b is provided by the radial waveguide illustrated in Fig. 2·15. In the $r\phi z$ polar coordinate system appropriate to regions of this type, the transverse cross sections are ϕz surfaces and transmission is in the direction of the radius r. Radial waveguides are encountered in many of the resonant cavities employed in ultra high-frequency oscillator tubes, filters, etc.; free space can also be regarded as a radial waveguide of infinite height. As stated in Sec. 1·7 the transverse electromagnetic field in radial waveguides cannot be represented, in general, as a superposition of transverse vector modes. There exists only a scalar representation that, for no H_z field, is expressible in terms of E-type modes and, for no E_z field, in terms of H-type modes.

---

[1] The delta function $\delta(x - x')$ is defined by the conditions that its integral be unity if the interval of integration includes the point x' and be zero otherwise.

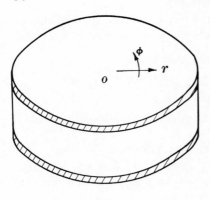
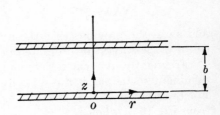

General view   Side view
FIG. 2·15.—Radial waveguide of cylindrical cross section.

E-type modes.—The field components of an E-type mode in the radial guide of Fig. 2.15 can be represented as

$$E_z = -V'_i \frac{\epsilon_n}{b} \cos \frac{n\pi}{b} z \, {\cos \atop \sin} m\phi,$$
$$E_\phi = \mp V'_i \frac{\epsilon_n}{b} \frac{m}{\kappa_n r} \frac{n\pi}{\kappa_n b} \sin \frac{n\pi}{b} z \, {\sin \atop \cos} m\phi, \quad (67a)$$
$$E_r = -j\zeta I'_i \frac{\epsilon_m}{2\pi r} \frac{n\pi}{kb} \sin \frac{n\pi}{b} z \, {\cos \atop \sin} m\phi,$$

$$H_z = 0,$$
$$H_\phi = I'_i \frac{\epsilon_m}{2\pi r} \cos \frac{n\pi}{b} z \, {\cos \atop \sin} m\phi, \quad (67b)$$
$$H_r = \pm j\eta V'_i \frac{\epsilon_n}{b} \frac{k}{\kappa_n} \frac{m}{\kappa_n r} \cos \frac{n\pi}{b} z \, {\sin \atop \cos} m\phi,$$

where

$$\kappa_n = \sqrt{k^2 - \left(\frac{n\pi}{b}\right)^2},$$
$$\epsilon_n = 1 \quad \text{if } n = 0,$$
$$\epsilon_n = 2 \quad \text{if } n \neq 0,$$
$$m = 0, 1, 2, 3, \cdots, \qquad n = 0, 1, 2, 3, \cdots.$$

The z dependence of the E-type modes is determined by the transmission-line behavior of the mode voltage V'_i and current I'_i. The latter quantities satisfy the transmission-line equations (1·64) with

$$Z'_i = \zeta \frac{b\epsilon_m}{2\pi r \epsilon_n} \frac{\kappa_n^2}{\kappa'_i k},$$
$$\kappa'_i = \sqrt{k^2 - \left(\frac{n\pi}{b}\right)^2 - \left(\frac{m}{r}\right)^2}, \quad (68)$$
$$\kappa_n = \sqrt{k^2 - \left(\frac{n\pi}{b}\right)^2}.$$

For either $m = 0$ or $n = 0$ the transverse electric and magnetic fields of each mode have only a single component. As is evident from the form of Eqs. (67) there exist two independent E_{mn}-type modes with different ϕ polarizations; the amplitudes of these degenerate modes are different from each other although this has not been explicitly indicated.

In a radial waveguide the concept of guide wavelength loses its customary significance because of the nonperiodic nature of the field variation in the transmission direction. Consequently the usual relation between guide wavelength and cutoff wavelength is no longer valid. However, the cutoff wavelength, defined as the wavelength at which $\kappa'_i = 0$, is still useful as an indication of the "propagating" or "nonpropagating" character of a mode. For an E-type mode the cutoff wavelength is

$$\lambda'_{ci} = \frac{1}{\sqrt{\left(\dfrac{n}{2b}\right)^2 + \left(\dfrac{m}{2\pi r}\right)^2}}, \tag{69}$$

and its dependence on r indicates that the mode is propagating in those regions for which $\lambda < \lambda'_{ci}$ and nonpropagating when $\lambda > \lambda'_{ci}$.

In terms of the rms mode voltage and current the total outward power flow in an E-type mode is Re $(V'_i I'^*_i)$. For computations of power on a matched line it should be noted that the input admittance of a matched radial line is not equal to its characteristic admittance [cf. Eq. (1·74)].

The dominant E-type mode in the radial waveguide of Fig. 2·15 is the $m = n = 0$ mode and is seen to be a transverse electromagnetic mode. The nonvanishing field components of this TEM-mode follow from Eqs. (67) as (omitting the mode designations)

$$\left. \begin{array}{l} E_z = -\dfrac{V(r)}{b}, \\[6pt] H_\phi = \dfrac{I(r)}{2\pi r}, \end{array} \right\} \tag{70a}$$

and the corresponding characteristic impedance and mode constant as

$$Z = \zeta \frac{b}{2\pi r} \quad \text{and} \quad k. \tag{70b}$$

Frequently the excitation and guide dimensions are such that the dominant mode characterizes the field almost everywhere. The total outward power carried by the dominant mode at any point r in a matched nondissipative radial guide is

$$P = \eta \frac{2\pi r}{b} \frac{2}{\pi k r h_0^2(kr)} |V(r)|^2, \tag{71}$$

as is evident from the power relation and Eqs. (1·74); the function $h_0(kr)$

is the amplitude of the zeroth-order Hankel function and is defined in Eq. (1·80).

H-type modes.—The field components of an *H*-type mode in the radial guide of Fig. 2·15 are given by

$$E_z = 0,$$
$$E_\phi = V_i'' \frac{\epsilon_m}{2\pi r} \sin \frac{n\pi}{b} z \, \frac{\sin}{\cos} m\phi,$$
$$E_r = \mp j\zeta I_i'' \frac{\epsilon_n}{b} \frac{k}{\kappa_n} \frac{m}{\kappa_n r} \sin \frac{n\pi}{b} z \, \frac{\cos}{\sin} m\phi,$$
$$\quad (72a)$$

$$H_z = I_i'' \frac{\epsilon_n}{b} \sin \frac{n\pi}{b} z \, \frac{\sin}{\cos} m\phi,$$
$$H_\phi = \pm I_i'' \frac{\epsilon_n}{b} \frac{m}{\kappa_n r} \frac{n\pi}{\kappa_n b} \cos \frac{n\pi}{b} z \, \frac{\cos}{\sin} m\phi,$$
$$H_r = -j\eta V_i'' \frac{\epsilon_m}{2\pi r} \frac{n\pi}{kb} \cos \frac{n\pi}{b} z \, \frac{\sin}{\cos} m\phi,$$
$$\quad (72b)$$

where

$$\kappa_n = \sqrt{k^2 - \left(\frac{n\pi}{b}\right)^2},$$

$$m = 0, 1, 2, \cdots. \qquad n = 1, 2, 3, \cdots.$$

The *z*-dependent mode voltage V_i'' and current I_i'' obey the radial transmission-line equations (1·64) with

$$Z_i'' = \zeta \frac{2\pi r \epsilon_n}{b\epsilon_m} \frac{\kappa_i'' k}{\kappa_n^2},$$
$$\kappa_i'' = \sqrt{k^2 - \left(\frac{n\pi}{b}\right)^2 - \left(\frac{m}{r}\right)^2}.$$
$$\quad (73)$$

The existence of two distinct H_{mn}-type modes with different ϕ polarizations is to be noted.

The cutoff wavelength of the H_{mn}-type mode is identical with that of the E_{mn}-type mode and, as in the latter case, is indicative of regions of propagation and nonpropagation. The total outward power flow in an *H*-type mode is given in terms of the rms mode voltage and current by Re $(V_i'' I_i''^*)$.

The dominant *H*-type mode in the radial guide of Fig. 2·15 is the $m = 0$, $n = 1$ mode. The nonvanishing field components of this mode can be written as

$$E_\phi = \frac{V}{2\pi r} \sin \frac{\pi}{b} z,$$
$$H_z = \frac{2I}{b} \sin \frac{\pi}{b} z,$$
$$H_r = -j\eta \frac{V}{2\pi r} \frac{\lambda}{2b} \cos \frac{\pi}{b} z,$$
$$\quad (74)$$

on omission of the distinguishing mode indices. The characteristic impedance and mode constant of the dominant H-type mode are

$$Z = \zeta \frac{4\pi r}{b} \frac{k}{\kappa} \quad \text{and} \quad \kappa = \sqrt{k^2 - \left(\frac{\pi}{b}\right)^2}. \tag{75}$$

The total outward power flow carried by the dominant H-type mode in a matched nondissipative radial line is

$$P = \zeta \frac{4\pi r}{b} \frac{k}{\kappa} \frac{2}{\pi \kappa r h_0^2(\kappa r)} |I(r)|^2. \tag{76}$$

b. *Cylindrical Sector Cross Sections.*—Another example of a radial waveguide is provided by the nonuniform region illustrated in Fig. 2·16.

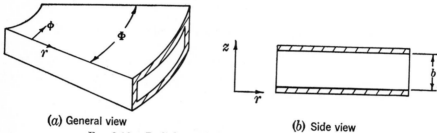

(a) General view (b) Side view

Fig. 2·16.—Radial waveguide of sectoral cross section.

In the $r\phi z$ coordinate system indicated therein the cross-sectional surfaces are cylindrical sectors of aperture Φ and height b.

E-type modes.—In the above type of radial waveguide the field components of an E-type mode are

$$\left. \begin{aligned} E_z &= -V_i' \frac{\epsilon_n}{b} \cos \frac{n\pi}{b} z \sin \frac{m\pi}{\Phi} \phi, \\ E_\phi &= V_i' \frac{\epsilon_n}{b} \frac{m\pi}{\kappa_n r\Phi} \frac{n\pi}{\kappa_n b} \sin \frac{n\pi}{b} z \cos \frac{m\pi}{\Phi} \phi, \\ E_r &= -j\zeta I_i' \frac{2}{r\Phi} \frac{n\pi}{kb} \sin \frac{n\pi}{b} z \sin \frac{m\pi}{\Phi} \phi, \end{aligned} \right\} \tag{77a}$$

$$\left. \begin{aligned} H_z &= 0, \\ H_\phi &= I_i' \frac{2}{r\Phi} \cos \frac{n\pi}{b} z \sin \frac{m\pi}{\Phi} \phi, \\ H_r &= -j\eta V_i' \frac{\epsilon_n}{b} \frac{k}{\kappa_n} \frac{m\pi}{\kappa_n r\Phi} \cos \frac{n\pi}{b} z \cos \frac{m\pi}{\Phi} \phi, \end{aligned} \right\} \tag{77b}$$

where

$$\kappa_n = \sqrt{k^2 - \left(\frac{n\pi}{b}\right)^2}$$

$$m = 1, 2, 3, \cdots, \quad n = 0, 1, 2, 3, \cdots.$$

The z dependence of this E_{mn}-type mode is determined by the transmission-line behavior of the mode voltage V_i' and current I_i', as given by Eqs. (1·64) with

$$\left. \begin{array}{l} Z_i' = \zeta \dfrac{2b}{r\Phi\epsilon_n} \dfrac{\kappa_n^2}{\kappa_i' k}, \\[6pt] \kappa_i' = \sqrt{k^2 - \left(\dfrac{n\pi}{b}\right)^2 - \left(\dfrac{m\pi}{r\Phi}\right)^2}. \end{array} \right\} \quad (78)$$

The cutoff wavelength of the E_{mn}-type mode is

$$\lambda_{ci}' = \dfrac{1}{\sqrt{\left(\dfrac{n}{2b}\right)^2 + \left(\dfrac{m}{2r\Phi}\right)^2}} \quad (79)$$

and as before is indicative of the regions of propagation and nonpropagation.

The dominant E-type mode in the radial waveguide of Fig. 2·16 is the $m = 1$, $n = 0$ mode. The nonvanishing field components of this mode are (omitting mode indices)

$$\left. \begin{array}{l} E_z = -\dfrac{V}{b} \sin \dfrac{\pi}{\Phi} \phi, \\[6pt] H_\phi = \dfrac{2I}{r\Phi} \sin \dfrac{\pi}{\Phi} \phi, \\[6pt] H_r = -j\eta \dfrac{V}{b} \dfrac{\lambda}{2r\Phi} \cos \dfrac{\pi}{\Phi} \phi. \end{array} \right\} \quad (80)$$

The characteristic impedance and mode constant are

$$Z = \zeta \dfrac{2b}{r\Phi} \dfrac{k}{\kappa'} \quad \text{and} \quad \kappa' = \sqrt{k^2 - \left(\dfrac{\pi}{r\Phi}\right)^2}, \quad (81)$$

and the transmission-line behavior is described in terms of the "standing waves"

$$J_{\pi/\Phi}(kr) \quad \text{and} \quad N_{\pi/\Phi}(kr). \quad (82)$$

In terms of the dominant-mode rms voltage V the total outward power in a matched nondissipative guide is

$$P = \eta \dfrac{r\Phi}{2b} \dfrac{\kappa'}{k} \dfrac{2}{\pi k r h^2_{\pi/\Phi}} |V(r)|^2. \quad (83)$$

H-type modes.—The field components of an H-type mode in the radial waveguide of Fig. 2·16 are

$$\left.\begin{aligned}
E_z &= 0, \\
E_\phi &= V_i'' \frac{\epsilon_m}{r\Phi} \sin \frac{n\pi}{b} z \cos \frac{m\pi}{\Phi} \phi, \\
E_r &= j\zeta I_i'' \frac{\epsilon_n}{b} \frac{k}{\kappa_n} \frac{m\pi}{\kappa_n r\Phi} \sin \frac{n\pi}{b} z \sin \frac{m\pi}{\Phi} \phi,
\end{aligned}\right\} \quad (84a)$$

$$\left.\begin{aligned}
H_z &= I_i'' \frac{\epsilon_n}{b} \sin \frac{n\pi}{b} z \cos \frac{m\pi}{\Phi} \phi, \\
H_\phi &= -I_i'' \frac{\epsilon_n}{b} \frac{m\pi}{\kappa_n r\Phi} \frac{n\pi}{\kappa_n b} \cos \frac{n\pi}{b} z \sin \frac{m\pi}{\Phi} \phi, \\
H_r &= -j\eta V_i'' \frac{\epsilon_m}{r\Phi} \frac{n\pi}{kb} \cos \frac{n\pi}{b} z \cos \frac{m\pi}{\Phi} \phi,
\end{aligned}\right\} \quad (84b)$$

where

$$\kappa_n = \sqrt{k^2 - \left(\frac{n\pi}{b}\right)^2}$$

$$m = 0, 1, 2, 3, \cdots, \qquad n = 1, 2, 3, \cdots.$$

The z-dependent mode voltage V_i'' and current I_i'' satisfy the radial transmission-line equations (1·64) with

$$\left.\begin{aligned}
Z_i'' &= \zeta \frac{r\Phi \epsilon_n}{b \epsilon_m} \frac{\kappa_i'' k}{\kappa_n^2}, \\
\kappa_i'' &= \sqrt{k^2 - \left(\frac{n\pi}{b}\right)^2 - \left(\frac{m\pi}{r\Phi}\right)^2}.
\end{aligned}\right\} \quad (85)$$

The cutoff wavelength is the same as that given in Eq. (79) for an E_{mn}-type mode. The total outward mode power is given by Re $(V_i'' I_i''^*)$.

The dominant H-type mode is the $(m = 0, n = 1)$-mode. The nonvanishing field components of the dominant H-type mode are (omitting mode indices)

$$\left.\begin{aligned}
E_\phi &= \frac{V}{r\Phi} \sin \frac{\pi}{b} z, \\
H_z &= \frac{2I}{b} \sin \frac{\pi}{b} z, \\
H_r &= -j\eta \frac{V}{r\Phi} \frac{\lambda}{2b} \cos \frac{\pi}{b} z.
\end{aligned}\right\} \quad (86)$$

The characteristic impedance and mode constant are

$$Z = \zeta \frac{r\Phi}{2b} \frac{k}{\kappa} \qquad \text{and} \qquad \kappa = \sqrt{k^2 - \left(\frac{\pi}{b}\right)^2}. \quad (87)$$

The transmission-line behavior of the dominant H-type mode is expressed in terms of the standing waves

$$J_0(\kappa r) \qquad \text{and} \qquad N_0(\kappa r).$$

The total outward power flow transported by this dominant mode in a matched nondissipative sectoral guide can be expressed in terms of the rms current I as

$$P = \zeta \frac{r\Phi}{2b} \frac{k}{\kappa} \frac{2}{\pi \kappa r h_0^2(\kappa r)} |I(r)|^2. \tag{88}$$

2·8. Spherical Waveguides.[1]

a. Fields in Free Space.—On introduction of a $r\theta\phi$ spherical coordinate system, as shown in Fig. 2·17, it is evident that free space may be regarded as a nonuniform transmission region or spherical waveguide. The transmission direction is along the radius r and the cross sections transverse thereto are complete spherical surfaces described by the coordinates θ and ϕ. In practice many spherical cavities may be conveniently regarded as terminated spherical guides.

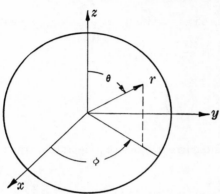

FIG. 2·17.—Spherical coordinate system for waves in space.

The $\mathbf{e'}$-mode functions characteristic of the E-modes in a spherical guide are derivable from scalar functions

$$\Phi_i = \Psi_i = \frac{1}{N_i} P_n^m(\cos\theta) \frac{\cos}{\sin} m\phi, \tag{89}$$

where

$$N_i^2 = \frac{4\pi}{\epsilon_m} \frac{n(n+1)}{2n+1} \frac{(n+m)!}{(n-m)!},$$
$$m = 0, 1, 2, \cdots n-1, n$$
$$n = 1, 2, 3, \cdots,$$
$$\epsilon_m = 1 \quad \text{if } m = 0,$$
$$\epsilon_m = 2 \quad \text{if } m \neq 0,$$

and $P_n^m(\cos\theta)$ is the associated Legendre function of order n and degree m. Typical Legendre functions of argument $\cos\theta$ are

$$\left.\begin{array}{ll} P_0 = 1, & \\ P_1 = \cos\theta, & P_1^1 = -\sin\theta, \\ P_2 = \frac{1}{2}(3\cos^2\theta - 1), & P_2^1 = -3\sin\theta\cos\theta, \quad P_2^2 = 3\sin^2\theta, \\ P_3 = \frac{1}{2}(5\cos^3\theta - 3\cos\theta), & P_3^1 = -\frac{3}{2}\sin\theta(5\cos^2\theta - 1), \\ & P_3^2 = 15\sin^2\theta\cos\theta, \quad P_3^3 = -15\sin^3\theta. \end{array}\right\} \tag{90}$$

---

[1] *Cf.* S. A. Schelkunoff, *Electromagnetic Waves*, Chaps. 10 and 11, Van Nostrand, 1943.

By Eqs. (1·3), (1·10), and (89) the field components of the E_{mn}-mode are

$$E_\theta = -\frac{V'_i}{r}\frac{1}{N_i}\frac{dP_n^m(\cos\theta)}{d\theta}\genfrac{}{}{0pt}{}{\cos}{\sin}m\phi,$$

$$E_\phi = \pm\frac{V'_i}{r}\frac{m}{N_i\sin\theta}P_n^m(\cos\theta)\genfrac{}{}{0pt}{}{\sin}{\cos}m\phi, \qquad (91a)$$

$$E_r = -j\frac{\zeta I'_i}{r^2}\frac{n(n+1)}{kN_i}P_n^m(\cos\theta)\genfrac{}{}{0pt}{}{\cos}{\sin}m\phi,$$

$$H_\theta = \mp\frac{I'_i}{r}\frac{m}{N_i\sin\theta}P_n^m(\cos\theta)\genfrac{}{}{0pt}{}{\sin}{\cos}m\phi,$$

$$H_\phi = -\frac{I'_i}{r}\frac{1}{N_i}\frac{dP_n^m(\cos\theta)}{d\theta}\genfrac{}{}{0pt}{}{\cos}{\sin}m\phi, \qquad (91b)$$

$$H_r = 0,$$

where the degeneracy of the E_{mn}-mode is indicated by the two possible polarizations in ϕ. The r dependence of the fields is determined by the mode voltage V'_i and current I'_i. These quantities obey the spherical transmission-line equations (1·94). The components of the mode functions $\mathbf{e}'_i$ normalized according to Eqs. (1·5) with $dS = \sin\theta\,d\theta\,d\phi$ are obtained from E_θ and E_ϕ of Eq. (91a) on omission of the amplitude factor V'_i/r. The outward power carried by the ith mode is Re $(V_i I_i^*)$.

The dominant mode in a spherical waveguide is the electric dipole mode $n = 1$. For the case of circular symmetry ($m = 0$) the non-vanishing components of this mode are

$$E_\theta = \frac{V}{r}\sqrt{\frac{3}{8\pi}}\sin\theta,$$

$$E_r = -j\frac{\zeta I}{kr^2}\sqrt{\frac{3}{2\pi}}\cos\theta, \qquad (92)$$

$$H_\phi = \frac{I}{r}\sqrt{\frac{3}{8\pi}}\sin\theta,$$

on omission of the mode designations. In terms of the rms voltage V the total power carried by an outward traveling dominant mode is (cf. Sec. 1·8)

$$P = \frac{\eta}{1 - \frac{1}{(kr)^2} + \frac{1}{(kr)^4}}|V(r)|^2. \qquad (93)$$

The $\mathbf{e}''_i$-mode functions characteristic of the H-modes in a spherical waveguide are derivable from the scalar functions shown in Eq. (89). The field components of the H_{mn}-mode follow from Eqs. (1·4), (1·10), and (89) as

$$E_\theta = \pm \frac{V_i''}{r} \frac{m}{N_i \sin \theta} P_n^m(\cos \theta) \frac{\sin}{\cos} m\phi,$$

$$E_\phi = \frac{V_i''}{r} \frac{1}{N_i} \frac{dP_n^m(\cos \theta)}{d\theta} \frac{\cos}{\sin} m\phi, \quad (94a)$$

$$E_r = 0,$$

$$H_\theta = -\frac{I_i''}{r} \frac{1}{N_i} \frac{dP_n^m(\cos \theta)}{d\theta} \frac{\cos}{\sin} m\phi,$$

$$H_\phi = \pm \frac{I_i''}{r} \frac{m}{N_i \sin \theta} P_n^m(\cos \theta) \frac{\sin}{\cos} m\phi, \quad (94b)$$

$$H_r = -j\zeta \frac{V_i''}{r^2} \frac{n(n+1)}{kN_i} P_n^m(\cos \theta) \frac{\cos}{\sin} m\phi.$$

The r dependence of the mode fields is determined by the spherical transmission-line behavior of the mode voltage V_i'' and current I_i''. The components of the normalized mode function are obtained from E_θ and E_ϕ of Eq. (94b) on omission of the amplitude factor V_i''/r.

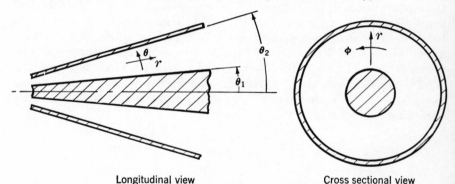

Longitudinal view   Cross sectional view
Fig. 2·18.—Conical waveguide.

As for the case of modes in a radial waveguide, the concepts of cutoff wavelength and guide wavelength lose their customary significance in a spherical guide because of the lack of spatial periodicity along the transmission direction. The cutoff wavelength

$$\lambda_{ci} = \frac{2\pi r}{\sqrt{n(n+1)}} \quad (95)$$

of both the E_{mn}- and H_{mn}-modes is, however, indicative of the regions wherein these modes are propagating or nonpropagating. For regions such that $\lambda < \lambda_{ci}$ the mode fields decay spatially like $1/r$ and hence may be termed "propagating"; conversely for $\lambda > \lambda_{ci}$ the mode fields decay faster than $1/r$ and may, therefore, be termed "nonpropagating."

b. Conical Waveguides.—A typical conical waveguide together with its associated spherical coordinate system $r\theta\phi$ is illustrated in Fig. 2·18.

The transmission direction is along the radius r, and the cross sections transverse thereto are $\theta\phi$ spherical surfaces bounded by cones of aperture θ_1 and θ_2. The conical waveguide is seen to bear the same relation to a spherical waveguide that a coaxial guide bears to a circular guide. Examples of conical guides are provided by tapered sections in coaxial guide, conical antennas, etc.

The dominant E-mode in the conical guide of Fig. 2·18 is a transverse electromagnetic mode whose nonvanishing components are

$$E_\theta = \frac{V}{r \ln \dfrac{\cot\,(\theta_1/2)}{\cot\,(\theta_2/2)}} \frac{1}{\sin\,\theta},$$
$$H_\phi = \frac{I}{2\pi r} \frac{1}{\sin\,\theta}. \qquad (96)$$

The r dependence of the dominant mode voltage V and current I is determined by the spherical transmission-line equations (1·94); for this case of the dominant ($n = 0$) mode these reduce to uniform transmission-line equations. The choice of normalization is such that the characteristic impedance and propagation wave number are

$$Z = \frac{\zeta}{2\pi} \ln \frac{\cot\,(\theta_1/2)}{\cot\,(\theta_2/2)} \quad \text{and} \quad \kappa = k. \qquad (97)$$

The total outward dominant-mode power flow in a matched nondissipative conical guide is correspondingly

$$P = \frac{\zeta}{2\pi} \ln \frac{\cot\,(\theta_1/2)}{\cot\,(\theta_2/2)} |I(r)|^2. \qquad (98)$$

The cutoff wavelength of the dominant mode is infinite.

The attenuation constant of the dominant mode in a conical guide is a function of r and is given by

$$\alpha = \left(\frac{\mathcal{R}_1/\zeta}{r \sin\,\theta_1} + \frac{\mathcal{R}_2/\zeta}{r \sin\,\theta_2}\right) \frac{1}{2 \ln \dfrac{\cot\,(\theta_1/2)}{\cot\,(\theta_2/2)}}, \qquad (99)$$

where $\mathcal{R}_1$ and $\mathcal{R}_2$ are the characteristic resistances (*cf.* Table 1·2) of the inner and outer metallic cones.

The $\mathbf{e}'_i$ mode functions characteristic of the E-modes in a conical guide are derivable by Eqs. (1·3) from scalar functions of the form

$$\Phi_i = [P_n^m(\cos\,\theta)P_n^m(-\cos\,\theta_1) - P_n^m(-\cos\,\theta)P_n^m(\cos\,\theta_1)]\genfrac{}{}{0pt}{}{\cos}{\sin} m\phi, \qquad (100)$$

where the indices n (nonintegral in general) are determined by the roots of

$$P_n^m(\cos \theta_2)P_n^m(-\cos \theta_1) - P_n^m(-\cos \theta_2)P_n^m(\cos \theta_1) = 0,$$

where $m = 0, 1, 2, 3, \cdots$. The $\mathbf{e}_i''$ mode functions characteristic of the H-modes are derivable by Eqs. (1.4) from scalar functions

$$\Psi_i = P_n^m(\cos \theta) \frac{dP_n^m(-\cos \theta_1)}{d\theta} - P_n^m(-\cos \theta) \frac{dP_n^m(\cos \theta_1)}{d\theta}, \quad (101)$$

where the indices n are the roots of

$$\frac{dP_n^m(\cos \theta_2)}{d\theta} \frac{dP_n^m(-\cos \theta_1)}{d\theta} - \frac{dP_n^m(-\cos \theta_2)}{d\theta} \frac{dP_n^m(\cos \theta_1)}{d\theta} = 0$$

where $m = 0, 1, 2, 3, \cdots$. The lack of adequate tabulations both of the roots n and of the fractional order Legendre functions does not justify a detailed representation of the field components of the higher modes. The special case, $\theta_2 = \pi - \theta_1$, of a conical antenna has been investigated in some detail by Schelkunoff (*loc. cit.*).

CHAPTER 3

MICROWAVE NETWORKS

3·1. Representation of Waveguide Discontinuities.—Waveguide structures are composite regions containing not only uniform or nonuniform waveguide regions but also discontinuity regions. The latter are regions wherein there exist discontinuities in cross-sectional shape; these discontinuities may occur within or at the junction of waveguide regions. As indicated in the preceding chapters the fields within each of the waveguide regions are usually completely described by only a single propagating mode. In contrast the complete description of the fields within a discontinuity region generally requires, in addition to the dominant propagating mode, an infinity of nonpropagating modes. Since a waveguide region can be represented by a single transmission line appropriate to the propagating mode, it might be expected that the representation of the discontinuity regions would require an infinity of transmission lines. This expectation is essentially correct but unnecessarily complicated. The nonpropagating nature of the higher-mode transmission lines restricts the complication in field description to the immediate vicinity of the discontinuity. Hence, the discontinuity fields can be effectively regarded as "lumped." The effect of these lumped discontinuities is to introduce corresponding discontinuities into the otherwise continuous spatial variation of the dominant-mode voltage and current on the transmission lines representative of the propagating modes in the over-all microwave structure. Such voltage-current discontinuities can be represented by means of lumped-constant equivalent circuits. The equivalent circuits representative of the discontinuities together with the transmission lines representative of the associated waveguides comprise a microwave network that serves to describe the fields almost everywhere within a general waveguide structure. The present chapter is principally concerned with the general nature and properties of the parameters that characterize such microwave networks.

The determination of the fields within a waveguide structure is primarily an electromagnetic-boundary-value problem. An electromagnetic-boundary-value problem involves the determination of the electric field **E** and magnetic field **H** at every point within a closed region of space. These fields are required to satisfy the Maxwell field equations and to assume prescribed values on the boundary surface enclosing the given region. According to a fundamental theorem a unique solution to this problem exists if the tangential component of *either* the electric field *or* the magnetic field is specified at the boundary surface. The reformulation of this field problem in terms of conventional network concepts will be illustrated for a general type of waveguide structure.

a. *Impedance Representation.*—A typical waveguide structure is depicted in Fig. 3·1. The over-all structure is composed of a discontinuity, or junction, region J and a number of arbitrary waveguide regions 1, . . . , N. The boundary conditions appropriate to this structure are

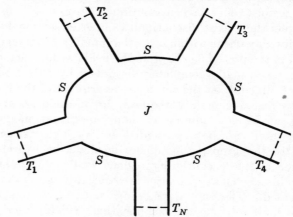

FIG. 3·1.—Junction of N waveguides.

that the electric-field components tangential to the metallic boundary surface S, indicated by solid lines, vanish and that the magnetic-field components tangential to the "terminal," or boundary, surfaces T_1, . . . , T_N, indicated by dashed lines, assume prescribed but arbitrary values. It is further assumed that the dimensions and frequency of excitation are such that only a single mode can be propagated in each of the waveguide regions although this is not a necessary restriction. It is thereby implied that the terminal surfaces T_1, . . . , T_N are so far removed from the junction region J that the fields at each terminal surface are of dominant-mode type. Consequently the tangential electric field $\mathbf{E}_t$ and magnetic field $\mathbf{H}_t$ at any terminal surface T_m may be completely characterized by the equations (*cf.* Sec. 1·3)

$$\left. \begin{array}{l} \mathbf{E}_t(x,y,z_m) = V_m\,\mathbf{e}_m, \\ \mathbf{H}_t(x,y,z_m) = I_m\,\mathbf{h}_m, \end{array} \quad \mathbf{h}_m = \mathbf{z}_{0m} \times \mathbf{e}_m, \right\} \quad (1)$$

where $\mathbf{e}_m$ and $\mathbf{h}_m$ are the vector mode functions indicative of the cross-sectional form of the dominant mode in the mth guide, where $\mathbf{z}_{0m}$ denotes the outward unit vector along the axes of the mth guide, and where the

Sec. 3·1] REPRESENTATION OF WAVEGUIDE DISCONTINUITIES

voltage V_m and current I_m denote the rms amplitudes of the respective fields at T_m; the normalization is such that $\operatorname{Re}(V_m I_m^*)$ represents the average power flowing in the mth guide toward the junction region.

The above-quoted uniqueness theorem states that the electric field within the space enclosed by the terminal surfaces is uniquely determined by the tangential magnetic fields or, equivalently, by the currents I_1, ..., I_N at the terminals $T_1, \ldots, T_N$. In particular the tangential electric fields or, equivalently, the voltages $V_1, \ldots, V_N$ at the terminal surfaces $T_1, \ldots, T_N$ are determined by the currents $I_1, \ldots, I_N$. The linear nature of the field equations makes it possible to deduce the form of the relations between the voltages and currents at the various terminals without the necessity of solving the field equations. By linearity it is evident that the voltages $V_1, \ldots, V_N$ set up by the current I_1, or I_2, ... or I_N acting alone must be of the form

$$\left.\begin{array}{llll}V_1 = Z_{11}I_1, & V_1 = Z_{12}I_2, & V_1 = Z_{1N}I_N \\ V_2 = Z_{21}I_1, & V_2 = Z_{22}I_2, & V_2 = Z_{2N}I_N, \\ \cdot & \cdot & \cdot \\ \cdot & \cdot & \cdot \\ \cdot & \cdot & \cdot \\ V_N = Z_{N1}I_1, & V_N = Z_{N2}I_2, & V_N = Z_{NN}I_N,\end{array}\right\} \quad (2)$$

where the Z_{mn} are proportionality factors, or impedance coefficients, indicative of the voltage set up at the terminal T_m by a unit current acting *only* at the terminal T_n. By superposition the voltages resulting from the simultaneous action of all the currents are given by

$$\left.\begin{array}{l}V_1 = Z_{11}I_1 + Z_{12}I_2 + \cdots + Z_{1N}I_N, \\ V_2 = Z_{21}I_1 + Z_{22}I_2 + \cdots + Z_{2N}I_N, \\ \cdot \quad \cdot \quad \cdot \quad \cdot \\ \cdot \quad \cdot \quad \cdot \quad \cdot \\ V_N = Z_{N1}I_1 + Z_{N2}I_2 + \cdots + Z_{NN}I_N.\end{array}\right\} \quad (3)$$

These so-called network equations, which completely describe the behavior of the propagating modes in the given microwave structure, are frequently characterized simply by the array of impedance coefficients

$$\mathbf{Z} = \begin{pmatrix} Z_{11} & Z_{12} & \cdots & Z_{1N} \\ Z_{21} & Z_{22} & \cdots & Z_{2N} \\ \cdot & \cdot & \cdots & \cdot \\ \cdot & \cdot & \cdots & \cdot \\ \cdot & \cdot & \cdots & \cdot \\ Z_{N1} & Z_{N2} & \cdots & Z_{NN} \end{pmatrix},$$

called the impedance matrix of the structure.

The foregoing analysis of an N terminal pair microwave structure is the exact analogue of the familiar Kirchhoff mesh analysis of an n terminal pair low-frequency electrical structure. As in the latter case, many properties of the impedance coefficients Z_{mn} may be deduced from general considerations without the necessity of solving any field equations. The more important of these properties are $Z_{mn} = Z_{nm}$ and $\mathrm{Re}\,(Z_{mn}) = 0$. With appropriate voltage-current definitions (cf. Chap. 2) the former of these relations are generally valid, whereas the latter pertain only to nondissipative structures. In addition to the above, many useful properties may be derived if certain geometrical symmetries exist in a waveguide structure. Such symmetries impose definite relations among the network parameters Z_{mn} (cf. Sec. 3·2)—relations, it is to be stressed, that can be ascertained without the necessity of solving any field equations. These relations reduce the number of unknown parameters and often yield important qualitative information about the properties of microwave structures.

The form of the network equations (3) together with the reciprocity relations $Z_{mn} = Z_{nm}$ imply the existence of a lumped-constant equivalent circuit which provides both a schematic representation and a structural equivalent of the relations between the voltages and currents at the terminals of the given microwave structure. This equivalent circuit, or network representation, provides no information not contained in the original network equations, but nevertheless serves the purpose of casting the results of field calculations in a conventional engineering mold from which information can be derived by standard engineering calculations. In view of this representation, the boundary-value problem of the determination of the relations between the far transverse electric and magnetic fields on the terminal surfaces is seen to be reformulated as a network problem of the determination of the impedance parameters Z_{mn}. These parameters may be determined either theoretically from the field equations or experimentally by standing-wave measurements on the structure. In either case it is evident that the impedance parameters provide a rigorous description of the dominant modes at the terminal surfaces and hence of the electromagnetic fields almost everywhere. This "far" description, of course, does not include a detailed analysis of the fields in the immediate vicinity of the discontinuities.

In the reformulation of the field description as a network problem the choice of terminal planes is seen to be somewhat arbitrary. This arbitrariness implies the existence of a variety of equivalent networks for the representation of a waveguide structure. Any one of these networks completely characterizes the far field behavior. No general criterion exists to determine which of the equivalent networks is most appropriate. This ambiguous situation does not prevail for the case of lumped low-frequency networks, because there is generally no ambiguity in the choice of terminals of a lumped circuit. However, even at low frequencies there are, in general, many circuits equivalent to any given one, but usually there is a "natural" one distinguished by having a minimum number of impedance elements of simple frequency variation. It is doubtful whether a corresponding "natural" network exists, in general, for any given waveguide structure. In special cases, however, the same criteria of a minimum number of network parameters, simple frequency dependence, etc., can be employed to determine the best network representation. These determinations are facilitated by the ability to transform from a representation at one set of terminal planes to that at another (*cf.* Sec. 3·3).

Various definitions of voltage and current may be employed as measures of the transverse fields in waveguide regions. The arbitrariness in definition introduces an additional source of flexibility in the network representation of waveguide structures. For example, if the voltages V_n and currents I_n employed in Eqs. (3) are transformed into a new set $\bar{V}_n$ and $\bar{I}_n$ by

$$V_n = \frac{\bar{V}_n}{\sqrt{N_n}}, \qquad I_n = \bar{I}_n \sqrt{N_n}, \tag{4a}$$

the transformed network equations retain the same form as Eqs. (3) provided the transformed impedance elements are given by

$$\bar{Z}_{mn} = Z_{mn} \sqrt{N_m N_n}. \tag{4b}$$

The new representation may possess features of simplicity not contained in the original representation. Because of this it is frequently desirable to forsake the more conventional definitions of voltage and current in order to secure a simplicity of circuit representation. It should be noted that the new definitions are equivalent to a change in the characteristic impedances of the terminal waveguides or, alternatively, to an introduction of ideal transformers at the various terminals.

b. Admittance Representation.—Although the preceding reformulation of the "far" field description of the microwave structure of Fig. 3·1 has been carried through an impedance basis, an equivalent reformulation on an admittance basis is possible. In the latter case the original boundary value problem is specified by indication of the transverse components of the electric rather than the magnetic field on the terminal surfaces $T_1, \ldots, T_N$. The introduction of voltages V_m and currents I_m on the terminal planes together with a Kirchhoff analysis on a node basis (i.e. V_m rather than I_m specified at T_m) leads in this case to network equations of the form

$$\left.\begin{aligned}I_1 &= Y_{11}V_1 + Y_{12}V_2 + \cdots + Y_{1N}V_N, \\ I_2 &= Y_{21}V_1 + Y_{22}V_2 + \cdots + Y_{2N}V_N, \\ &\;\;\vdots \\ I_N &= Y_{N1}V_1 + Y_{N2}V_2 + \cdots + Y_{NN}V_N,\end{aligned}\right\} \quad (5)$$

where the admittance elements Y_{mn} possess the same general properties $Y_{mn} = Y_{nm}$ and in the nondissipative case $\mathrm{Re}\,(Y_{mn}) = 0$ as the impedance elements of Eqs. (3). In this case the admittance element Y_{mn} represents the current set up at the terminal T_m by a unit voltage applied *only* at the terminal T_n. As an alternative to Eqs. (5) the array of admittance coefficients

$$\mathbf{Y} = \begin{pmatrix} Y_{11} & Y_{12} & \cdots & Y_{1N} \\ Y_{21} & Y_{22} & \cdots & Y_{2N} \\ \vdots & \vdots & & \vdots \\ Y_{N1} & Y_{N2} & \cdots & Y_{NN} \end{pmatrix},$$

called the admittance matrix of the waveguide structure, is sometimes employed to characterize the dominant-mode behavior of the given structure.

The statements relative to the arbitrariness in choice of terminal planes and voltage-current definitions apply equally well to the admittance description. However, the equivalent network representation of the network equations (5) is dual rather than identical with the network representation of Eqs. (3).

c. Scattering Representation.—An alternative description of the fields within the waveguide structure of Fig. 3·1 stems from a reformulation of the associated field problem as a scattering problem. Accordingly, in addition to the general requirement of the vanishing of the electric-field components tangential to the metallic surfaces, the original boundary-value problem is defined by specification of the amplitudes of the waves incident on the terminal planes $T_1, \ldots, T_N$. In this scattering type of description the dominant-mode fields at any point in the waveguide regions are described by the amplitudes of the incident and reflected (scattered) waves at that point. In particular the fields at the terminal plane T_m are described by

$$\left.\begin{aligned}\mathbf{E}_t(x,y,z_m) &= (a_m + b_m)\mathbf{e}_m, \\ \mathbf{H}_t(x,y,z_m) &= (a_m - b_m)\mathbf{h}_m,\end{aligned} \quad \mathbf{h}_m = \mathbf{z}_{0m} \times \mathbf{e}_m,\right\} \quad (6)$$

where $\mathbf{e}_m$ and $\mathbf{h}_m$ are vector mode functions characteristic of the transverse form of the dominant mode in the mth guide and a_m and b_m are, respectively, the complex amplitudes of the electric field in the incident and reflected wave components of the dominant mode field at T_m. The normalization of the mode functions is such that the total inward power flow at any terminal T_m is given by $|a_m|^2 - |b_m|^2$; this corresponds to a choice of unity for the characteristic impedance of the mth guide.

The fundamental existence theorem applicable to the scattering formulation of a field problem states that the amplitudes of the scattered waves at the various terminals are uniquely related to the amplitudes of the incident waves thereon. As in the previous representations the form of this relation is readily found by adduction of the linear nature of the electromagnetic field. Because of linearity the amplitudes of the reflected waves set up at the terminal planes $T_1, \ldots, T_N$ by a single incident wave a_1 at T_1, or a_2 at $T_2, \ldots$ are

$$\left.\begin{aligned}
b_1 &= S_{11}a_1, & b_1 &= S_{12}a_2, & b_1 &= S_{1N}a_N, \\
b_2 &= S_{21}a_1, & b_2 &= S_{22}a_2, & b_2 &= S_{2N}a_N, \\
&\vdots & &\vdots & &\vdots \\
b_N &= S_{N1}a_1, & b_N &= S_{N2}a_2, & b_n &= S_{NN}a_N.
\end{aligned}\right\} \quad (7)$$

Therefore, by superposition the amplitudes of the scattered waves arising from the simultaneous incidence of waves of amplitudes $a_1, \ldots, a_N$ are

$$\left.\begin{aligned}
b_1 &= S_{11}a_1 + S_{12}a_2 + \cdots + S_{1N}a_N, \\
b_2 &= S_{21}a_1 + S_{22}a_2 + \cdots + S_{2N}a_N, \\
&\vdots \\
b_N &= S_{N1}a_1 + S_{N2}a_2 + \cdots + S_{NN}a_N
\end{aligned}\right\} \quad (8)$$

where the proportionality factor, or scattering coefficient, S_{mn} is a measure of the amplitude of the wave scattered into the mth guide by an incident wave of unit amplitude in the nth guide. In particular, therefore, the coefficient S_{mm} represents the reflection coefficient at the terminal T_m when all other terminals are "matched." For brevity it is frequently desirable to characterize the scattering properties of a waveguide structure by the array of coefficients

$$\mathbf{S} = \begin{pmatrix} S_{11} & S_{12} & \cdots & S_{1N} \\ S_{21} & S_{22} & \cdots & S_{2N} \\ \vdots & \vdots & & \vdots \\ S_{N1} & S_{N2} & \cdots & S_{NN} \end{pmatrix}, \quad (9)$$

called the scattering matrix, rather than by Eqs. (8).

The elements S_{mn} of the scattering matrix **S** may be determined either theoretically or experimentally. The values so obtained are dependent on the choice of terminal planes and the definitions of incident and scattered amplitudes. Certain general properties of scattering coefficients may be deduced from general considerations. For example, with the above definitions [Eqs. (6)] of the amplitudes a_n and b_n, it can be shown that

1. The reciprocity relations $S_{mn} = S_{nm}$,

2. The unitary relations $\sum_{\beta=1}^{N} S^{*}_{\beta m} S_{\beta n} = \delta_{mn} = \begin{cases} 1 \text{ if } m = n, \\ 0 \text{ if } m \neq n, \end{cases}$ (10)

are valid; the latter apply only to nondissipative structures.

If the given structure possesses geometrical symmetries, it is possible to derive corresponding symmetry relations among the scattering coefficients. These relations, derivable without the necessity of solving field equations or performing measurements, are identical with those for the elements of the impedance or admittance matrices of the same structure.

The reformulation of field problems either as network problems or as scattering problems provides fully equivalent and equally rigorous descriptions of the far field in a microwave structure. The choice of one or the other type of description is difficult to decide in many cases. In favor of the impedance or admittance descriptions are the following facts: (1) The descriptions are in close accord with conventional low-frequency network descriptions; (2) they can be schematically represented by equivalent circuits; (3) they lead to simple representations of many series or shunt combinations of discontinuities and junctions. In favor of the scattering description are the facts: (1) It is particularly simple and intuitive when applied to the important case of matched or nearly matched microwave structures; (2) reference-plane transformations can be effected quite simply by phase shifts of the scattering coefficients.

For the most part impedance or admittance descriptions are employed throughout the present volume since it is desired to stress the connection between microwave network analysis and the conventional low-frequency network analysis. For interrelations among the various descriptions the reader is referred to *Principles of Microwave Circuits* by C. G. Montgomery and R. Dicke, Vol. 8 of this series.

3·2. Equivalent Circuits for Waveguide Discontinuities.—The Kirchhoff analysis of the far fields within a general N-terminal pair microwave structure can be expressed in terms of $N(N + 1)/2$ complex parameters [*cf.* Eqs. (3) and (5)] and represented by a general N-terminal-pair equivalent network. If the structure possesses geometrical symmetries, it is possible to reduce the number of unknown network parameters and correspondingly simplify the form of the equivalent network by means of a Kirchhoff analysis that utilizes these symmetries. Symmetrical structures are characterized by the existence of two or more terminal planes looking into any one of which the structure appears electrically identical. As outlined in the preceding section, a Kirchhoff analysis of the response due to current excitation at one of these symmetrical terminal planes is described by impedance coefficients given by one of the columns in Eqs. (2). The columns describing the responses due to current excitation at the other symmetrical terminal planes can be expressed in terms of these same impedance coefficients, but in different order. It is thus evident that the symmetry properties of the given structure can serve to reduce the number of unknown impedance coefficients. Results of analyses utilizing structural symmetries will be tabulated in this section for several microwave discontinuities.

The Kirchhoff analysis of a symmetrical microwave structure can be effected on either an impedance or an admittance basis. The choice of analysis is generally dictated (at least for the $N > 3$ terminal pair structures) by the type of geometrical symmetry possessed by the structure. It is not implied hereby that only one type of description is possible in a given case. An impedance or an admittance description is always possible. In a structure with a certain type of symmetry the impedance description, for example, may be found most desirable since the parameters of the equivalent circuit for the structure may be simply related to the elements of the impedance matrix but not to those of the admittance matrix. The possible existence of another equivalent circuit whose parameters are simply related to the elements of the admittance rather than of the impedance matrix is not excluded. However, the two equivalent circuits will not, in general, be equally simple. The preferred description is that based on the simplest equivalent circuit.

In the following the equivalent circuits together with the corresponding impedance, or admittance, representations of several general classes of microwave structures will be presented. No detailed effort will be made either to show how the symmetries of the structure delimit the form of the matrix and circuit representations or to discuss the reasons for the choice of a particular representation. The consistency of a representation with the symmetry of a structure can be readily verified on application of a Kirchhoff analysis both to the given structure and to the equivalent circuit. These analyses lead, of course, to the same matrix representation. Incidentally the recognition of the applicability of conventional Kirchhoff analyses to microwave structures constitutes an important engineering asset, for one can thereby set up and delimit the impedance or admittance matrix or, alternatively, the equivalent circuit representation thereof and derive much information about the behavior of a given microwave structure without the necessity of solving any field equations.

For a given type of geometrical symmetry the equivalent circuit information will be seen to apply equally well to a variety of waveguide structures of which only a few will be pointed out. Since no specific choice either of terminal planes or of voltage-current definitions will be made, the representations to be presented are of a quite general form and can be considerably simplified by a judicious choice of these factors (*cf.* Sec. 3·3). However, the positive directions of voltage and current will be indicated since the form of the impedance or admittance matrix

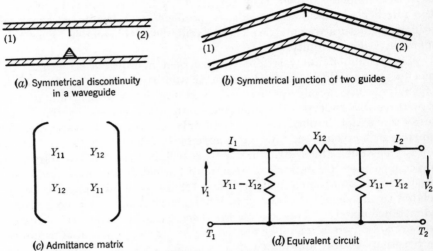

Fig. 3·2.—Symmetrical two-terminal-pair waveguide structures.

(though not the equivalent circuit) depends on this choice. Furthermore the location of the terminal planes, though arbitrary, must be in accord with the symmetry of the given structure. It is assumed throughout that only the dominant mode can be propagated in each of the waveguides, this unnecessary restriction being employed only for the sake of simplicity

a. *Two-terminal-pair Networks.*—Typical two-terminal-pair waveguide structures of arbitrary cross section are illustrated in Fig. 3·2a and b. For a symmetrical choice of the terminals T_1 and T_2 relative to the central plane, the symmetry of the structure imposes a corresponding symmetry on the admittance matrix and equivalent circuit representation of the over-all structure. The general representation of a two-terminal-pair structure is thereby reduced to that shown in Fig. 3·2c and d. The positive directions of voltage and current have been so chosen as to obtain positive off-diagonal elements in the admittance matrix.

The simplification in circuit description resulting from the symmetry of the above two-terminal-pair structures can be taken into account equally well on an impedance basis. In this case the circuit representation is expressed in terms of a symmetrical T-circuit rather than of the π-circuit employed in the admittance description.

   b. *Three-terminal-pair Networks.*—An arbitrary junction of three waveguides may be represented by either the impedance or the admittance matrix shown in Fig. 3·3a or b. The equivalent circuits correspond-

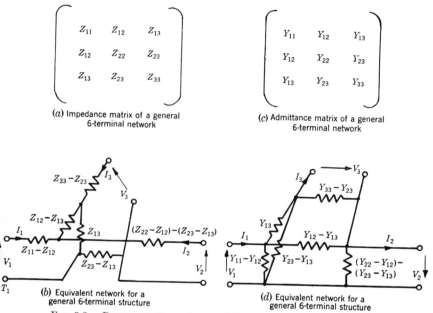

Fig. 3·3.—Representations of general three-terminal-pair structure.

ing to these matrices are dual to one another and can be represented as indicated in Fig. 3·3c and d. These representations can be considerably simplified for the case of symmetrical structures.

   An important class of symmetrical three-terminal-pair structures is that in which geometrical symmetry exists with respect to a plane. Such symmetry implies that the symmetry plane bisects one of the guides, the so-called stub guide, and is centrally disposed relative to the remaining two guides, the latter being designated as main guides. Structures with this planar symmetry may possess either E- or H-plane symmetry, depending on the type and relative orientation of the propagating modes in the main and stub guides. E-plane symmetry obtains when symmetrical electric-field excitation in the main guides results in no coupling to the stub guide. On the other hand, H-plane symmetry implies that antisymmetrical electric-field excitation in the main guides produces no excitation of the stub guide. It should be noted that such properties are not present if modes other than the dominant can be propagated in the stub guide.

STRUCTURES WITH E-PLANE SYMMETRY.—Two junctions with E-plane symmetry are illustrated and represented in Fig. 3·4. When formed of guides with rectangular cross section, such junctions are characterized

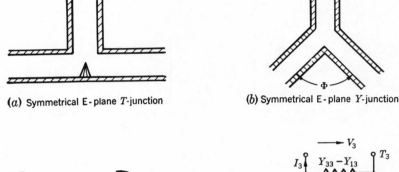

(a) Symmetrical E-plane T-junction

(b) Symmetrical E-plane Y-junction

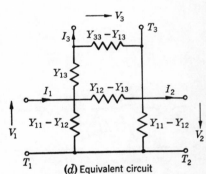

(c) Admittance matrix

(d) Equivalent circuit

FIG. 3·4.—Symmetrical three-terminal-pair structures—E-plane symmetry.

by the fact that the far electric field is everywhere parallel to the plane of the above figures. The indicated admittance matrix and equivalent circuit representations of such structures depend on a symmetrical choice of terminal planes in guides (1) and (2).

For the special case of a Y junction with $\Phi = 120°$, it follows from the added symmetry that $Y_{12} = Y_{13}$ and $Y_{11} = Y_{33}$, provided the terminal plane in guide (3) is selected in the same symmetrical manner as those in guides (1) and (2). The equivalent circuit of Fig. 3·4d therefore reduces to that shown in Fig. 3·5a.

For the case of a Y junction with $\Phi = 0$, the so called E-plane bifurcation, the sum of the terminal voltages is zero and consequently $Y_{ij} = \infty$ ($i, j = 1, 2,$ or 3) if the terminal planes are all chosen at the plane of the bifurcation. Although the admittance matrix is singular in this case, differences of the matrix elements are finite and the equivalent circuit of Fig. 3·4d reduces to that shown in Fig. 3·5b. For a bifurcation with a

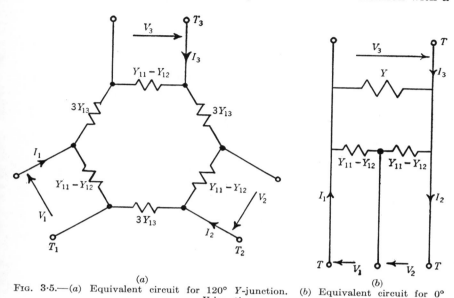

Fig. 3·5.—(a) Equivalent circuit for 120° Y-junction. (b) Equivalent circuit for 0° Y-junction.

dividing wall of arbitrary thickness

$$Y = Y_{12} + Y_{33} - 2Y_{13},$$

and for a wall of zero thickness

$$Y = -\frac{Y_{11} - Y_{12}}{2}.$$

STRUCTURES WITH H-PLANE SYMMETRY.—The sectional views of the junctions illustrated in Fig. 3·4a and b apply as well to junctions with H-plane symmetry. In the latter case, for guides of rectangular cross section, the far magnetic field is everywhere parallel to the plane of the figures. Coaxial T and Y junctions, though not possessing the same geometrical structure, have the same type of field symmetry; junctions of this type are illustrated in Fig. 3·6a and b. The associated impedance matrix and equivalent circuit representations shown in Fig. 3·6 correspond to a symmetrical choice of terminal planes in guides (1) and (2).

If $\Phi = 120°$ in the H-plane Y junctions of Fig. 3·4b, the higher degree of symmetry implies that $Z_{12} = Z_{13}$ and $Z_{11} = Z_{33}$, provided the terminal plane in guide (3) is chosen symmetrically with those in guides (1) and (2).

In this case the equivalent network of Fig. 3·6d becomes completely symmetrical and is composed of a common shunt arm of impedance Z_{13} and identical series arms of impedance $Z_{11} - Z_{12}$.

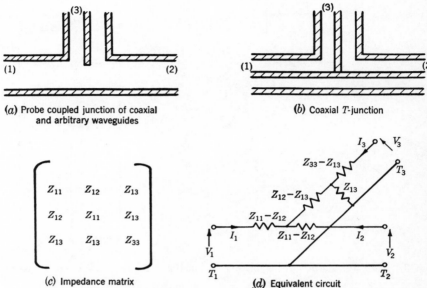

FIG. 3·6.—Symmetrical three-terminal-pair structures—H-plane symmetry.

c. *Four-terminal-pair Networks.* JUNCTIONS WITH E-PLANE SYMMETRY.—Junctions of four rectangular guides with E-plane symmetry are indicated in Fig. 3·7a and b. Since there exist two symmetry planes, either guides (1) and (2) or guides (3) and (4) can be designated as the main guides or as the stub guides. The designation E-plane is consistent with the fact that the far electric-field intensity is everywhere parallel to the sectional plane indicated in the figure. If guides (1), (2) and guides (3), (4) are identical and the terminal planes in identical guides are chosen symmetrically, the admittance matrix and equivalent circuit representations of the structure are shown in Fig. 3·7c and d.

The indicated equivalent circuit applies to the junction in Fig. 3·7b only if the thickness of the dividing wall is sufficiently large to make negligible the E-mode coupling (i.e., the coupling resulting when the normal electric field is a maximum at the aperture). If this situation does not prevail, as is the case when the thickness of the dividing wall is small, the $+45°$ diagonal elements of the admittance matrix of Fig. 3·7c should be changed from Y_{13} to Y_{14} in order to take account of both E- and H-modes of coupling through the aperture. For the case of four identical guides, a dividing wall of zero thickness, and all terminal planes

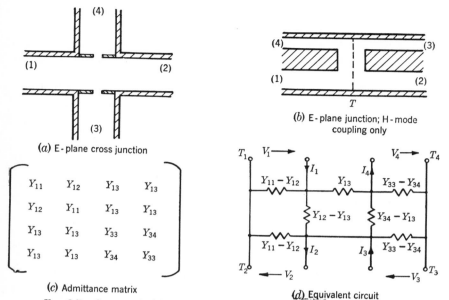

FIG. 3·7.—Symmetrical four-terminal-pair structures—H-plane symmetry.

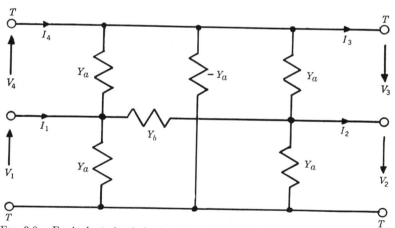

FIG. 3·8.—Equivalent circuit for junction of Fig. 3·7b with wall of zero thickness.

chosen coincident at the central reference plane T, the elements of the admittance matrix become infinite but differ from one another by a finite amount. The equivalent circuit corresponding to the resulting singular matrix is illustrated in Fig. 3·8 where

$$Y_a = 2(Y_{11} - Y_{12}) = 2(Y_{13} - Y_{14}) = 2(Y_{33} - Y_{3'}),$$
$$Y_b = 2(Y_{12} - Y_{13}) = 2(Y_{34} - Y_{13}).$$

JUNCTIONS WITH H-PLANE SYMMETRY.—The sectional views of Fig. 3·7a and b apply equally well to junctions with H-plane symmetry; in such junctions the far magnetic field intensity is everywhere parallel (i.e., the far electric field is everywhere perpendicular) to the plane of the indicated sectional view. The coaxial guide junctions shown in Fig. 3·9a and b also possess the same field symmetry as the H-plane junctions.

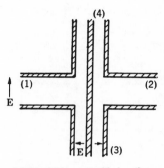

(a) Coaxial to waveguide junction

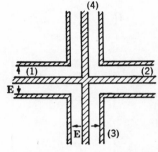

(b) Cross junction of two coaxial guides

$$\begin{pmatrix} Z_{11} & Z_{12} & Z_{13} & Z_{13} \\ Z_{12} & Z_{11} & Z_{13} & Z_{13} \\ Z_{13} & Z_{13} & Z_{33} & Z_{34} \\ Z_{13} & Z_{13} & Z_{34} & Z_{33} \end{pmatrix}$$

(c) Impedance matrix

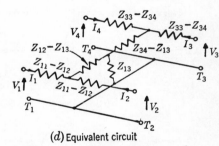

(d) Equivalent circuit

FIG. 3·9.—Symmetrical four-terminal-pair structures—H-plane symmetry.

If the terminal planes T_1 and T_2 are chosen symmetrically (as likewise T_3 and T_4), the equivalent circuit and impedance matrix representations of this class of structures are shown in Fig. 3·9c and d.

If the dividing wall in the H-plane junction of Fig. 3·7b is of small thickness, the +45° diagonal elements of the impedance matrix are to be changed from Z_{13} to Z_{14}. The corresponding equivalent circuit is shown in Fig. 3·10a. The special case of identical guides, a dividing wall of zero thickness, and all terminal planes coinciding at the central reference plane T is represented by the equivalent circuit of Fig. 3·10b.

MAGIC T-JUNCTIONS.—Two typical magic T-junctions are depicted in Fig. 3·11a and b. In Fig. 3·11a a symmetrical junction of four rectangular guides is illustrated in which guide (3) is the H-plane stub and guide (4) is the E-plane stub. Figure 3·11b is a symmetrical junction of one

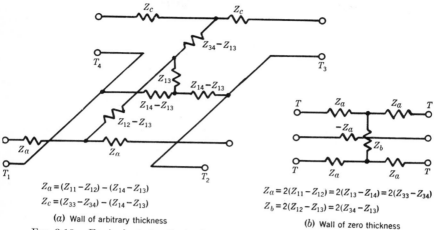

$Z_a = (Z_{11} - Z_{12}) - (Z_{14} - Z_{13})$
$Z_c = (Z_{33} - Z_{34}) - (Z_{14} - Z_{13})$

(a) Wall of arbitrary thickness

$Z_a = 2(Z_{11} - Z_{12}) = 2(Z_{13} - Z_{14}) = 2(Z_{33} - Z_{34})$
$Z_b = 2(Z_{12} - Z_{13}) = 2(Z_{34} - Z_{13})$

(b) Wall of zero thickness

FIG. 3·10.—Equivalent circuits for junction of Fig. 3·7b—H-plane symmetry.

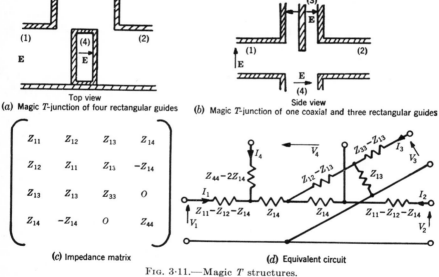

(a) Magic T-junction of four rectangular guides

(b) Magic T-junction of one coaxial and three rectangular guides

(c) Impedance matrix

(d) Equivalent circuit

FIG. 3·11.—Magic T structures.

coaxial and three rectangular guides; in this figure the coaxial guide (3) is the H-plane stub. If the terminal planes in the identical guides (1) and (2) are chosen symmetrically, the impedance matrix and equivalent circuit representations of these junctions are given in Figs. 3·11c and d.

3·3. Equivalent Representations of Microwave Networks.—Many of the equivalent circuits indicated in the preceding sections may be unsuitable in practice either because of difficulties in carrying out network computations or because of the complexity in the measurement and frequency dependence of the circuit parameters. By an appropriate choice both of voltage-current definitions and of reference planes, alternative circuits can be devised in which such difficulties are minimized. Several equivalent representations obtained in this manner will be described in the present section. Since symmetrical N-terminal-pair representations can often be reduced by symmetry analyses (bisection

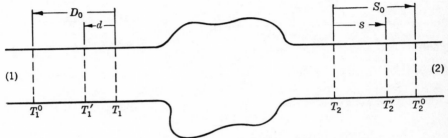

Fig. 3·12.—General two-terminal-pair waveguide discontinuity.

theorems) to a number of two-terminal-pair networks or less, equivalent representations of the basic two-terminal-pair structures will be considered first.

The arbitrary discontinuity at a junction of two different guides illustrated in Fig. 3·12 is an example of a general two-terminal-pair microwave structure. The over-all structure may be represented by transmission lines of characteristic impedances $Z_1 = 1/Y_1$ and $Z_2 = 1/Y_2$, connected at the terminal planes T_1 and T_2 by either the T or π equivalent

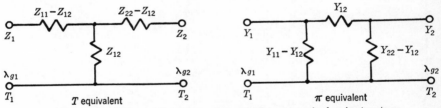

Fig. 3·13a.—Circuit representations of a general two-terminal-pair structure.

circuit indicated in Fig. 3·13a. The relations among the circuit parameters of the T and π representations at the terminals T_1 and T_2 are

$$Z_{11} - Z_{12} = \frac{Y_{22} - Y_{12}}{|Y|}, \qquad Y_{11} - Y_{12} = \frac{Z_{22} - Z_{12}}{|Z|},$$

$$Z_{12} = \frac{Y_{12}}{|Y|}, \qquad Y_{12} = \frac{Z_{12}}{|Z|}, \qquad (11)$$

$$Z_{22} - Z_{12} = \frac{Y_{11} - Y_{12}}{|Y|}, \qquad Y_{22} - Y_{12} = \frac{Z_{11} - Z_{12}}{|Z|},$$

where

$$|Y| = Y_{11}Y_{22} - Y_{12}^2 = \frac{1}{|Z|}, \qquad |Z| = Z_{11}Z_{22} - Z_{12}^2 = \frac{1}{|Y|}.$$

The relation between the input impedance Z_{in} (or input admittance Y_{in}) at T_1 and the output impedance Z_{out} (or output admittance Y_{out}) at T_2 is given by

$$Z_{in} = Z_{11} - \frac{Z_{12}^2}{Z_{22} + Z_{out}}, \quad \left(\text{or } Y_{in} = Y_{11} - \frac{Y_{12}^2}{Y_{22} + Y_{out}} \right). \quad (12)$$

At the same terminals T_1 and T_2 alternative representations of the above discontinuity are provided by the series-shunt circuits of Fig. 3·13b.

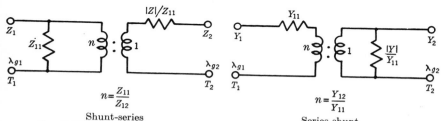

Shunt-series Series-shunt
FIG. 3·13b.—Circuit representations of a general two-terminal-pair structure.

These dual circuits are equivalent to the T and π-circuits shown in Fig. 3·13a. The primary-secondary turns ratio of the ideal transformer is denoted by $n/1$; the corresponding *impedance* ratio at the transformer terminals is $n^2/1$. It is evident that for a structure in which the determinant $|Z|$ or $|Y|$ vanishes, the equivalent circuit becomes either purely shunt or purely series, respectively. In this special case the ideal transformer can be omitted if the characteristic impedance of the output line is changed to $n^2 Z_2$; i.e., if the voltage-current definitions in the output guide are changed.

A variety of other equivalent representations for two-terminal-pair structures can be found by employing transmission lines as circuit elements.

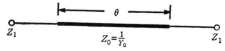

FIG. 3·14.—Transmission-line representation of a symmetrical two-terminal-pair structure.

Thus, as shown in Fig. 3·14, a transmission line of length $\theta = \kappa l$ and characteristic impedance Z_0 can be employed to represent a *symmetric* discontinuity structure with $Z_{11} = Z_{22}(Y_{11} = Y_{22})$. In terms of the parameters of the circuit representations of Fig. 3·13a and b, the transmission-line parameters are

$$Z_0 = \sqrt{Z_{11}^2 - Z_{12}^2}, \qquad Y_0 = \sqrt{Y_{11}^2 - Y_{12}^2},$$
$$\tan\frac{\theta}{2} = \sqrt{\frac{Z_{12} - Z_{11}}{Z_{12} + Z_{11}}}, \qquad \tan\frac{\theta}{2} = \sqrt{\frac{Y_{12} - Y_{11}}{Y_{12} + Y_{11}}}.$$

The consideration of the corresponding representations for asymmetrical structures will be deferred until the closely related question of the transformation of reference planes is treated.

Transformations of Reference Planes.—Equivalent circuit representations of a waveguide discontinuity may be considerably simpler at one set of terminal planes than at another. The investigation of simplifications of this type requires the ability to determine the equivalent circuit parameters at one set of reference planes from the knowledge of the parameters at any other set. For the case of the structure shown in Fig. 3·12, shifts of the input terminals from T_1 to T_1', a distance d away from the junction, and of the output terminals from T_2 to T_2', a distance s away from the junction, can be accomplished in several ways. A straightforward way of effecting this shift involves the addition of transmission lines (or their equivalent circuits) of lengths d and s to the input and output terminals, respectively; the characteristic impedance and propagation wave number of the input and output lines being $Z_1 = 1/Y_1$, $\kappa_1 = 2\pi/\lambda_{g1}$, and $Z_2 = 1/Y_2$, $\kappa_2 = 2\pi/\lambda_{g2}$. The computation of the "shifted" parameters can be carried out by standard circuit techniques. Though somewhat laborious, this method has the virtue of being applicable to N-terminal-pair structures involving both uniform and nonuniform transmission lines. Phase shift of the scattering matrix of a microwave structure provides an alternative method of reference-plane transformations, but this will not be discussed herein.

For the particular case of uniform lines there is another way of effecting the desired transformation. This method is based on the fact that an arbitrary two-terminal-pair network can be represented as an ideal transformer at certain "characteristic" reference planes. Since reference plane transformations to and from these "characteristic" terminals can be readily accomplished, a simple means of carrying out arbitrary transformations is thereby provided.

The existence of an ideal transformer representation of the two-terminal-pair structure of Fig. 3·12 follows from the fact that at the terminals T_1 and T_2 the input-output relations of Eqs. (12) can be rewritten in terms of three new parameters D_0, S_0, γ as

$$\tan \kappa_1 (D - D_0) = \gamma \tan \kappa_2 (S - S_0) \tag{13}$$

if the change of variables

$$\left.\begin{array}{ll} Z_{\text{in}} = -jZ_1 \tan \kappa_1 D, & Y_{\text{in}} = +jY_1 \cot \kappa_1 D, \\ Z_{\text{out}} = +jZ_2 \tan \kappa_2 S, & Y_{\text{out}} = -jY_2 \cot \kappa_2 S \end{array}\right\} \tag{14}$$

is made. The relations between the parameters D_0, S_0, γ of the tangent relation (13) and the parameters of the T or π representation of Fig. 3·13 are given either as

$$\left.\begin{aligned} a &= -\frac{\alpha\beta + \gamma}{\beta - \alpha\gamma}, \\ c &= \frac{1 + \alpha\beta\gamma}{\beta - \alpha\gamma}, \\ b &= \frac{\alpha - \beta\gamma}{\beta - \alpha\gamma}, \end{aligned}\right\} \quad (15)$$

or conversely as

$$\left.\begin{aligned} \alpha &= \frac{1 + c^2 - a^2 - b^2}{2(a - bc)} \pm \sqrt{\left[\frac{1 + c^2 - a^2 - b^2}{2(a - bc)}\right]^2 + 1}, \\ \beta &= \frac{1 + a^2 - c^2 - b^2}{2(c - ba)} \mp \sqrt{\left[\frac{1 + a^2 - c^2 - b^2}{2(c - ba)}\right]^2 + 1}, \\ -\gamma &= \frac{1 + a^2 + c^2 + b^2}{2(b + ac)} \pm \sqrt{\left[\frac{1 + a^2 + c^2 + b^2}{2(b + ac)}\right]^2 - 1}, \end{aligned}\right\} \quad (16)$$

where for the

T Representation $\qquad\qquad\qquad \pi$ Representation

$$\left.\begin{aligned} \alpha &= \tan\frac{2\pi}{\lambda_{g1}} D_0 & \alpha &= -\cot\frac{2\pi}{\lambda_{g1}} D_0 \\ \beta &= \tan\frac{2\pi}{\lambda_{g2}} S_0 & \beta &= -\cot\frac{2\pi}{\lambda_{g2}} S_0 \\ a &= -j\frac{Z_{11}}{Z_1} & a &= -j\frac{Y_{11}}{Y_1} \\ c &= -j\frac{Z_{22}}{Z_2} & c &= -j\frac{Y_{22}}{Y_2} \\ b &= \frac{Z_{11}Z_{22} - Z_{12}^2}{Z_1 Z_2} & b &= \frac{Y_{11}Y_{22} - Y_{12}^2}{Y_1 Y_2} \end{aligned}\right\} \quad (17)$$

The relations (15) are determined by expansion and identification of terms in Eqs. (12) and (13); Eqs. (16) follow from Eqs. (15) by inversion. Equations (15) are not valid for the degenerate case $\alpha = \beta = 0$, as is to be expected from the corresponding degeneracy in the impedance representation of an ideal transformer. The $\pm$ signs in Eqs. (16) indicate the existence of two sets of α, β, γ equivalent to a, b, c; these sets are positive or negative reciprocals of each other. For each value of γ given by Eqs. (16), the corresponding set of values for α and β may be obtained from

$$\alpha = \frac{b + \gamma}{c + a\gamma}, \qquad \beta = \frac{\gamma(b + ac) + (1 + a^2)}{c - ab},$$

or
$$\alpha = -\frac{a + c\gamma}{1 + b\gamma}, \qquad \beta = -\frac{\gamma(b^2 + c^2) + (b + ac)}{\gamma(c - ab)}.$$

For each value of α, the corresponding set of values of β and γ are

$$\beta = \frac{b - \alpha c}{a + \alpha}, \qquad \gamma = \frac{b - \alpha c}{\alpha a + 1},$$

or

$$\beta = \frac{1 - \alpha a}{c + \alpha b}, \qquad \gamma = -\frac{\alpha + a}{c + \alpha b}.$$

For nondissipative structures with purely reactive output impedances it is evident from Eqs. (14) that both D and S are real. The quantity D is then the distance from the terminal T_1 to a voltage node in the input line and is counted positive in the direction away from the junctions;

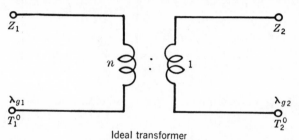

Ideal transformer

FIG. 3·15.—Ideal transformer representation of a nondissipative two-terminal-pair structure at characteristic reference planes.

correspondingly, S is the distance from T_2 to a voltage node in the output line and is also positive in the direction away from the junction. Thus if γ is written as $-n^2 Z_2/Z_1$, Eq. (13) states that "characteristic" terminals T_1^0 and T_2^0 exist, distant D_0 and S_0 away from T_1 and T_2, at which the input impedance is a constant n^2 times the output impedance.[1] Therefore at the terminals T_1^0 and T_2^0 the equivalent circuit of the nondissipative waveguide junction shown in Fig. 3·12 is the ideal transformer depicted in Fig. 3·15.

The equivalence between the transformer representation at T_1^0, T_2^0 and the T or π representation at T_1, T_2 can be rephrased as an equivalence at the same set of terminals. For example, if lengths D_0 and S_0 of input and output transmission lines are added, respectively, to the terminals T_1 and T_2 of the T or π representation of Figs. 3·13, a representation is obtained at T_1^0 and T_2^0 that is equivalent to the transformer representation of Fig. 3·15. Conversely, if lengths $-D_0$ and $-S_0$ of input and output lines are added to the terminals T_1^0, T_2^0 of the transformer representation,

---

[1] *Cf.* A. Weissfloch, *Hochfreq. u. Elektro.*, vol. 60, 1942, pp. 67 *et seq.*

a representation is obtained at T_1, T_2 that is equivalent to the T or π representation.

The transformer representation embodied in Eq. (13) provides a relatively simple means of determining the parameters a', b', and c' [cf. Eqs. (17)] of a network representation at any terminals T_1', T_2' from the corresponding parameters a, b, and c of a representation at the terminals T_1, T_2. Let it be assumed, as indicated in Fig. 3·12, that the reference planes T_1' and T_2' are located at distances d and s, respectively, away from T_1 and T_2. The form of the tangent relation relative to the new reference planes T_1' and T_2' can be readily obtained from that at the reference planes T_1 and T_2 by rewriting Eq. (13) as

$$\tan \kappa_1[(D - d) - (D_0 - d)] = \gamma \tan \kappa_2[(S - s) - (S_0 - s)]. \quad (18)$$

Comparison of Eqs. (13) and (18) indicates that relative to the new terminals T_1' and T_2' the parameters α', β', and γ', as defined in Eqs. (17), are given by

$$\left.\begin{aligned} \alpha' &= \tan \kappa_1(D_0 - d) = \frac{\alpha - \alpha_0}{1 + \alpha\alpha_0}, \\ \beta' &= \tan \kappa_2(S_0 - s) = \frac{\beta - \beta_0}{1 + \beta\beta_0}, \\ \gamma' &= \gamma, \end{aligned}\right\} \quad (19)$$

where

$$\left.\begin{aligned} \alpha_0 &= \tan \kappa_1 d = \tan \frac{2\pi}{\lambda_{g1}} d, \\ \beta_0 &= \tan \kappa_2 s = \tan \frac{2\pi}{\lambda_{g2}} s. \end{aligned}\right\} \quad (20)$$

At the new terminals T_1', T_2' the relations between a', b', c' and α', β', γ' are the same (except for the prime) as those between a, b, c and α, β, γ given in Eqs. (15). The elimination of α', β', γ' from the primed relations by means of Eqs. (19), followed by the use of Eqs. (15), leads to the desired relations [Eqs. (21)] between the shifted and original network parameters.

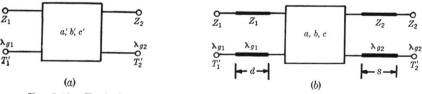

FIG. 3·16.—Equivalent representations of shifted two-terminal-pair network.

On transformation to new terminals T_1' and T_2', located at distances d and s from T_1 and T_2, the two-terminal-pair networks indicated in Figs. 3·13a and b can be schematically represented as in either Fig. 3·16a or b. The boxes represent networks of the T or π type or any of their equivalents; the heavy lines represent lengths of transmission lines. The relations between the parameters a, b, c of the original $T(\pi)$ representation and the parameters a', b', c' of the transformed $T(\pi)$ representation are[1]

$$\left.\begin{aligned} a' &= \frac{a + \alpha_0 + \beta_0 b - \alpha_0\beta_0 c}{1 - \alpha_0 a - \beta_0 c - \alpha_0\beta_0 b}, \\ c' &= \frac{c + \alpha_0 b + \beta_0 - \alpha_0\beta_0 a}{1 - \alpha_0 a - \beta_0 c - \alpha_0\beta_0 b}, \\ b' &= \frac{b - \alpha_0 c - \beta_0 a - \alpha_0\beta_0}{1 - \alpha_0 a - \beta_0 c - \alpha_0\beta_0 b}, \end{aligned}\right\} \quad (21)$$

where

$$\alpha_0 = \tan\frac{2\pi}{\lambda_{g1}} d, \qquad \beta_0 = \tan\frac{2\pi}{\lambda_{g2}} s.$$

Equations (21) apply as well to the case where a', b', c' are parameters of a $T(\pi)$ representation and a, b, c are parameters of a $\pi(T)$ representation, provided the relation between α_0, β_0 and d, s is

$$\alpha_0 = -\cot\frac{2\pi}{\lambda_{g1}} d, \qquad \beta_0 = -\cot\frac{2\pi}{\lambda_{g2}} s.$$

It is to be noted that the two distinct sets of transformation relations distinguished by the parentheses in the preceding sentences are dual to each other.

As an illustration of the use of Eqs. (21) let it be required to determine the shifts d and s of the input and output terminals of the waveguide structure of Fig. 3·12 in order to transform the representations of Fig.

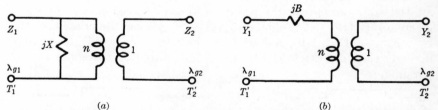

Fig. 3·17.—Equivalent representations by shift of terminal planes. (a) Shunt representation of arbitrary two-terminal-pair network shown in Figs. 3·13a and b. (b) Series representation of arbitrary two-terminal-pair network shown in Figs. 3·13a and b.

3·13a and b into the pure shunt (series)[1] representation of Fig. 3·17a and b. Let a, b, c be the parameters, as defined in Eqs. (17), of the original representation, and correspondingly let a', b', c' be the impedance (admittance) parameters of the transformed representation. Since for a shunt

---

[1] The following statements apply to the cases *either* within *or* without parentheses, respectively.

(series) representation $b' = 0$, it follows from Eqs. (21) that for an arbitrary α_0

$$\beta_0 = \frac{b - \alpha_0 c}{a + \alpha_0}, \tag{22a}$$

where

$$\alpha_0 = \tan \frac{2\pi}{\lambda_{g1}} d, \qquad \beta_0 = \tan \frac{2\pi}{\lambda_{g2}} s \tag{22b}$$

for a $T(\pi)$ to shunt (series) representation, or

$$\alpha_0 = -\cot \frac{2\pi}{\lambda_{g1}} d, \qquad \beta_0 = -\cot \frac{2\pi}{\lambda_{g2}} s \tag{22b'}$$

for a $T(\pi)$ to series (shunt) representation. On substitution of Eq. (22a) into (21) the parameters of the shunt (series) representation become

$$\left. \begin{aligned} a' &= \frac{(a + \alpha_0)^2 + (b - \alpha_0 c)^2}{(1 - \alpha_0^2)(a - bc) + \alpha_0(1 + c^2 - a^2 - b^2)}, \\ \frac{a'}{c'} &= \frac{(a + \alpha_0)^2 + (b - \alpha_0 c)^2}{(1 + \alpha_0^2)(b + ac)}, \end{aligned} \right\} \tag{22c}$$

where for the

Shunt Representation	Series Representation
$a' = \dfrac{X}{Z_1}$	$a' = \dfrac{B}{Y_1}$
$\dfrac{a'}{c'} = n^2 \dfrac{Z_2}{Z_1}$	$\dfrac{a'}{c'} = \dfrac{1}{n^2} \dfrac{Y_2}{Y_1}$

Both the shunt impedance jX and series admittance jB of the transformed representation are shown in Fig. 3·17a and b (also cf. Fig. 3·13b). A further simplification of the transformed representation is obtained on removal of the ideal transformer by modification of the output characteristic admittance, a procedure indicated previously.

A useful special case of the above transformation occurs when the original network is symmetrical ($a = c$) and the input and output terminals are shifted by equal amounts $d = s$. The shift $d = s$ required to secure a shunt (series) representation is given by Eq. (22a) as

$$\alpha_0 = \beta_0 = -a + \sqrt{b + a^2}.$$

Therefore, Eqs. (22c) become

$$\left. \begin{aligned} a' &= \frac{\sqrt{b + a^2}}{1 - b}, \\ \frac{a'}{c'} &= 1. \end{aligned} \right\} \tag{23}$$

These expressions assume greater significance when one notes that $\sqrt{b + a^2}$ represents the shunt reactance (series susceptance) of the original symmetrical T (π) network.

Another simple case obtains when $d = 0$. From Eqs. (22) it is evident that for a $T(\pi)$ to shunt (series) transformation

$$\left. \begin{aligned} \beta_0 &= \tan \frac{2\pi}{\lambda_{g2}} s = \frac{b}{a}, \\ a' &= \frac{a^2 + b^2}{a - bc}, \\ \frac{a'}{c'} &= \frac{a^2 + b^2}{b + ac}, \end{aligned} \right\} \quad (24)$$

whereas for a $T(\pi)$ to series (shunt) transformation

$$\left. \begin{aligned} \beta_0 &= -\cot \frac{2\pi}{\lambda_{g2}} s = -c, \\ a' &= \frac{1 + c^2}{bc - a}, \\ \frac{a'}{c'} &= \frac{1 + c^2}{b + ac}. \end{aligned} \right\} \quad (25)$$

N-terminal-pair Structures.—Equivalent representations of an N-terminal-pair waveguide structure can be obtained either at a given set of reference planes or at shifted reference planes. In the former case representations of the type depicted in Figs. 3·13a and b can be employed to secure equivalent representations of the two-terminal-pair networks that compose the over-all N-terminal-pair network. In addition multiwinding ideal transformers can be usefully employed. Since no impedance or admittance description of an ideal transformer exists, its description must necessarily be phrased in terms of terminal voltage and currents. For the case of a three-winding ideal transformer, illustrated in Fig. 3·18a, the network equations are

$$\left. \begin{aligned} V_1 I_1 + V_2 I_2 + V_3 I_3 &= 0, \\ \frac{V_1}{n_1} = \frac{V_2}{n_2} &= \frac{V_3}{n_3}, \end{aligned} \right\} \quad (26)$$

where n_1, n_2, and n_3 are proportional to the number of turns on the various windings. The relation between the input admittance Y_{in} at the terminals T_3 and the output admittances Y_1 and Y_2 at the terminals T_1 and T_2 follows from Eqs. (26) as

$$Y_{\text{in}} = \left(\frac{n_1}{n_3}\right)^2 Y_1 + \left(\frac{n_2}{n_3}\right)^2 Y_2. \quad (27)$$

The three-winding ideal transformer is evidently a natural generalization of the familiar two-winding ideal transformer.

A network utilizing the ideal transformer of Fig. 3·18a is shown in Fig. 3·18b. The terminals T_4 can be regarded as terminals of an output line or of a lumped-constant circuit element. Accordingly Fig. 3·18b

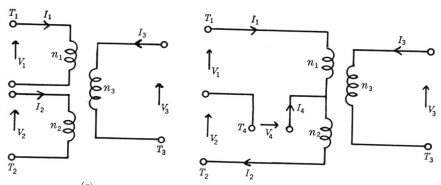

FIG. 3·18.—(a) Three-winding ideal transformer. (b) Network with three-winding ideal transformer.

represents either a four- or a three-terminal-pair network. The corresponding network equations follow from Eqs. (26) as

$$(V_1 - V_4)I_1 + (V_2 + V_4)I_2 + V_3 I_3 = 0,$$
$$\frac{V_1 - V_4}{n_1} = \frac{V_2 + V_4}{n_2} = \frac{V_3}{n_3}, \quad (28)$$
$$I_4 = I_2 - I_1.$$

The relation between the input admittance Y_{in} at terminals T_3 and output admittances Y_1, Y_2, and Y_4 at terminals T_1, T_2, and T_4 is given by

$$Y_{in} = \frac{(n_1 + n_2)^2 Y_1 Y_2 + n_1^2 Y_1 Y_4 + n_2^2 Y_2 Y_4}{n_3^2(Y_1 + Y_2 + Y_4)}. \quad (29a)$$

The input admittance Y_{in} at T_4 in terms of output admittances Y_1, Y_2, and Y_3 at T_1, T_2, and T_3 is

$$Y_{in} = \frac{(n_1 + n_2)^2 Y_1 Y_2 + n_3^2(Y_1 + Y_2)Y_3}{n_1^2 Y_1 + n_2^2 Y_2 + n_3^2 Y_3}. \quad (29b)$$

The input admittance Y_{in} at T_2 in terms of output admittances Y_1, Y_3, and Y_4 at T_1, T_3, and T_4 is

$$Y_{in} = \frac{n_1^2 Y_1 Y_4 + n_3^2(Y_1 + Y_4)Y_3}{(n_1 + n_2)^2 Y_1 + n_3^2 Y_3 + n_2^2 Y_4}. \quad (29c)$$

The special case $n_1 = n_2$ describes a hybrid coil, a network frequently employed to represent a magic T (cf. Fig. 3·11) at appropriate reference planes. For this case it is apparent from Eqs. (29a) and (b) that when $Y_1 = Y_2$, the input admittance at $T_3(T_4)$ is independent of the output admittance at $T_4(T_3)$. This important property forms the basis for many applications of the magic T in bridge circuits, etc. By the use of additional elements in the network of Fig. 3·18b it is possible to obtain

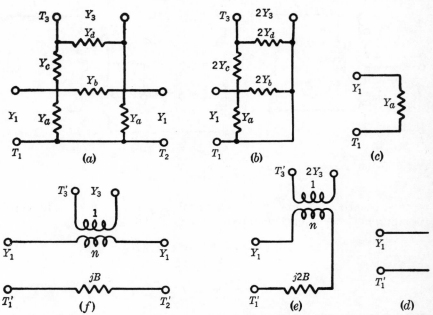

Fig. 3·19.—Steps in reference-plane transformation of representation in (a) to representation in (f).

equivalent representation of arbitrary three- or more terminal-pair networks.

Equivalent representations of an N-terminal-pair structure can also be obtained by transformation of reference planes. Reference plane shifts can be effected quite simply if the over-all network representation can be reduced to a number of two- (or less) terminal-pair networks. Such reductions, for the case of symmetrical N-terminal-pair networks, can be accomplished by the use of symmetry analyses (bisection theorems). The reduced networks are fully equivalent to the over-all network in that the former compose the latter and conversely the latter reduce to the former. If the reduced networks are two-terminal-pair networks, the transformation equations (21) can be employed to secure new representations at other terminal planes. With the knowledge of the new representations of the reduced networks the over-all network can be composed at the new reference planes.

For illustration a transformation process will be employed to obtain simplified representations of the three-terminal-pair structures depicted in Figs. 3·4 and 3·6. The over-all network representations at the terminals T_1, T_2, and T_3 are reproduced in Figs. 3·19a and 3·20a with a somewhat different notation. On bisection of the over-all networks by placement

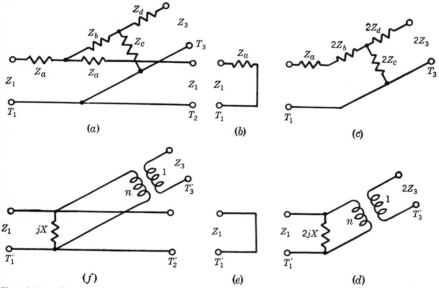

Fig. 3·20.—Steps in reference-plane transformation of representation in (a) to representation in (f).

of a short or open circuit at the electrical centers, the reduced networks indicated in Fig. 3·19b or c and Fig. 3·20b or c are obtained. Simplified representations of these reduced networks can easily be found. On appropriate shift of the terminal T_1 a distance d away from the junction, the reduced network of Fig. 3·19c becomes the open circuit indicated in Fig. 3·19d; and correspondingly, the network of Fig. 3·20b becomes the short circuit of Fig. 3·20e. In addition the terminal T_3 can be shifted a distance s away from the junction so as to transform the reduced networks of Figs. 3·19b and 3·20c into the series and shunt representations of Figs. 3·19e and 3·20d. The over-all network representation at the new terminals T'_1, T'_2, and T'_3, as shown in Figs. 3·19f and 3·20f, is then composed by recombination of the transformed reduced networks. The relations between the parameters a, b, and c of the original and a', b', and c' of the transformed networks are given by Eqs. (22a) and (c), where in this case

For Fig. 3·19

$$\alpha_0 = \tan \frac{2\pi}{\lambda_{g1}} d = \frac{jY_a}{Y_1}$$

$$\beta_0 = \tan \frac{2\pi}{\lambda_{g3}} s$$

$$a = -j \frac{Y_a + 2Y_b + 2Y_c}{Y_1}$$

$$c = -j \frac{2Y_d + 2Y_c}{2Y_3}$$

$$b = -ac - \frac{4Y_c^2}{2Y_1 Y_3}$$

$$a' = \frac{2B}{Y_1}$$

$$\frac{a'}{c'} = \frac{1}{n^2} \frac{2Y_3}{Y_1}$$

For Fig. 3·20

$$\alpha_0 = \tan \frac{2\pi}{\lambda_{g1}} d = \frac{jZ_a}{Z_1}$$

$$\beta_0 = \tan \frac{2\pi}{\lambda_{g3}} s$$

$$a = -j \frac{Z_a + 2Z_b + 2Z_c}{Z_1}$$

$$c = -j \frac{2Z_d + 2Z_c}{2Z_3}$$

$$b = -ac - \frac{4Z_c^2}{2Z_1 Z_3}$$

$$a' = \frac{2X}{Z_1}$$

$$\frac{a'}{c'} = n^2 \frac{2Z_3}{Z_1}$$

(30)

As mentioned above the representations in Figs. 3·19f and 3·20f can be further simplified through removal of the ideal transformer by a suitable modification of the characteristic impedance of the output line at terminal T'_3.

3·4. Measurement of Network Parameters.—The experimental determination of the $N(N+1)/2$-network parameters that characterize an N-terminal-pair waveguide structure involves the placement of known impedances at $N-1$ "output" terminals and measurement of the resulting impedance at the remaining "input" terminal. A variable length of short-circuited line provides a convenient form of output impedance. A standing-wave detector or its equivalent provides a means for the measurement of input impedance (cf. Vol. 11 of this series). Input impedance measurements must be performed for $N(N+1)/2$ arbitrary but independent sets of output terminations. The determinations of the network parameters from these measurements can be considerably simplified by a judicious choice and placement of the output impedances. For example, the placement of arbitrary but fixed output impedances at $N-2$ terminals reduces the over-all network to a two-terminal-pair network, the parameters of which can be readily measured. A proper choice of the fixed output impedances gives rise to two-terminal-pair networks from whose measured parameters the unknown $N(N+1)/2$ parameters are easily determined. The proper choice of output impedances is generally apparent from the form of the equivalent circuit for the over-all structure. Since the measurement of the parameters of an N-terminal-pair network can be reduced to the measurement of the parameters of two-terminal-pair networks, only the latter will be considered in this section.

The three network parameters characteristic of an arbitrary two-terminal-pair waveguide structure can be measured by various methods. The conventional network method involves the measurement of input impedance for three particular values of output impedance. This method has the advantage of being applicable to dissipative structures and to both uniform and nonuniform lines. For nondissipative structures it is desirable to employ pure reactive output impedances (short-circuited lines), since they give rise to an infinite standing-wave ratio in the input line. Under these conditions the output impedance is a simple function of the length of the short-circuited line, and the input impedance is a correspondingly simple function of the distance to the minimum in the input line. It is convenient to employ any three of the following pairs of measured values for input impedance Z_{in} and corresponding output impedance Z_{out},

	1	2	3	4	5
Z_{in}	z_0	z_∞	z	∞	0
Z_{out}	0	∞	Z	Z_∞	Z_0

(31)

for the determination of the unknown network parameters. The quantity z_0 represents the input impedance set up by a zero output impedance, etc. In terms of the above values the impedance elements for a T representation (cf. Fig. 3·13a) of a two-terminal-pair network can be expressed as in any of the following columns:

	1, 2, 3	1, 2, 4	1, 2, 5
Z_{12}	$\pm \sqrt{\dfrac{(z_\infty - z)(z_\infty - z_0)Z}{(z - z_0)}}$	$\pm \sqrt{(z_0 - z_\infty)Z_\infty}$	$\pm \sqrt{\dfrac{z_\infty Z_0 (z_0 - z_\infty)}{z_0}}$
Z_{11}	z_∞	z_∞	z_∞
Z_{22}	$\dfrac{Z_{12}^2}{z_\infty - z_0}$	$-Z_\infty$	$-\dfrac{z_\infty Z_0}{z_0}$

where the numbers at the head of each column indicate the particular set of three measured values in (31) on which the equations are based. The admittance elements of a π representation (cf. Fig. 3·13a) follow from the above expressions on the duality replacement of impedances by admittances. It is to be noted that either of two elements, $+$ or $-Z_{12}(Y_{12})$, can be employed for the representation of the input-output impedance measurements. This ambiguity can be resolved by a measurement of transfer impedance or of any other quantity that yields the relative phase at the input and output terminals. In many cases the correct sign may be ascertained theoretically.

For the case of uniform lines the transformer representation discussed in Sec. 3·3 provides a basis for an alternative method of measurement of a two-terminal-pair structure. As illustrated in Figs. 3·12 and 3·15 a nondissipative two-terminal-pair structure can be represented at characteristic terminals T_1^0 and T_2^0 by an ideal transformer of impedance ratio n^2 connecting input and output lines of characteristic impedances Z_1 and Z_2. The transformer ratio n^2 and the location of the characteristic terminals are readily determined from the following typical measurements of

1. The standing-wave ratio and position of the minimum in the input line for a matched load ($Z_{out} = Z_2$) in the output line.
2. The position of a short circuit in the output line such that the corresponding minimum in the input line coincides with that in measurement 1.

From the transformation properties of the ideal transformer it is evident that the standing-wave ratio in measurement 1 is equal to $n^2 Z_2/Z_1$, which as shown in Sec. 3·3, is denoted by $-\gamma$; the locations of the terminals T_1^0 and T_2^0 are given by the position of the maximum in measurement 1 and of the short circuit in measurement 2. As in Fig. 3·12, the distances from the input and output terminals to T_1^0 and T_2^0 are designated as D_0 and S_0, respectively. The parameters of a T or π representation (cf. Fig. 3·13a) are then expressed in terms of D_0, S_0, γ by means of Eqs. (15) and (17).

The accuracy of the two preceding methods of measurement is difficult to ascertain because of the uncertainty in individual standing-wave measurements. This difficulty can be partially removed by averaging a large number of such measurements. For the case of nondissipative two-terminal-pair structures in uniform lines, a more systematic procedure may be employed if more accuracy is required. This precision method involves a plot of the measured values of the positions of the input minima vs. the corresponding positions of the output short circuits. An analysis of this plot with the aid of the previously considered tangent relation in the form

$$\tan 2\pi(D' - D_0') = \gamma \tan 2\pi(S' - S_0') \tag{32}$$

yields the data required for the determination of the network parameters. Equation (32) is just Eq. (13) rewritten with $D' = D/\lambda_{g1}$, $D_0' = D_0/\lambda_{g1}$, $S' = S/\lambda_{g2}$, and $S_0' = S_0/\lambda_{g2}$.

As mentioned in Sec. 3·3, the tangent relation provides a representation of the input-output impedance relation that is particularly well suited for measurements in nondissipative waveguides. This is evident from Eqs. (14), which indicate that S is identically the distance measured from the output terminals to a short circuit in the output line and D is the distance from the input terminals to the corresponding minimum position in the input line. The essence of the present method is the determination of a set of parameters D_0, S_0, and γ that, on insertion into Eq. (32), provide a curve of D vs. S which best reproduces the experimental curve of D vs. S. By Eqs. (15) and (17) it is apparent that the parameters D_0, S_0, and γ are equivalent to the network parameters of a T or π representation. A virtue of the present method is that a comparison of the computed and measured curves of D vs. S indicates immediately the average accuracy of the final set of parameters D_0, S_0, and γ.

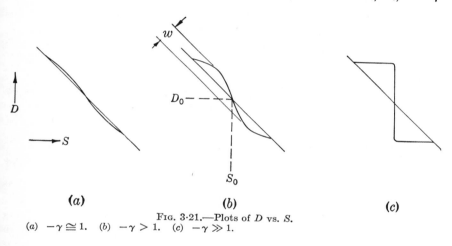

Fig. 3·21.—Plots of D vs. S.
(a) $-\gamma \cong 1$. (b) $-\gamma > 1$. (c) $-\gamma \gg 1$.

In practice it has been possible to obtain, at wavelengths of about 3 cm, an average difference of less than $0.0005\lambda_g$ between the experimental and computed curves of D vs. S. An accuracy of this magnitude implies that the limitations in the accuracy of equivalent circuit measurements lie not in the standing-wave measurements but rather in the mechanical measurements required to locate the input and output terminals.

The details of a successive approximation procedure for the precise determination of the parameters D_0, S_0, and γ will now be outlined. The measured values of D' and S' when plotted yield curves of the form indicated in Fig. 3·21.[1] From Eq. (32) one sees that this curve should be repetitive with a period of a half wavelength in both D' and S' and symmetrical about a line of slope -1. The curve intersects the line of slope -1 at points of maximum and minimum slope. The point of intersection at the maximum slope is $D' = D'_0$, $S' = S'_0$. The maximum slope is γ, and the minimum slope is $1/\gamma$. First approximations to D_0 and S_0 are

[1] A. Weissfloch, *Hochfreq. u. Elektro.*, Vol. 60, 1942, pp. 67 *et seq.*

obtained from the locations of the points of maximum and minimum slopes by suitable averages. A first approximation to γ is given by the average of the maximum and the reciprocals of the minimum slopes. An additional value for the average is given in terms of w', the width in guide wavelengths between points of slope -1, as

$$-\gamma = \cot^2 2\pi \left(\frac{1}{8} - \frac{\sqrt{2}}{4} w' \right). \tag{33a}$$

If the guide wavelengths in the input and output lines are unequal, the value of w' may be determined from a plot of D' vs. S'. In practice, however, it is most convenient to plot the absolute values of D and S as in Fig. 3·21; in terms of the maximum spread w of this curve

$$-\gamma = \cot^2 2\pi \left(\frac{1}{8} - \frac{w}{4} \sqrt{\frac{1}{\lambda_{g1}^2} + \frac{1}{\lambda_{g2}^2}} \right) \tag{33b}$$

It is to be noted that the slope of the symmetry axis of this curve is $-\lambda_{g1}/\lambda_{g2}$ and hence the maximum slope of this curve is $\gamma \lambda_{g1}/\lambda_{g2}$, etc.

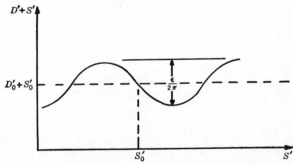

FIG. 3·22.—Plot of $D' + S'$ vs. S' for $-\gamma \cong 1$.

Almost "matched" two-terminal-pair structures have a $-\gamma$ value of approximately unity and consequently give rise to a D vs. S curve from which it is difficult to evaluate and locate the points of maximum slope. Since in such cases $\gamma = -1 - \epsilon (\epsilon \ll 1)$, Eq. (32) may be rewritten in the approximate form

$$D' + S' \cong D_0' + S_0' - \frac{\epsilon}{4\pi} \sin 4\pi (S' - S_0'). \tag{34}$$

Thus, if the experimental data is plotted in the form $D' + S'$ vs. S', the curve shown in Fig. 3·22 is obtained. The values of $D_0' + S_0'$, S_0', and $\epsilon/2\pi$ can be easily read from this curve and furnish first approximations to the required parameters D_0, S_0, and γ.

The knowledge of the first approximations to D_0, S_0, and γ is suffi ciently accurate in many cases. More accurate values can be obtained if a theoretical curve of D vs. S, as computed by means of Eq. (32) and the first approximation values, is compared with the measured curve. A convenient mode of comparison is a plot of the difference between the experimental and computed values of D vs. those values of S corresponding to the experimental points. This difference curve of $\Delta D' = D'_{\text{exp}} - D'_{\text{comp}}$ vs. S' may or may not possess regularity. If regularity is exhibited, the first approximations to D'_0, S'_0, and γ are inaccurate

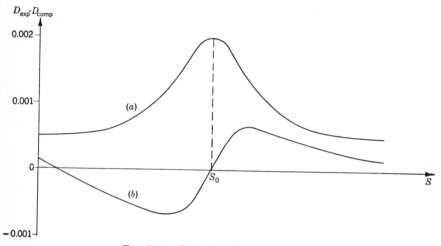

Fig. 3·23.—Plot of typical error curves.
(a) $\gamma = -2$; $\Delta S_0 = 0.001$, $\Delta D_0 = \Delta\gamma = 0$. (b) $\gamma = -2$; $\Delta\gamma = 0.01$, $\Delta D_0 = \Delta S_0 = 0$

and the difference curve should be analyzed to obtain corrections $\Delta D'_0$, $\Delta S'_0$, and $\Delta\gamma$ to the first approximations. To determine these corrections it is necessary to know the expected form of the curve of $\Delta D'$ vs. S' arising from variations $\Delta D'_0$, $\Delta S'_0$, and $\Delta\gamma$ in Eq. (32). The differential form of Eq. (32), namely,

$$\Delta D' = \Delta D'_0 - \frac{\gamma\,\Delta S'_0 - \dfrac{\sin 4\pi(S' - S'_0)}{4\pi}\,\Delta\gamma}{\cos^2 2\pi(S' - S'_0) + \gamma^2 \sin^2 2\pi(S' - S'_0)} \qquad (35)$$

is the theoretical equation for the difference curve.

The actual difference, or error, curve of $\Delta D' = D'_{\text{exp}} - D'_{\text{comp}}$ vs. S' arises from errors $\Delta D'_0$, $\Delta S'_0$, and $\Delta\gamma$ in the choice of D'_0, S'_0, and γ by the procedure described above. This curve is plotted in Fig. 3·23 for typical values of the various parameters. If, for example, the amplitudes of the actual error curve at $S' - S'_0 = 0, \frac{1}{8}$, and $\frac{1}{4}$ are designated as Δ_0, $\Delta_{1/8}$, and $\Delta_{1/4}$, then from Eq. (35) the required corrections to be *added* to D'_0, S'_0, and γ are found to be

$$\left.\begin{aligned}\Delta D_0' &= \frac{1}{\gamma^2 - 1}(\gamma^2 \Delta_{1/4} - \Delta_0), \\ \Delta S_0' &= \frac{\gamma}{\gamma^2 - 1}(\Delta_{1/4} - \Delta_0), \\ \Delta \gamma' &= 2\pi[(1 + \gamma^2)\Delta_{1/8} - \Delta_0 - \gamma^2 \Delta_{1/4}].\end{aligned}\right\} \quad (36)$$

Other methods of analysis of the error curve can be employed depending on the value of γ. The corrected values for D_0', S_0', and γ usually suffice to describe *all* the measured data to within experimental accuracy. The accuracy itself can be estimated by plotting another difference curve employing the second-approximation values of D_0', S_0', and γ. This curve should possess no regularity; its average deviation provides a measure of the average error in the "electrical" measurement.

The electrical error in the standing-wave measurements is to be distinguished from the "mechanical" error in the measurement of distance to the terminal planes. Since the evaluation of γ and the location of the point of maximum slope can be obtained merely from relative values of D and S, it is evident that these determinations involve only the electrical error. However, the absolute evaluation of D_0' and S_0' necessary for the determination of the network parameters may involve the measurement of the distances from the point of maximum slope to the input and output terminals. Because of difficulties in maintaining accurate mechanical tolerances in microwave structures, the latter measurement is usually the largest source of error.

From the error curve, etc., it is possible to estimate the over-all experimental errors $\delta D_0'$, $\delta S_0'$, and $\delta\gamma/\gamma$ in D_0', S_0', and γ. The corresponding relative errors $\delta a/a$, $\delta c/c$, and $\delta b/b$ in the network parameters a, b and c, [cf. Eqs. (15)] arising from the experimental errors may be expressed as

$$\left.\begin{aligned}\frac{\delta a}{a} &= \sqrt{\left(\frac{4\pi A_1}{\sin 4\pi D_0'}\delta D_0'\right)^2 + \left(\frac{4\pi A_2}{\sin 4\pi S_0'}\delta S_0'\right)^2 + \left(A_2 \frac{\delta\gamma}{\gamma}\right)^2}, \\ \frac{\delta c}{c} &= \sqrt{\left(\frac{4\pi A_3}{\sin 4\pi D_0'}\delta D_0'\right)^2 + \left(\frac{4\pi A_4}{\sin 4\pi S_0'}\delta S_0'\right)^2 + \left(A_3 \frac{\delta\gamma}{\gamma}\right)^2}, \\ \frac{\delta b}{b} &= \sqrt{\left(\frac{4\pi A_5}{\sin 4\pi D_0'}\delta D_0'\right)^2 + \left(\frac{4\pi A_5}{\sin 4\pi S_0'}\delta S_0'\right)^2 + \left(A_6 \frac{\delta\gamma}{\gamma}\right)^2},\end{aligned}\right\} \quad (37)$$

where

$$A_1 = \frac{\alpha(\beta^2 + \gamma^2)}{(\beta - \alpha\gamma)(\alpha\beta + \gamma)}, \qquad A_4 = \frac{\beta(1 + \alpha^2\gamma^2)}{(\alpha\gamma - \beta)(1 + \alpha\beta\gamma)},$$

$$A_2 = \frac{\beta\gamma(1 + \alpha^2)}{(\alpha\gamma - \beta)(\alpha\beta + \gamma)}, \qquad A_5 = \frac{\alpha\beta(1 - \gamma^2)}{(\alpha - \beta\gamma)(\beta - \alpha\gamma)},$$

$$A_3 = \frac{\alpha\gamma(1 + \beta^2)}{(1 + \alpha\beta\gamma)(\beta - \alpha\gamma)}, \qquad A_6 = \frac{\gamma(\alpha^2 - \beta^2)}{(\alpha - \beta\gamma)(\beta - \alpha\gamma)},$$

SEC. 3·4]   MEASUREMENT OF NETWORK PARAMETERS   137

and α, β, γ, a, b, and c are defined in Eqs. (15) and (17). The sensitivity of the network parameters a, b, and c to errors in D_0, S_0, and γ evidently depends on the choice of terminals as well as on the type of microwave structure. For a desired accuracy in the values of the network parameters, the above equations furnish from approximate values for D_0, S_0, and γ the precision necessary in the determination of the latter quantities. Instead of the relative error $\delta b/b$ it is frequently necessary to know the relative error

$$\frac{\delta \sqrt{b+ac}}{\sqrt{b+ac}} = \frac{b\frac{\delta b}{b} + ac\left(\frac{\delta a}{a} + \frac{\delta c}{c}\right)}{2(b+ac)}. \tag{37a}$$

Distance Invariant Representations.—As is evident from Eqs. (37) a precise determination of the equivalent-circuit parameters of a microwave structure requires a precise measurement of the distances D_0 and S_0, between the characteristic and the prescribed terminal planes. Accurate measurements of these distances may require an absolute mechanical accuracy at $\lambda = 3$ cm, for example, of a mil in a distance of a few inches; this is exceedingly difficult to attain—particularly when the prescribed terminal planes are relatively inaccessible. Inaccuracies in mechanical measurements may result in disproportionately large errors in the circuit parameters. For instance, inaccurate distance measurements on a symmetrical structure may lead to an asymmetrical circuit representation (i.e., $a \neq c$) despite the certainty of structural symmetry. All such difficulties arise because of the dependence of the circuit representation on "mechanical" measurements of distances to prescribed terminal planes. The difficulties can be avoided by use of a representation in which the type of circuit is prescribed and in which the locations of the terminal planes are not prescribed but rather ascertained from the measurements. The values of the circuit parameters in such a representation may be made highly accurate since they may be ascertained solely from the "electrical" measurement of the maximum slope γ of the D vs. S curve. Inaccuracies in absolute distance measurements have no effect on the determination of γ and hence are manifest only as proportionate inaccuracies in the locations of terminal planes. The determination of distance invariant representations of this type will now be discussed.

The equivalent circuit representative of a nondissipative microwave structure is dependent on the determination of three parameters. As noted above, the tangent parameters D_0, S_0, and γ form a convenient set. To secure a representation in which the circuit parameters are independent of measurements of the two distance parameters D_0 and S_0, it is necessary to prescribe two bits of information about the desired representation. For example, the circuit representation may be prescribed to be both symmetrical and shunt. The impedance parameters of the desired representation will therefore satisfy the prescribed conditions

$$a = c \quad \text{and} \quad b = 0. \tag{38}$$

By Eqs. (15) this implies

$$\alpha\beta = -1, \quad \alpha = \beta\gamma, \tag{39a}$$

and hence α and β are prescribed to be

$$\alpha = \pm \sqrt{-\gamma}, \quad \beta = \mp \frac{1}{\sqrt{-\gamma}}. \tag{39b}$$

The relative reactance of the shunt element is therefore, by Eqs. (39) and (15),

$$a = c = \mp \frac{\sqrt{-\gamma}}{1 + \gamma} \tag{40}$$

and is manifestly dependent on only the γ of the D vs. S curve of the given structure. The locations of the terminal planes for this shunt representation follow from measurements of the locations of the characteristic reference planes and from the values [by Eqs. (39b) and (17)] of D_0 and S_0, the distances between the characteristic and the desired terminal planes.

There exist a variety of other distance invariant representations of a microwave structure. For example, the representation may be prescribed to be purely series; the admittance parameters of this representation are then identical with the impedance parameters of Eqs. (40). The transformer representation illustrated in Fig. 3·15 also belongs to this category. Of the various possible representations the most desirable is usually the one in which the associated terminal planes are located in closest proximity to the physical terminal planes of the structure.

3·5. Theoretical Determination of Circuit Parameters.[1]—The presence of a discontinuity structure in a waveguide results in discontinuities in the propagating mode fields at the "terminals" of the given structure. As noted in previous sections such field or, equivalently, voltage-current discontinuities can be schematically represented by means of a lumped-constant equivalent circuit. The theoretical determination of the equivalent-circuit parameters requires mathematical methods that do not properly lie within the realm of microwave network engineering. Instead, such determinations generally involve the solution of so-called boundary value or field problems. The present section is primarily intended to

---

[1] A comprehensive account of the theory of guided waves is in preparation by J. Schwinger and the author.

SEC. 3·5] *DETERMINATION OF CIRCUIT PARAMETERS* 139

sketch those field theoretical techniques, devised largely by J. Schwinger, which have been employed to obtain the equivalent-circuit results presented in Chaps. 4 to 8.

The field problems to be discussed are concerned with the behavior of electromagnetic fields not everywhere within a region but rather only in those regions relatively "far" from a discontinuity structure; the behavior in the latter is, of course, just that of the propagating modes. The solution of such field problems presupposes the ability to determine the electric and magnetic fields set up in a waveguide by electric currents (i.e., tangential magnetic fields) on obstacle-type discontinuities and by magnetic currents (i.e., tangential electric fields) on aperture-type discontinuities. The field representations summarized in Eqs. (1·6) are of the desired form provided the mode functions $\mathbf{e}_i$, the mode voltages V_i, and the mode currents I_i can be determined. The mode functions are so determined that the mode fields possess in the waveguide cross section the transverse xy behavior dictated both by the field equations (1·1) and by the requirement of vanishing tangential electric field on the nondissipative guide walls. Explicit evaluations of the mode functions $\mathbf{e}_i$ for a variety of waveguide cross sections are presented in Chap. 2. The corresponding evaluation of the mode voltages and mode currents then follows from the requirements that the mode fields possess the longitudinal z dependence dictated by the field equations, and in addition that the total fields satisfy the boundary conditions imposed by the presence of the discontinuity and the nature of the excitation in the waveguide. As shown in Sec. 1·2, the determination of the longitudinal z dependence of the mode amplitudes V_i and I_i constitutes a conventional transmission-line problem and is described implicitly by the transmission-line equations (1·8). These transmission-line considerations are a necessary preliminary to a major source of difficulty: the explicit evaluation of the electric or magnetic currents set up on discontinuity surfaces by the given excitation in the waveguide. These discontinuity currents must be so determined that the total fields satisfy prescribed boundary conditions on the discontinuity surfaces. Once the discontinuity currents are found, the various mode voltages and currents follow by straightforward transmission-line considerations. In particular there follow the dominant-mode voltage-current relations at the terminals of the discontinuity, and hence the equivalent-circuit parameters characteristic thereof.

The preceding paragraph has sketched in only qualitative detail the salient features of a general method for the determination of equivalent-circuit parameters. The methods employed in Chaps. 4 to 8 are basically of a similar nature and differ mostly in their technique of successive approximation to the desired rigorous results. These methods have been classified as

1. The variational method.
2. The integral equation method.
3. The equivalent static method.
4. The transform method.

The particular method employed in the derivation of the equivalent circuit results in Chaps. 4 to 8 has always been indicated under *Restrictions* in each section therein. The above methods will be briefly illustrated in this section. The field problems presented by the *Asymmetrical Capacitive Window* described in Sec. 5·1b (cf. Fig. 5·1-2) and the *E-plane Bifurcation* described in Sec. 6·4 (cf. Fig. 6·4-1) have been chosen for simplicity

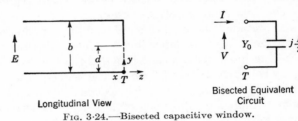

Longitudinal View   Bisected Equivalent Circuit

FIG. 3·24.—Bisected capacitive window.

of illustration; the methods to be discussed are, however, of quite general applicability.

The equivalent circuit for a capacitive window in a rectangular guide in which only the dominant H_{10}-mode can be propagated is, in general, a four-terminal network. However, if the window is formed by an obstacle of zero thickness and if the input and output terminal planes are chosen coincident with the plane of the window, the equivalent network becomes pure shunt, since the electric field and hence the dominant mode voltage are continuous at the terminal plane. The associated field problem, whose solution is necessary for the determination of the relevant equivalent-circuit parameter, need be concerned with the field behavior in only the input half of the structure. This is a consequence of choosing the arbitrary excitation in the input and output guides such that the inward-flowing dominant-mode currents at the input and output terminals are equal. Under these circumstances the tangential magnetic and electric fields in the aperture plane are zero and a maximum, respectively, with a consequent symmetry of the field structure about the terminal plane. A sketch of the bisected structure and its associated equivalent circuit is shown in Fig. 3·24. It is first necessary to find in the input region $z < 0$ a solution of the field equations (1·1) subject to the boundary conditions of

1. Vanishing of the electric field tangential to the guide walls at $y = 0, b$.

SEC. 3·5] DETERMINATION OF CIRCUIT PARAMETERS 141

2. Vanishing of the electric field tangential to the obstacle surface at $z = 0$, $d < y < b$.
3. Vanishing of the magnetic field tangential to the aperture surface at $z = 0$, $0 < y < d$.
4. Excitation by the dominant-mode current I at the terminal plane T.

This field problem will be solved not for the case of a rectangular wave guide region but rather for the simpler case of a parallel-plate waveguide; the uniformity of the structure in the x direction implies that the equivalent circuit results for the parallel-plate guide (with principal mode incident) become identical with those for the rectangular guide on replacement of the space wavelength λ in the former by the guide wavelength λ_g in the latter.

The transverse electric and magnetic fields in the input region $z < 0$ may be represented in accordance with Eqs. (1·6) as

$$\left. \begin{array}{l} \mathbf{E}_t(y,z) = \sum V_n(z)\mathbf{e}_n(y), \\ \mathbf{H}_t(y,z) = \sum I_n(z)\mathbf{h}_n(y), \end{array} \right\} \quad (41)$$

where $\mathbf{h}_n = \mathbf{z}_0 \times \mathbf{e}_n$.

By virtue of principal-mode excitation and uniformity of the field structure in the x direction there is no z component of magnetic field; hence only E-modes are necessary in the representation. From Eqs. (2·15) the E-mode functions in a parallel-plate guide of height b are given by

$$\left. \begin{array}{l} \mathbf{e}_n = -\sqrt{\dfrac{\epsilon_n}{b}} \cos \dfrac{n\pi y}{b} \, \mathbf{y}_0 = e_n(y)\mathbf{y}_0, \\ \mathbf{h}_n = +\sqrt{\dfrac{\epsilon_n}{b}} \cos \dfrac{n\pi y}{b} \, \mathbf{x}_0 = h_n(y)\mathbf{x}_0, \end{array} \right\} \quad (42)$$

where $n = 0, 1, 2, 3, \ldots$.

Imposition of the boundary conditions 2 and 3 at the terminal plane $z = 0$ results in

$$0 = \sum_0^\infty V_n e_n(y), \quad d < y < b, \quad (43a)$$

$$0 = \sum_0^\infty I_n h_n(y), \quad 0 < y < d, \quad (43b)$$

where V_n and I_n represent the mode amplitudes at $z = 0$. From the orthogonality [Eqs. (1·5)] of the mode functions it follows from Eqs. (43) that

$$V_n = \int_{\text{ap}} E(y)e_n(y)\,dy, \tag{44a}$$

$$I_n = \int_{\text{ob}} H(y)h_n(y)\,dy, \tag{44b}$$

the integrals being extended over the aperture and obstacle surfaces, respectively, and the functions $E(y)$ and $H(y)$ being defined by

$$\mathbf{E}_t(y,o) = E(y)\mathbf{y}_0$$

and

$$\mathbf{H}_t(y,o) = H(y)\mathbf{x}_0.$$

The mode amplitudes V_n and I_n are not unrelated. The nonpropagating nature of the higher modes implies that the higher mode transmission lines are "matched," and hence for $n > 0$ (noting the convention that impedances are positive in the direction of increasing z)

$$I_n = -Y_n V_n \quad \text{or} \quad V_n = -Z_n I_n, \tag{45}$$

where by Sec. 2·2c the characteristic admittance Y_n and the propagation wave number κ_n of the nth E-mode transmission line are given by

$$Y_n = \frac{1}{Z_n} = \frac{\omega\epsilon}{\kappa_n}, \quad \kappa_n = \sqrt{k^2 - \left(\frac{n\pi}{b}\right)^2}. \tag{46}$$

Equation (43b) may be rewritten by means of Eqs. (45) in the form

$$I\,h(y) = \sum_{1}^{\infty} Y_n V_n h_n(y), \quad 0 < y < d, \tag{47}$$

where hereafter for simplicity of notation the o subscript shall be omitted from all the dominant-mode quantities. By the use of the expressions (44a) for the mode amplitudes V_n, Eq. (47) can be rewritten as an integral equation for the determination of the electric field $E(y)$ in the aperture, namely, as

$$I\,h(y) = -\int_{\text{ap}} G(y,y')E(y')\,dy', \quad 0 < y < d, \tag{48}$$

where

$$G(y,y') = \sum_{1}^{\infty} Y_n h_n(y) h_n(y').$$

The interchange of the order of summation and integration involved in the transition from Eqs. (47) to (48) is permissible provided the boundary condition (43b) is understood to relate to the plane $z = 0 -$. After solu-

Sec. 3·5] DETERMINATION OF CIRCUIT PARAMETERS 143

tion of the integral equation (48) for $E(y)$ the circuit parameter B is obtained from the relation

$$j\frac{B}{2} = \frac{I}{V} = \frac{I}{-\int_{\text{ap}} E(y)h(y)\,dy}, \qquad (49)$$

a result which is independent of I by virtue of the linear relation between the aperture field $E(y)$ and the exciting current I. An alternative expression for the shunt susceptance B can be derived on an obstacle basis rather than the aperture basis considered above; the derivation on an obstacle basis proceeds via Eqs. (43b), (45), and (44b) by determination of the relation between the obstacle field $H(y)$ and the exciting voltage. However, the evaluations of the circuit parameter B to be discussed below will be confined to the aperture treatment based on Eq. (49).

a. The Variational Method.[1]—Although a rigorous method of evaluation of the susceptance of the capacitive window requires the solution of the integral equation (48) for $E(y)$, it is possible to avoid this by a variety of approximate methods. The variational method is based on the following expression for the susceptance:

$$j\frac{B}{2} = \frac{I}{V} = \sum_1^\infty Y_n \left(\frac{V_n}{V}\right)^2 = \sum_1^\infty Y_n \left(\frac{\int_{\text{ap}} E(y)h_n(y)\,dy}{\int_{\text{ap}} E(y)h(y)\,dy}\right)^2 \qquad (50)$$

derivable on multiplication of Eq. (48) by $E(y)$ and integration over the aperture region $0 < y < d$. The importance of Eq. (50) lies in its stationary character with respect to variations of $E(y)$ about the correct aperture field. More explicitly, if a field $E(y)$ correct only to the first order is inserted into Eq. (50), the susceptance determined therefrom is correct to the second order. Moreover for the case of the above-mentioned capacitive window, wherein only E-modes are exited, the susceptance determined from the so-called variational expression (50) is a minimum for the correct field $E(y)$. A judicious choice of a trial field is thus capable of giving reasonably correct values of susceptance without the necessity of solving the integral equation (48). One also notes that the variational expression is dependent only on the form of a trial field $E(y)$ and not on its amplitude.

As an example let us employ the trial field $E(y) = 1$ in the variational expression. Evaluation of the resulting integrals in Eqs. (50) and use of Eqs. (46) then lead to the approximate expression for the relative susceptance

[1] A monograph on the use of variational principles in diffraction problems, etc., is being prepared by H. Levine and J. Schwinger (to be published by John Wiley and Son, New York).

$$\frac{B}{Y} = \frac{8b}{\lambda} \sum_{1}^{\infty} \frac{1}{\sqrt{n^2 - \left(\frac{2b}{\lambda}\right)^2}} \left(\frac{\sin\frac{n\pi d}{b}}{\frac{n\pi d}{b}}\right)^2 \qquad (51)$$

The relative susceptance may be evaluated numerically by direct summation of the series in Eq. (51). For example, when $d/b = 0.5$ and $2b/\lambda \ll 1$, the addition of a few terms of the series yields

$$\frac{B}{Y} = \frac{8b}{\lambda}(0.42),$$

whereas the rigorous result given in Sec. 5·1b (if $\lambda \to \lambda_g$) yields

$$\frac{B}{Y} = \frac{8b}{\lambda}(0.35),$$

the approximate result being about 20 per cent larger. For other values of d/b, for example, when $d/b \ll 1$, direct summation of the series may become prohibitively tedious because of poor convergence. In this range it is desirable to employ an alternative method of summation. To this end Eq. (51) may be rewritten as

$$\frac{B}{Y} = \frac{8b}{\lambda}\left[\sum_{1}^{\infty}\frac{\sin^2 n\alpha}{n^3\alpha^2} + \frac{1}{2}\left(\frac{2b}{\lambda}\right)^2\sum_{1}^{\infty}\frac{\sin^2 n\alpha}{n^5\alpha^2} + \cdots\right] \qquad (52)$$

for $2b/\lambda < 1$ and with $\alpha = \pi d/b$. Equation (52) can be summed in a variety of ways. For example, if the sums in Eq. (52) are designated as

$$F(\alpha) = \sum_{1}^{\infty}\frac{\sin^2 n\alpha}{n^3} \quad \text{and} \quad G(\alpha) = \sum_{1}^{\infty}\frac{\sin^2 n\alpha}{n^5},$$

then twofold differentiation with respect to α yields

$$F''(\alpha) = 2\sum_{1}^{\infty}\frac{\cos 2n\alpha}{n} = -2\ln 2|\sin \alpha| \approx -2\ln 2\alpha, \qquad (53a)$$

$$G''(\alpha) = 2\sum_{1}^{\infty}\frac{\cos 2n\alpha}{n^3} = 2\zeta(3) - 4F(\alpha), \qquad (53b)$$

with

$$F(o) = F'(o) = G(o) = G'(o) = 0,$$

$$\zeta(3) = \sum_{1}^{\infty}\frac{1}{n^3} = 1.202.$$

Hence on twofold integration of Eqs. (53) to order α^2

$$F(\alpha) \approx -\alpha^2 \ln\left(\frac{2\alpha}{e^{\frac{3}{2}}}\right) = \alpha^2 \ln\left(\frac{2.23b}{\pi d}\right)$$

$$G(\alpha) \approx \zeta(3)\alpha^2 = 1.202 \left(\frac{\pi d}{b}\right)^2.$$

Therefore on insertion of these results into Eq. (52) the approximate value of the relative susceptance in the range $d/b \ll 1$ and $2b/\lambda < 1$ becomes

$$\frac{B}{Y} = \frac{8b}{\lambda}\left[\ln\left(\frac{2.23b}{\pi d}\right) + \frac{1}{2}\left(\frac{2b}{\lambda}\right)^2 1.202 \cdots\right]. \tag{54a}$$

This is to be compared with the rigorous value given in Sec. 5·1b to the corresponding order and in the same range (note $\lambda \to \lambda_g$):

$$\frac{B}{Y} = \frac{8b}{\lambda}\left[\ln\left(\frac{2b}{\pi d}\right) + \frac{1}{2}\left(\frac{2b}{\lambda}\right)^2 \cdots\right]. \tag{54b}$$

It is to be noted that the approximate value is again somewhat larger than the correct one.

Although a variational procedure can provide a reasonably good approximation to the correct values of circuit parameters, its accuracy is dependent on a judicious choice of the trial field. A more refined choice of field is embodied in the so-called Rayleigh-Ritz procedure wherein the trial field is represented in terms of a set of independent (and usually orthogonal) functions in the aperture domain. To illustrate this procedure for the case of the capacitive window let us represent the trial field in the form

$$E(y) = 1 + \sum_{1}^{N} a_m \cos \frac{m\pi y}{d}, \tag{55}$$

where, in accord with the properties of the variational expression (50), the coefficients a_m must be so chosen as to make the susceptance stationary and in particular a minimum. The substitution of Eq. (55) into (50) yields

$$\frac{B}{Y} = \frac{8b}{\lambda}\left[D_{00} + 2\sum_{1}^{N} D_{0l}a_l + \sum_{l,m=1}^{N} D_{lm}a_l a_m\right] \tag{56}$$

where, in view of Eq. (46) and the orthogonality of $\cos(m\pi y/d)$ and $h(y)$ in the aperture,

$$D_{lm} = D_{ml} = \sum_{n=1}^{\infty} \frac{1}{\sqrt{n^2 - \left(\frac{2b}{\lambda}\right)^2}} \frac{\int_{\text{ap}} \cos\frac{l\pi y}{d} h_n(y)\, dy \int_{\text{ap}} \cos\frac{m\pi y}{d} h_n(y)\, dy}{\left(\int_{\text{ap}} h(y)\, dy\right)^2}.$$

The imposition of the stationary requirements on B/Y, namely,

$$\frac{\partial \frac{B}{Y}}{\partial a_m} = \frac{16b}{\lambda} \left(D_{0m} + \sum_{l=1}^{N} D_{lm} a_l \right) = 0 \qquad (57)$$
$$m = 1, 2, 3, \cdots, N$$

leads to a set of N linear simultaneous equations for the determination of the unknown coefficients a_l. On use of these equations in Eq. (56) one obtains for the relative susceptance the simple result

$$\frac{B}{Y} = \frac{8b}{\lambda} \left[D_{00} + \sum_{1}^{N} D_{0l} a_l \right] \qquad (58)$$

which represents an upper bound for the correct result. For $N = 0$ Eq. (58) reduces to the value

$$\frac{B}{Y} = \frac{8b}{\lambda} D_{00}$$

previously obtained in Eq. (51). For $N \to \infty$ the representation of $E(y)$ in Eq. (55) becomes complete and Eq. (58) generally converges to the rigorous result.

It should be pointed out that a corresponding variational procedure can be developed in terms of the obstacle current rather than the aperture field; in the case of the capacitive window this leads to approximate values of susceptance that provide a lower bound to the correct value. A combination of these two procedures can quite accurately determine the true value of susceptance.

b. The Integral-equation Method.—The susceptance of the capacitive window can be obtained by direct solution of the integral equation (48) for the aperture field $E(y)$. It is not generally feasible to solve this integral equation exactly because of the termwise method of solution. However, an approximate integral-equation solution for $E(y)$ coupled with its subsequent use in the variational expression (50) provides a highly accurate procedure.

A general method of solution of the integral equation (48) is based on a diagonal representation of the kernel $G(y,y')$ in terms of a set of functions orthogonal in the aperture domain. The relevant functions are

Sec. 3·5]   DETERMINATION OF CIRCUIT PARAMETERS   147

the eigenfunctions of the kernel regarded as an integral operator in the aperture domain. It is difficult to obtain such a representation for the dynamic kernel $G(y,y')$. However, the desired representation can be obtained for the static ($k^2 \to 0$) kernel $G_s(y,y')$, where

$$G_s(y,y') = \sum_1^\infty Y_{ns}\, h_n(y) h_n(y')$$

$$= \frac{2j\omega\epsilon}{\pi} \sum_1^\infty \frac{\cos\frac{n\pi y}{b} \cos\frac{n\pi y'}{b}}{n} = -j\frac{\omega\epsilon}{\pi} \ln 2\left|\cos\frac{\pi y}{b} - \cos\frac{\pi y'}{b}\right| \quad (59)$$

where the static characteristic admittance Y_{ns} of the nth E-mode is

$$Y_{ns} = \frac{j\omega\epsilon}{\frac{n\pi}{b}} = \frac{1}{Z_{ns}}. \quad (60)$$

On introduction of the change of variable

$$\cos\frac{\pi y}{b} = \sin^2\frac{\pi d}{2b} \cos\theta + \cos^2\frac{\pi d}{2b}, \quad (61)$$

which transforms the y domain 0 to d into the θ domain 0 to π, one obtains the desired diagonal representation of the static kernel as

$$G_s(y,y') = -j\frac{\omega\epsilon}{\pi} \ln\left(2 \sin^2\frac{\pi d}{2b} |\cos\theta - \cos\theta'|\right)$$

$$= -j\frac{\omega\epsilon}{\pi}\left[\ln\sin^2\frac{\pi d}{2b} - 2\sum_1^\infty \frac{\cos n\theta \cos n\theta'}{n}\right] \quad (62)$$

in terms of the orthogonal functions $\cos n\theta$ in the transformed aperture domain 0 to π.

To utilize the representation (62) the integral equation (48) is written in the form

$$\hat{I}\, h(y) + \sum_1^\infty \hat{I}_n h_n(y) = -\int_0^d G_s(y,y') E(y')\, dy' \quad (63)$$

with

$$\left.\begin{array}{l}\hat{I} = I, \\ \hat{I}_n = (Y_{ns} - Y_n) V_n.\end{array}\right\} \quad (63a)$$

In view of the linearity of this integral equation the solution $E(y)$ can be written as

$$E(y) = \hat{I}\,\mathcal{E}(y) + \sum_{1}^{\infty} \hat{I}_n \mathcal{E}_n(y), \tag{64}$$

whence by Eq. (44a) it follows that

$$\left.\begin{aligned} V &= Z_{00}\hat{I} + \sum_{1}^{\infty} Z_{0n}\hat{I}_n, \\ V_m &= Z_{m0}\hat{I} + \sum_{1}^{\infty} Z_{mn}\hat{I}_n, \end{aligned}\right\} \quad (m = 1,\,2,\,3,\,\cdots) \tag{65}$$

on use of the definitions

$$Z_{mn} = -\int_0^d \mathcal{E}_n(y) h_m(y)\,dy. \tag{65a}$$

Equations (65) represent in network terms the coupling of the static modes excited by the capacitive discontinuity. In view of Eqs. (63a), which represent the terminal conditions on the higher-mode lines, the desired dominant-mode admittance I/V can be obtained from Eqs. (65) by a conventional network calculation if the impedance parameters Z_{mn} are known. The latter may be found on determination of the partial fields $\mathcal{E}_n(y)$ from the integral equations ($n = 0,\,1,\,2,\,\cdots$),

$$h_n(y) = -\int_0^d G_s(y,y')\mathcal{E}_n(y')\,dy', \quad 0 < y < d, \tag{66}$$

obtained by substitution of Eq. (64) into (63) and equating the coefficients of I_n. It is readily shown with the aid of Eqs. (66) and (65a) that the parameters Z_{mn} obey the reciprocity relations $Z_{mn} = Z_{nm}$.

To illustrate the approximate solution of the integral equation (63) let us place $\hat{I}_n = 0$ for $n \geqslant 2$. The determination of the impedance parameters for this case first requires the solution of the integral equation (66) for $n = 0$ and $n = 1$. For $n = 0$ Eq. (66) becomes, on introduction of the representation (62) and the change of variable (61),

$$\sqrt{\frac{1}{b}} = -j\frac{2\omega\epsilon}{\pi}\int_0^\pi \mathcal{E}(\theta')\left(\ln \csc \frac{\pi d}{2b} + \sum_{1}^{\infty} \frac{\cos n\theta \cos n\theta'}{n}\right) d\theta', \tag{67}$$

where for simplicity $\mathcal{E}(\theta) = \mathcal{E}(y)\,dy/d\theta$. On equating the coefficients of $\cos n\theta$ on both sides of Eq. (67), one obtains

SEC. 3·5]   DETERMINATION OF CIRCUIT PARAMETERS   149

$$\begin{aligned}\int_0^\pi \mathcal{E}(\theta')\, d\theta' &= -\frac{\frac{1}{\sqrt{b}}}{j\frac{2\omega\epsilon}{\pi}\ln\csc\frac{\pi d}{2b}}, \\ \int_0^\pi \mathcal{E}(\theta')\cos n\theta'\, d\theta' &= 0, \qquad n \geq 1.\end{aligned} \qquad (68)$$

Hence, by Eqs. (44a) and (61) with ($\alpha = \sin^2 \pi d/2b$, $\beta = \cos^2 \pi d/2b$), there follows

$$\begin{aligned}Z_{00} &= -\frac{1}{\sqrt{b}}\int_0^\pi \mathcal{E}(\theta)\, d\theta = \frac{1}{\frac{2j\omega\epsilon b}{\pi}\ln\csc\frac{\pi d}{2b}}, \\ Z_{10} &= -\sqrt{\frac{2}{b}}\int_0^\pi \mathcal{E}(\theta)(\alpha\cos\theta + \beta)\, d\theta = \frac{\sqrt{2}\,\beta}{\frac{2j\omega\epsilon b}{\pi}\ln\csc\frac{\pi d}{2b}}, \\ Z_{20} &= -\sqrt{\frac{2}{b}}\int_0^\pi \mathcal{E}(\theta)(\alpha^2\cos 2\theta + 4\alpha\beta\cos\theta + (1-3\alpha)\beta)\, d\theta \\ &= \frac{\sqrt{2}\,(1-3\alpha)\beta}{\frac{2j\omega\epsilon b}{\pi}\ln\csc\frac{\pi d}{2b}},\end{aligned} \qquad (69)$$

the parameters Z_{n0} being obtained in a like manner. Similarly for $n = 1$. Equation (66) becomes on use of Eqs. (62) and (61)

$$\sqrt{\frac{2}{b}}(\alpha\cos\theta + \beta) = -j\frac{2\omega\epsilon}{\pi}\int_0^\pi \mathcal{E}_1(\theta')\left(\ln\csc\frac{\pi d}{2b} + \sum_1^\infty \frac{\cos n\theta \cos n\theta'}{n}\right) d\theta'; \qquad (70)$$

whence

$$\begin{aligned}\int_0^\pi \mathcal{E}_1(\theta')\, d\theta' &= -\frac{\sqrt{\frac{2}{b}}\,\beta}{j\frac{2\omega\epsilon}{\pi}\ln\csc\frac{\pi d}{2b}}, \\ \int_0^\pi \mathcal{E}_1(\theta')\cos\theta'\, d\theta' &= -\frac{\sqrt{\frac{2}{b}}\,\alpha}{j\frac{2\omega\epsilon}{\pi}}, \\ \int_0^\pi \mathcal{E}_1(\theta')\cos n\theta'\, d\theta' &= 0, \qquad n \geq 2,\end{aligned} \qquad (71)$$

and, as in Eqs. (69) with $\mathcal{E}(\theta)$ replaced by $\mathcal{E}_1(\theta)$, one obtains by Eqs. (44a) and (61)

$$Z_{01} = \frac{\sqrt{2}\,\beta}{j\dfrac{2\omega\epsilon b}{\pi}\ln\csc\dfrac{\pi d}{2b}},$$

$$Z_{11} = \frac{1}{j\dfrac{\omega\epsilon b}{\pi}}\left(\alpha^2 + \frac{\beta^2}{\ln\csc\dfrac{\pi d}{2b}}\right),$$ (72)

$$Z_{21} = \frac{\beta}{j\dfrac{\omega\epsilon b}{\pi}}\left[4\alpha^2 + \frac{(1-3\alpha)\beta}{\ln\csc\dfrac{\pi d}{2b}}\right],$$

etc., for Z_{n1}. As a check the verification of the reciprocity relation $Z_{10} = Z_{01}$ is to be noted.

With the knowledge of the relevant impedance parameters the desired discontinuity susceptance can now be obtained from Eqs. (65) by a simple "network" calculation. For in view of Eqs. (63a) and $\hat{I}_2 = \hat{I}_3 = \cdots = 0$, one finds that

$$\frac{V}{I} = \left(Z_{00} - \frac{Z_{10}^2}{Z_{11} + \dfrac{1}{Y_1 - Y_{1s}}}\right),$$

and thus on substitution of Eqs. (60), (69), and (72) there follows after some algebraic manipulation

$$\frac{B}{Y} = \frac{8b}{\lambda}\left[\ln\csc\frac{\pi d}{2b} + \frac{Q\cos^4\dfrac{\pi d}{2b}}{1 + Q\sin^4\dfrac{\pi d}{2b}}\right],$$ (73)

where

$$Q = \frac{1}{\sqrt{1 - \left(\dfrac{2b}{\lambda}\right)^2}} - 1.$$

Equation (73) provides an approximate value for the relative susceptance that is seen to agree with the first two terms of the more accurate value quoted in Sec. 5·1b (if $\lambda \to \lambda_g$).

Since only the first two E-modes have been treated correctly, the aperture field $E(y)$ determined by the above integral-equation method of solution is approximate. However, as stated above, the accuracy of the equivalent-circuit calculation can be improved by use of this approximate field as a trial field in the variational expression (50). To investigate the limits of validity of such a procedure let us first rewrite Eq. (50) in the form

$$j\frac{B}{2} = \sum_{1}^{\infty} Y_{ns}\left(\frac{V_n}{V}\right)^2 + \sum_{1}^{\infty} (Y_n - Y_{ns})\left(\frac{V_n}{V}\right)^2. \tag{74}$$

The field $E(y)$ defined by the integral equation (63) will be employed as a trial field, but now the $\hat{I}_n$ are to be regarded as unknown coefficients that are to be determined so as to make stationary (minimum in this case) the susceptance B of Eq. (74). On multiplication of Eq. (63) by $E(y)$ and integration over the aperture, there is obtained on use of Eq. (44a)

$$\hat{I} V + \sum_{1}^{\infty} \hat{I}_n V_n = \sum_{1}^{\infty} Y_{ns} V_n^2. \tag{75}$$

Substitution of this result into Eq. (74) and use of the network relations (65) then yield

$$j\frac{B}{2} = \frac{\sum_{m,n=0}^{\infty} Z_{mn}\hat{I}_m\hat{I}_n + \sum_{m=1}^{\infty} (Y_m - Y_{ms})\left(\sum_{n=0}^{\infty} Z_{mn}\hat{I}_n\right)^2}{\left(\sum_{m=0}^{\infty} Z_{0m}\hat{I}_m\right)^2}, \tag{76}$$

where one notes that $\hat{I}_0 = \hat{I}$. The imposition of the stationary conditions

$$\frac{\partial B}{\partial \hat{I}_n} = 0, \qquad n = 0, 1, 2, 3, \cdots$$

for the determination of the coefficients $\hat{I}_m$ leads to the set of equations (for all n)

$$jB\left(\sum_{0}^{\infty} Z_{0m}\hat{I}_m\right) Z_{0n} = 2Z_{0n}\hat{I} + 2\sum_{m=1}^{\infty} Z_{mn}[\hat{I}_m - (Y_{ms} - Y_m)V_m]$$

from which, on noting the identity of the left side and the first term of the right side of these equations, one readily obtains the solutions

$$\hat{I}_m = (Y_{ms} - Y_m)V_m, \qquad m = 1, 2, 3, \cdots. \tag{77}$$

Thus the choice of coefficients $\hat{I}_m$ that makes the susceptance B stationary is exactly the choice required to make $E(y)$ an exact solution of the integral equation (63). It is not generally feasible to satisfy all of the conditions (77). As an approximation one may choose $\hat{I}_m = 0$ for all $m > N$. Under these circumstances the network equations (65), which determine V_m and hence $\hat{I}_m$, become

$$V = Z_{00}\hat{I} + \sum_{1}^{N} Z_{0n}\hat{I}_n, \\ V_m = Z_{m0}\hat{I} + \sum_{1}^{N} Z_{mn}\hat{I}_n, \Biggr\} \quad (78)$$

with

$$\hat{I} = I, \quad \hat{I}_m = (Y_{ms} - Y_m)V_m, \quad m = 1, 2, \cdots, N.$$

The use of these network equations in the variational expression (76) then leads to the final variational result

$$j\frac{B}{2} = j\frac{B_0}{2} + \sum_{N+1}^{\infty} (Y_m - Y_{ms})\left(\frac{V_m}{V}\right)^2, \quad (79)$$

where

$$j\frac{B_0}{2} = \frac{I}{V} = \frac{\hat{I}}{\sum_{0}^{N} Z_{0m}\hat{I}_m} \quad (79a)$$

is the approximate discontinuity admittance obtained by the integral-equation method of solution with $\hat{I}_m = 0$ for $m > N$. The sum in Eq. (79) is thus seen to act as a variational correction to the integral-equation result.

As an example of the use of Eq. (79) let us consider the case $N = 1$ for which the integral equation result B_0/Y has already been derived in Eq. (73). Since the correction series in Eq. (79) converges relatively rapidly, only the first term thereof will be evaluated and this only to order $(2b/\lambda)^3$. Thus, on use of Eqs. (78), (46), and (60), the first term may be expressed relative to the characteristic admittance of the dominant mode as

$$\frac{Y_2 - Y_{2s}}{Y}\left(\frac{V_2}{V}\right)^2 \approx j\frac{1}{16}\left(\frac{2b}{\lambda}\right)^3 \left[\frac{Z_{20}Z_{11} - Z_{12}Z_{10} + \dfrac{Z_{20}}{Y_1 - Y_{1s}}}{Z_{00}Z_{11} - Z_{10}^2 + \dfrac{Z_{00}}{Y_1 - Y_{1s}}}\right]^2$$

which, on evaluation of the wavelength independent terms within the bracket by Eqs. (69) and (72), becomes

$$j\frac{1}{8}\left(\frac{2b}{\lambda}\right)^3\left(1 - 3\sin^2\frac{\pi d}{2b}\right)^2 \cos^4\frac{\pi d}{2b}. \quad (80)$$

The substitution of Eqs. (73) and (80) into (79) then yields as the variational result for the relative susceptance

$$\frac{B}{Y} = \frac{8b}{\lambda}\left[\ln\csc\frac{\pi d}{2b} + \frac{Q\cos^4\dfrac{\pi d}{2b}}{1 + Q\sin^4\dfrac{\pi d}{2b}} + \frac{1}{16}\left(\frac{2b}{\lambda}\right)^2\left(1 - 3\sin^2\frac{\pi d}{2b}\right)^2\cos^4\frac{\pi d}{2b}\right]. \tag{81}$$

This is identically the result quoted in Sec. 5·1b (if $\lambda \to \lambda_g$); it is estimated to be in error by less than 5 per cent for $2b/\lambda < 1$ and by less than 1 per cent for $4b/\lambda < 1$.

c. *The Equivalent Static Method.*—The equivalent static method of solving the capacitive window problem of Sec. 5·1b resembles strongly the integral-equation method just described. As in the latter, the original dynamic problem with only a dominant mode incident is reduced to a static, parallel-plate problem with an infinity of modes incident. However, the task of finding the fields produced by each of the incident static modes is now regarded as an electrostatic problem to be solved by conformal mapping of the original problem into a geometrically simpler problem for which the potential (i.e., static) solution can be found by means of complex function theory. The static problem for the case of only the lowest, principal mode incident constitutes a conventional electrostatic problem with a d-c voltage applied across the guide plates; the static problems with a higher mode incident, although less conventional electrostatic problems, are nevertheless solved in a manner similar to that employed for the lowest mode. The use of conformal mapping in the solution of the electrostatic problems implies a limitation of the equivalent static method to microwave problems that are essentially two dimensional. As in the previous method, no attempt is made in practice to obtain the formally possible exact solution; a variational procedure is, however, employed to improve the accuracy of an approximate solution.

As a preliminary to the electrostatic solution of the integral equation (63) subject to Eqs. (63a), the unknown aperture field is now represented as

$$E(y) = V\mathcal{E}(y) + \sum_{1}^{\infty}\hat{I}_n\mathcal{E}_n(y), \tag{82}$$

where the proportionality to V rather than to $\hat{I}$, as in Eq. (64), is more convenient for conformal mapping purposes. It then follows by Eqs. (44) that

$$-\int_{\text{ap}} \mathcal{E}(y)h(y)\,dy = 1, \qquad \int_{\text{ap}} \mathcal{E}_n(y)h(y)\,dy = 0$$

and also that

$$\begin{aligned}\hat{I} &= Y_{00}V - \sum_{1}^{\infty} T_{0n}\hat{I}_n, \\ V_m &= T_{m0}V + \sum_{1}^{\infty} Z_{mn}\hat{I}_n,\end{aligned} \Bigg\} \quad m = 1, 2, 3, \cdots, \quad (84)$$

where the significance of the proportionality factors Y_{00}, T_{0n} will be evident below, while by definition

$$\begin{aligned}T_{m0} &= -\int_{\text{ap}} \mathcal{E}(y) h_m(y)\, dy, \\ Z_{mn} &= -\int_{\text{ap}} \mathcal{E}_n(y) h_m(y)\, dy.\end{aligned} \Bigg\} \quad (85)$$

In contrast to Eqs. (65) the network equations (84) are of a "mixed" type; the "network" parameters comprise the admittance Y_{00}, the impedances Z_{mn}, and the transfer coefficients T_{0n}, T_{n0}. From the knowledge of these parameters the desired discontinuity admittance $\hat{I}/V$ can be found by straightforward solution of the network equations (84) subject to the terminal conditions (63a). To determine the network parameters it is first necessary to find the partial fields $\mathcal{E}_n(y)$. On substituting Eq. (82) into the integral equation (63) and equating coefficients, one finds that the partial fields are determined by the set of integral equations ($n = 1, 2, 3, \ldots$)

$$Y_{00}\, h(y) = -\int_0^d G_s(y,y') \mathcal{E}(y')\, dy', \quad (86a)$$

$$0 < y < d,$$

$$-T_{0n}\, h(y) + h_n(y) = -\int_0^d G_s(y,y') \mathcal{E}_n(y')\, dy', \quad (86b)$$

from which it is apparent that these partial fields differ from those encountered in the previous method. From Eqs. (85) and (86) one derives the reciprocity relations

$$T_{0n} = T_{n0} \quad \text{and} \quad Z_{mn} = Z_{nm}.$$

To find the partial field $\mathcal{E}(y)$ by conformal mapping one observes that the integral equation (86a) is characteristic of an electrostatic distribution with a y-component electric field and an x-component magnetic field (or stream function) given, when $z < 0$, by

DETERMINATION OF CIRCUIT PARAMETERS

$$\left.\begin{aligned}\mathcal{E}(y) &= e(y) + \sum_{1}^{\infty} T_{0n} e_n(y) e^{n\pi z/b}, \\ \mathcal{H}(y) &= (Y_{00} - j\omega\epsilon z)h(y) - \sum_{1}^{\infty} T_{0n} Y_{ns} h_n(y) e^{n\pi z/b},\end{aligned}\right\} \quad (88a)$$

for the individual terms in Eqs. (88a) are solutions of Laplace's equation, and as is evident from Eqs. (85) and (87), the integral equation (86a) is obtained from the boundary condition that $\mathcal{H}(y) = 0$ at $z = 0, 0 < y < d$. Equations (88a) are characteristic of an electrostatic distribution corresponding to a unit d-c voltage across the guide plates, the term Y_{00} being proportional to the "excess" static capacitance arising from the presence of the discontinuity at $z = 0$. Similarly the field distribution

$$\left.\begin{aligned}\mathcal{E}_n(y) &= Z_{ns} \sinh \frac{n\pi z}{b} e_n(y) + \sum_{m=1}^{\infty} Z_{mn} e_m(y) e^{m\pi z/b}, \\ \mathcal{H}_n(y) &= -T_{0n} h(y) + \cosh \frac{n\pi z}{b} h_n(y) - \sum_{m=1}^{\infty} Y_{ms} Z_{mn} h_m(y) e^{m\pi z/b},\end{aligned}\right\} \quad (88b)$$

associated with the integral equation (86b) corresponds to an electrostatic distribution with only the nth mode incident.

The parameters Y_{00}, T_{n0}, and Z_{mn} can be found either by solving the integral equations (86) or by finding the electrostatic field distributions (88a) and (88b). The latter involves the determination of a stream function $\psi(y,z)$ (i.e., a magnetic field) satisfying Laplace's equation and the boundary conditions appropriate to the parallel-plate waveguide geometry depicted in Fig. 3·25a. It is simpler to solve this problem not in the actual waveguide but rather in a geometrically transformed waveguide in which the boundary conditions on ψ are simple. From the theory of analytic complex functions a solution of Laplace's equation in the transformed guide is likewise a solution in the actual guide if the transformation is conformal. The desired transformation is one which conformally maps the original waveguide region in the $\zeta = z + jy$ plane of Fig. 3·25a into the upper half of the t-plane of Fig. 3·25b and thence into the waveguide region in the $\zeta' = z' + jy'$ plane shown in Fig. 3·25c. The corresponding points in the ζ, t, and ζ'-planes are so chosen that the aperture surface $0 < y < d$, $z = 0$ in the ζ-plane is transformed into the terminal surface $0 < y' < b$, $z' = 0$ in the ζ'-plane.

With the aid of the Schwartz-Christoffel transformation the function that maps the guide periphery in the ζ-plane of Fig. 3·25a into the real axis of the t-plane of Fig. 3·25b is found to be

$$\cosh \frac{\pi \zeta}{b} = t, \tag{89a}$$

while that which maps the guide periphery in the ζ'-plane of Fig. 3·25c into the real axis of the t-plane is

$$\alpha \cosh \frac{\pi \zeta'}{b} + \beta = t, \tag{89b}$$

where $\alpha = \sin^2 \pi d/2b$ and $\beta = \cos^2 \pi d/2b$ [note the significant connection

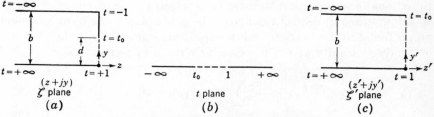

FIG. 3·25.—Conformal mapping from ζ to ζ' planes.
$\frac{\partial \psi}{\partial n} = 0$ on ——————————
$\psi = 0$ on ----------

between Eqs. (89b) and (61)]. The over-all mapping function for the transformation from the ζ- to the ζ'-planes is thus

$$\cosh \frac{\pi \zeta}{b} = \alpha \cosh \frac{\pi \zeta'}{b} + \beta. \tag{90}$$

The solutions of electrostatic problems in the transformed waveguide of Fig. 3·25c can be readily found, since the terminal conditions therein correspond to a simple open circuit ($\psi = 0$) at the terminal plane $z' = 0$.

The complex stream function

$$\Psi = \psi + j\phi \tag{91}$$

that corresponds to only the principal mode incident (i.e., an applied d-c voltage) in the waveguide of Fig. 3·25c is

$$\Psi = A\zeta', \tag{92a}$$

where the constant amplitude A (regarded as real) can be selected arbitrarily. The associated solution in the actual waveguide is therefore by Eq. (90)

$$\Psi = A \frac{b}{\pi} \cosh^{-1} \left(\frac{\cosh \frac{\pi \zeta}{b} - \beta}{\alpha} \right), \tag{92b}$$

or equivalently as a series in powers of $e^{\pi \zeta/b}$ as

$$\Psi = \frac{Ab}{\pi}\left[\ln\frac{1}{\alpha} - \frac{\pi\zeta}{b} - 2\beta e^{\pi\zeta/b} - \beta(1-3\alpha)e^{2\pi\zeta/b} + \cdots\right]. \quad (93)$$

If A is now set equal to $j\omega\epsilon/\sqrt{b}$, the stream function ψ follows from Eqs. (91) and (92a) as

$$\psi = j\frac{\omega\epsilon b}{\pi}\frac{1}{\sqrt{b}}\left[\ln\frac{1}{\alpha} - \frac{\pi}{b}z - 2\beta\cos\frac{\pi y}{b}e^{\pi z/b} - \beta(1-3\alpha)\cos\frac{2\pi y}{b}e^{2\pi z/b}\right.$$
$$\left. + \cdots\right] \quad (94)$$

and evidently represents the static magnetic field produced by a principal mode incident in the actual guide. On comparison of Eqs. (94) and (88a) one finally obtains

$$\left.\begin{array}{l} Y_{00} = j\dfrac{2\omega\epsilon b}{\pi}\ln\csc\dfrac{\pi d}{2b}, \\ T_{01} = \sqrt{2}\,\beta, \\ T_{02} = \sqrt{2}\,(1-3\alpha)\beta, \cdots . \end{array}\right\} \quad (95)$$

The corresponding complex stream function for only the $(n = 1)$-mode incident in the waveguide of Fig. 3·25c is simply

$$\Psi = B\sinh\frac{\pi\zeta'}{b}, \quad (96a)$$

whence by Eq. (90) the complex stream function in the actual guide is

$$\Psi = \frac{2B}{\alpha}\sinh^2\frac{\pi\zeta}{2b}\sqrt{1 + \frac{\alpha}{\sinh^2\dfrac{\pi\zeta}{2b}}}. \quad (96b)$$

The latter may be expressed as a series in powers of $e^{\pi\zeta/b}$ as

$$\Psi = \frac{B}{2\alpha}[e^{-\frac{\pi\zeta}{b}} - 2\beta + (1-2\alpha^2)e^{\frac{\pi\zeta}{b}} + \cdots], \quad (96c)$$

from which, on setting $B = \alpha\sqrt{2/b}$ and taking the real part, one obtains the real stream function

$$\psi = \sqrt{\frac{2}{b}}\left[-\beta + \cosh\frac{\pi z}{b}\cos\frac{\pi y}{b} - \alpha^2 e^{\pi z/b}\cos\frac{\pi y}{b} + \cdots\right] \quad (97)$$

characteristic of a static magnetic-field distribution in the actual guide with only the $n = 1$-mode incident. On comparison of Eqs. (97) and (88b) one finds that

$$T_{10} = \sqrt{2}\,\beta, \\ Z_{11} = \frac{\alpha^2}{j\dfrac{\omega\epsilon b}{\pi}}, \cdots \quad \Biggr\} \tag{98}$$

With the knowledge of the network parameters in Eqs. (95) and (98) it is now possible to obtain the solution of the capacitive window problem in the approximation wherein only the two lowest modes are treated exactly. The "mixed" network equations (84) reduce in this case to

$$\hat{I} = Y_{00}V - T_{01}\hat{I}_1, \\ V_1 = T_{10}V + Z_{11}\hat{I}_1, \Biggr\} \tag{99}$$

with the "terminal" conditions

$$\hat{I} = I \quad \text{and} \quad \hat{I}_1 = (Y_{1s} - Y_1)V_1,$$

whence

$$\frac{I}{V} = Y_{00} + \frac{T_{10}^2}{Z_{11} + \dfrac{1}{Y_1 - Y_{1s}}}.$$

On insertion of the values of the parameters given in Eqs. (95) and (98), one obtains for the discontinuity susceptance

$$\frac{B}{Y} = \frac{2I}{jYV} = \frac{8b}{\lambda}\left[\ln \csc \frac{\pi d}{2b} + \frac{Q\cos^4 \dfrac{\pi d}{2b}}{1 + Q\sin^4 \dfrac{\pi d}{2b}}\right], \tag{100}$$

the result obtained previously in Eq. (73) by the integral-equation method and in agreement with the first two terms of the value quoted in Sec. 5·1b (if $\lambda \to \lambda_g$).

The accuracy of an approximate solution of the above electrostatic problem can be improved by utilization of the aperture field $E(y)$ found thereby as a trial field in the variational expression (50). The basis for such a procedure may be developed in a manner similar to that in method b. As a trial field one first considers that electrostatic field $E(y)$ defined by the integral equation (63) in which the unknown coefficients $\hat{I}_n$ are to be so chosen as to make the variational susceptance B stationary. Writing the variational expression in the form (74) and utilizing Eqs. (75) and (84), one then obtains

$$j\frac{B}{2} = Y_{00} + \frac{\displaystyle\sum_{m,n=1}^{\infty} Z_{mn}\hat{I}_m\hat{I}_n + \sum_{m=1}^{\infty}(Y_m - Y_{ms})\left(T_{m0}V + \sum_{n=1}^{\infty}Z_{mn}\hat{I}_n\right)^2}{V^2}. \tag{101}$$

Sec. 3·5] DETERMINATION OF CIRCUIT PARAMETERS

Imposition of the stationary conditions

$$\frac{\partial B}{\partial \hat{I}_n} = 0, \qquad n = 1, 2, 3, \cdots, \infty,$$

with the arbitrary V being held constant, leads to an infinite set of equations

$$0 = \sum_{m=1}^{\infty} Z_{mn}[\hat{I}_m - (Y_{ms} - Y_m)V_m],$$

whose solution is readily seen to be

$$\hat{I}_m = (Y_{ms} - Y_m)V_m, \qquad m = 1, 2, 3, \cdots, \infty. \tag{102}$$

This variational choice for $\hat{I}_m$ is thus exactly that required to make the trial $E(y)$ rigorously correct. As a practical approximation let only the first N of Eqs. (101) be satisfied, the remaining $\hat{I}_m$ being set equal to zero. The network equations (84), which represent the solution of the electrostatic problem in this case, then become

$$\left.\begin{aligned}\hat{I} &= Y_{00}V - \sum_{1}^{N} T_{0n}\hat{I}_n, \\ V_m &= T_{m0}V + \sum_{1}^{N} Z_{mn}\hat{I}_n,\end{aligned}\right\} \quad m = 1, 2, 3, \cdots, N \tag{103}$$

subject to

$$\hat{I} = I \quad \text{and} \quad \hat{I}_m = (Y_{ms} - Y_m)V_m.$$

By means of Eqs. (103) the variational expression (101) may be rewritten as

$$j\frac{B}{2} = j\frac{B_0}{2} + \sum_{N+1}^{\infty} (Y_m - Y_{ms})\left(\frac{V_m}{V}\right)^2, \tag{104}$$

where we have defined

$$j\frac{B_0}{2} = Y_{00} - \sum_{1}^{N} \frac{T_{m0}\hat{I}_m}{V}, \tag{104a}$$

the latter being the approximate electrostatic result obtained by treating only the N lowest modes correctly. As in the corresponding, but not identical, variational expression (79) the sum in Eq. (104) represents a variational correction to the electrostatic result.

As an example, the use of Eq. (104) for the case $N = 1$ will be considered. The approximate electrostatic result B_0/Y is then given by Eq. (100). As an approximation the correction series in (104) will be replaced by its first term. By means of Eqs. (103) this term may be expressed in the form

$$\frac{Y_2 - Y_{2s}}{jY}\left(\frac{V_2}{V}\right)^2 = \left(\frac{1}{\sqrt{4 - \left(\frac{2b}{\lambda}\right)^2}} - \frac{1}{2}\right)\left(\frac{2b}{\lambda}\right)\left(T_{20} + \frac{Z_{21}T_{10}}{Z_{11} + \frac{1}{Y_1 - Y_{1s}}}\right)^2$$

$$\approx \frac{1}{16}\left(\frac{2b}{\lambda}\right)^3 (T_{20})^2 = \frac{1}{8}\left(\frac{2b}{\lambda}\right)^3 (1 - 3\alpha)^2\beta^2, \qquad (105)$$

the approximation being correct to order $(2b/\lambda)^3$. Insertion of this correction into Eq. (104) yields for B/Y the same result as obtained in Eq. (81) by the integral-equation method and quoted in Sec. 5·1b.

Although identical results have been obtained for the susceptance of capacitive window by both the integral-equation and the equivalent-static methods, it should be recognized that the latter method generally is the more powerful for two-dimensional problems. This is a consequence of the power of conformal mapping to solve Laplace's equation in rather complicated geometrical structures.

d. The Transform Method.—As a final illustration we shall consider the equivalent-circuit problem presented by a six-terminal microwave structure—the E-plane bifurcation of a rectangular guide treated in Sec. 6·4. Although the various methods described above are likewise applicable to the solution of this problem, we shall confine ourselves to an integral-equation method. However, since the E-plane bifurcation can be regarded advantageously as either an aperture or obstacle discontinuity of infinite extent, the associated integral equation will be of the so-called Wiener-Hopf type. The solution of this integral equation differs in detail from the integral-equation method discussed above and moreover is rigorous.

A longitudinal view of the E-plane bifurcation together with the choice of coordinate axes and dimensions is shown in Fig. 3·26a. For simplicity of analytical discussion it is assumed that only the H_{10}-mode can be propagated and that the bifurcation bisects the structure (note that the height at the bifurcation is b rather than $b/2$ as in Sec. 6·4). By a special choice of excitation the symmetry about the plane of the bifurcation can be utilized so as to require solution of the field problem in only half of the structure. Thus, an antisymmetrical choice of current excitation in guides 1 and 2 will lead to a correspondingly antisymmetrical magnetic-field distribution in which the tangential magnetic field in the "aperture" plane, $y = b$, $z > 0$, vanishes. The resulting field problem has then to be solved only in the reduced domain, shown in Fig. 3·26b, obtained on bisection of the structure about the plane of the bifurcation; the correspondingly reduced equivalent circuit shown in Fig. 3·26c is found on (open-circuit) bisection of the over-all equivalent circuit of Fig. 6·4-2. The two-terminal nature of the reduced circuit implies that no propagating mode is present in guide 3 under the assumed antisymmetrical excitation. A further simplification resulting from the uniformity of

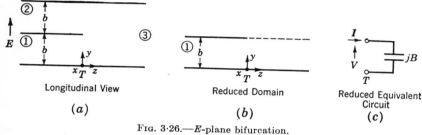

Longitudinal View
(a)

Reduced Domain
(b)

Reduced Equivalent Circuit
(c)

Fig. 3·26.—E-plane bifurcation.

the structure in the x direction permits the solution of the circuit problem in a parallel-plate rather than a rectangular guide; the parallel-plate results go over into those of the rectangular guide on replacement of the space wavelength λ in the former by the guide wavelength λ_g in the latter.

Because of the uniformity of the structure in the x direction and principal-mode excitation of guide 1 the fields are derivable from a single x-component, $\psi(y,z)$, of magnetic field. The field problem in the reduced domain of Fig. 3·26b is hence the scalar problem of determining a function ψ satisfying the wave equation

$$\left(\frac{\partial^2}{\partial y^2} + \frac{\partial^2}{\partial z^2} + k^2\right)\psi = 0 \tag{106}$$

and subject to the following boundary conditions (with $kb < \pi/2$):

1. $\dfrac{\partial \psi}{\partial y} = 0 \quad$ for $y = 0$ and for $y = b$, $z < 0$.
2. $\psi = 0 \quad$ for $y = b$, $z > 0$.
3. $\psi \to I \cos kz - jYV \sin kz \quad$ for $z \to -\infty$.
4. $\psi \to e^{-|\kappa_1'|z} \quad$ for $z \to +\infty$ where $|\kappa_1'| = \sqrt{\left(\dfrac{\pi}{2b}\right)^2 - k^2}$.

The quantities I and V represent the dominant-mode current and voltage at the terminal plane $T(z = 0)$, and Y is the characteristic admittance of guide 1. Condition 4 states that the dominant field in guide 3 is that of the lowest nonpropagating mode therein. It is required to solve the field problem specified in Eqs. (106) for the normalized susceptance

$$\frac{B}{Y} = \frac{I}{jYV}. \tag{107}$$

The representation of the field ψ which satisfies Eq. (106) will be determined by a Green's function method rather than by the equivalent modal method previously employed. To this end one utilizes the two-dimensional form of Green's theorem:

$$\iint (\psi \nabla^2 G - G \nabla^2 \psi) \, dy' \, dz' = \int \left(\psi \frac{\partial G}{\partial n'} - G \frac{\partial \psi}{\partial n'} \right) ds', \tag{108}$$

where

$$\nabla^2 = \frac{\partial^2}{\partial y'^2} + \frac{\partial^2}{\partial z'^2}.$$

The double integral with respect to the running variables y', z' is taken over the entire reduced domain in Fig. 3·26b; the line integral with respect to ds' is taken over the periphery thereof; and $\mathbf{n}'$ represents the outward normal direction at the periphery. The Green's function

$$G = G(y,z;y',z') = G(y'z';y,z)$$

will be defined by the inhomogeneous wave equation

$$\left(\frac{\partial^2}{\partial y^2} + \frac{\partial^2}{\partial z^2} + k^2 \right) G = -\delta(y - y')\delta(z - z') \tag{109}$$

and subject to the following boundary conditions:

1. $\dfrac{\partial G}{\partial y} = 0 \quad$ for $y = 0, b$.
2. $Im \, G = 0, \quad G$ bounded for $z \to -\infty$.
3. $G \to e^{-|\kappa_1|z} \quad$ for $z \to +\infty$ where $|\kappa_1| = \sqrt{\left(\dfrac{\pi}{b}\right)^2 - k^2}$.

The delta function is defined by

$$\delta(x) = 0 \quad \text{if } x \neq 0, \quad \int \delta(x - x') \, dx' = 1,$$

the interval of integration including the point x. Explicit representations of the Green's function defined by Eqs. (109) are readily found either in the form

SEC. 3·5]   DETERMINATION OF CIRCUIT PARAMETERS   163

$$G(y,z;y',z') = \frac{\sin k(z-z') - \sin k|z-z'|}{2kb}$$

$$+ \frac{2}{b} \sum_{1}^{\infty} \frac{e^{-\sqrt{(n\pi/b)^2 - k^2}|z-z'|}}{2\sqrt{\left(\frac{n\pi}{b}\right)^2 - k^2}} \cos \frac{n\pi y}{b} \cos \frac{n\pi y'}{b} \quad (110a)$$

or in the form

$$G(y,z;y',z') = -\frac{1}{2\pi} \int_{-\infty+j\eta}^{+\infty+j\eta} \frac{\cos \sqrt{k^2 - \zeta^2}\, y \cos \sqrt{k^2 - \zeta^2}\,(b-y')}{\sqrt{k^2 - \zeta^2}\, \sin \sqrt{k^2 - \zeta^2}\, b} e^{j\zeta(z-z')}\, d\zeta,$$

$$y < y',$$

$$= -\frac{1}{2\pi} \int_{-\infty+j\eta}^{+\infty+j\eta} \frac{\cos \sqrt{k^2 - \zeta^2}\,(b-y) \cos \sqrt{k^2 - \zeta^2}\, y'}{\sqrt{k^2 - \zeta^2}\, \sin \sqrt{k^2 - \zeta^2}\, b} e^{j\zeta(z-z')}\, d\zeta,$$

$$y > y', \quad (110b)$$

with

$$0 < \eta < \sqrt{\left(\frac{\pi}{b}\right)^2 - k^2}.$$

In conformity with the defining equation and boundary conditions in Eqs. (109) one notes that the representation (110a) satisfies the wave equation except at z', where it possesses a discontinuity in the z derivative of magnitude $-\delta(y - y')$; correspondingly, the representation (110b) satisfies the wave equation save at y', where it has a discontinuity in the y derivative of magnitude $-\delta(z - z')$. Equations (110a) and (110b) are seen to be representations in terms of modes transverse to the z- and y-axes, respectively.

On use of the defining equations and boundary conditions (106) and (109), one finds by Green's theorem (108) that ψ can be represented everywhere in the reduced domain as

$$\psi(y,z) = \int_0^\infty G(y,z;b,z') E(z')\, dz', \quad (111)$$

where for brevity we have set

$$E(z') \equiv \frac{\partial}{\partial y'} \psi(y',z') \bigg]_{y'=b}.$$

The unknown function $E(z')$, which is proportional to the z-component of electric field in the aperture plane, may be determined on imposition of the boundary condition 2 in Eq. (106). This condition leads to the homogeneous integral equation

$$0 = \int_0^\infty K(z - z')E(z')\,dz', \qquad z > 0, \tag{112}$$

where $K(z - z') \equiv G(b,z;b,z')$. The solution of Eq. (112) for $E(z)$ then determines $\psi(y,z)$ everywhere; in particular at $z \to -\infty$ we note from Eqs. (111) and (110a) that

$$\psi(y,z) = \frac{1}{kb} \int_0^\infty E(z') \sin k(z - z')\,dz',$$

whence, on comparison with boundary condition 3 of Eq. (106),

$$I = -\frac{1}{kb} \int_0^\infty E(z') \sin kz'\,dz',$$

$$jYV = -\frac{1}{kb} \int_0^\infty E(z') \cos kz'\,dz'.$$

In view of Eq. (107) the desired circuit parameter is seen to be

$$\frac{B}{Y} = \frac{\displaystyle\int_0^\infty E(z') \sin kz'\,dz'}{\displaystyle\int_0^\infty E(z') \cos kz'\,dz'} = -\tan\varphi, \tag{113}$$

where

$$\varphi = \text{phase of } \int_0^\infty E(z')e^{-jkz'}\,dz'.$$

If, in harmony with its interpretation as an electric field, we set the function $E(z')$ equal to zero on the "obstacle" plane $z < 0$, the desired angle φ becomes the phase of the Fourier transform of $E(z)$ at the "frequency" k.

To determine the Fourier transform of $E(z)$ it is suggestive to rewrite Eq. (112) as the Wiener-Hopf integral equation

$$H(z) = \int_{-\infty}^{+\infty} K(z - z')E(z')\,dz' \tag{114}$$

with

$$H(z) = 0, \qquad z > 0,$$
$$E(z) = 0, \qquad z < 0,$$

where, as is evident on comparison with Eq. (111), $H(z)$ is the magnetic-field function ψ on the obstacle plane $y = b$, $z < 0$. The diagonal representation of $K(z - z')$ given in Eq. (110b) (with $y = y' = b$) permits a simple algebraic representation of the integral equation (114) as

$$\mathcal{H}(\zeta) = \mathcal{K}(\zeta)\mathcal{E}(\zeta), \quad \text{where} \quad \mathcal{K}(\zeta) = -\frac{\cot \sqrt{k^2 - \zeta^2}\,b}{\sqrt{k^2 - \zeta^2}}, \tag{115}$$

SEC. 3·5]   DETERMINATION OF CIRCUIT PARAMETERS   165

in terms of the Fourier transforms

$$\begin{aligned}
\mathcal{H}(\zeta) &= \int_{-\infty}^{+\infty} H(z) e^{-j\zeta z}\, dz, & Im\,\zeta &> 0, \\
\mathcal{E}(\zeta) &= \int_{-\infty}^{+\infty} E(z) e^{-j\zeta z}\, dz, & Im\,\zeta &< \kappa_1, \\
\mathcal{K}(\zeta) &= \int_{-\infty}^{+\infty} K(z) e^{-j\zeta z}\, dz, & 0 &< Im\,\zeta < \kappa_1' < \kappa_1
\end{aligned} \quad (116)$$

of $H(z)$, $E(z)$, and $K(z)$, respectively. The validity of the transform relation (115), which is an expression of the so-called Faltung theorem for Fourier transforms, presupposes a common domain of regularity for the various transforms. The existence of a common domain $0 < Im\,\zeta < \kappa_1$ is manifest from the regularity domains indicated in Eqs. (116), the latter domains being determined by the infinite z behavior of the various functions. An additional boundary condition on the field is implied by the mere existence of the first two transforms in Eqs. (116); namely, the behavior of $E(z)$, for example, must be such that $E(z)$ is integrable about the singularity point $z = 0$. Thus it will be assumed that

$$E(z) \sim z^{-\alpha} \qquad \text{as } z \to 0 \tag{117a}$$

with $\alpha < 1$; hence by Eq. (116) it follows that

$$\mathcal{E}(\zeta) \sim (j\zeta)^{\alpha-1} \qquad \text{as } \zeta \to \infty. \tag{117b}$$

Although both the transforms $\mathcal{H}(\zeta)$ and $\mathcal{E}(\zeta)$ in Eq. (115) are unknown, they may nevertheless be evaluated by means of Liouville's theorem. The theorem states that, if an analytic function has no zeros or poles in the entire complex plane and is bounded at infinity, the function must be a constant. To utilize the theorem let it first be supposed that the transform $\mathcal{K}(\zeta)$ can be factored as

$$\mathcal{K}(\zeta) = \frac{\mathcal{K}_-(\zeta)}{\mathcal{K}_+(\zeta)} = -\frac{\cot \sqrt{k^2 - \zeta^2}\, b}{\sqrt{k^2 - \zeta^2}}, \tag{118}$$

where $\mathcal{K}_-(\zeta)$ has no poles or zeros in a lower half of the ζ-plane while $\mathcal{K}_+(\zeta)$ has no poles or zeros in an upper half of the ζ-plane. Equation (115) can then be rewritten in the form

$$\mathcal{K}_+(\zeta)\mathcal{H}(\zeta) = \mathcal{K}_-(\zeta)\mathcal{E}(\zeta). \tag{119a}$$

The left-hand member then has no zeros or poles in an upper half plane while the right-hand member has no zeros or poles in a lower half of the ζ-plane. If all the terms in Eq. (119a) have a common domain of regularity, the function which is equal to the left-hand member in the upper half plane and which by analytic continuation is equal to the right-hand member in the lower half plane must therefore have no poles or zeros in the entire plane. If, as is already implied, the analytically continued function is bounded at infinity, then by Liouville's theorem this function and hence both members of Eq. (119a) must be equal to some constant A. It thus follows that

$$\mathcal{E}(\zeta) = \frac{A}{\mathcal{K}_-(\zeta)} \quad \text{and} \quad \mathcal{K}(\zeta) = \frac{A}{\mathcal{K}_+(\zeta)}, \tag{119b}$$

whence from Eq. (113)

$$\tan^{-1}\left(\frac{B}{Y}\right) = - \text{ phase of } \left(\frac{A}{\mathcal{K}_-(k)}\right). \tag{120a}$$

or independently of A

$$\frac{B}{Y} = \frac{1}{j} \frac{\mathcal{K}_-(k) - \mathcal{K}_-(-k)}{\mathcal{K}_-(k) + \mathcal{K}_-(-k)}. \tag{120b}$$

To factor $\mathcal{K}(\zeta)$ in the regular manner desired in Eq. (118), one employs the known infinite product expansion of the cotangent function and obtains

$$\mathcal{K}(\zeta) = -\frac{1}{(k^2 - \zeta^2)b} \frac{\prod_1^\infty \left[1 - \left(\frac{2kb}{n\pi}\right)^2 + \left(\frac{2\zeta b}{n\pi}\right)^2\right]}{\left\{\prod_1^\infty \left[1 - \left(\frac{kb}{n\pi}\right)^2 + \left(\frac{\zeta b}{n\pi}\right)^2\right]\right\}^2}$$

or on factoring

$$\mathcal{K}(\zeta) = -\frac{1}{(k^2 - \zeta^2)b} \frac{\Pi(2b,\zeta)\Pi(2b,-\zeta)}{[\Pi(b,\zeta)]^2[\Pi(b,-\zeta)]^2} = \frac{\mathcal{K}_-(\zeta)}{\mathcal{K}_+(\zeta)}, \tag{121}$$

where the function

$$\Pi(b,\zeta) = \prod_1^\infty \left[\sqrt{1 - \left(\frac{kb}{n\pi}\right)^2} + j\frac{\zeta b}{n\pi}\right] e^{-j\frac{\zeta b}{n\pi}} \tag{122}$$

is manifestly regular and has no zeros in the lower half plane

$$Im\,\zeta < \sqrt{\left(\frac{\pi}{b}\right)^2 - k^2}.$$

If we now identify in Eq. (121)

$$\mathcal{K}_-(\zeta) = \frac{\Pi(2b,\zeta)}{[\Pi(b,\zeta)]^2} e^{\chi(\zeta)}, \tag{123a}$$

$$\mathcal{K}_+(\zeta) = - \frac{(k^2 - \zeta^2)b[\Pi(b,-\zeta)]^2}{\Pi(2b,-\zeta)} e^{\chi(\zeta)}, \tag{123b}$$

we see that $\mathcal{K}_-(\zeta)$ is regular and has no zeros in the lower half plane $Im\,\zeta < \sqrt{\left(\frac{\pi}{2b}\right)^2 - k^2}$ while $\mathcal{K}_+(\zeta)$ is regular and has no zeros in the upper

DETERMINATION OF CIRCUIT PARAMETERS

half plane $Im\ \zeta > 0$. The argument $\chi(\zeta)$ of the exponential is to be so selected that the behavior of $\mathcal{K}_+(\zeta)$ and $\mathcal{K}_-(\zeta)$ at infinity is algebraic. This requirement is necessary, in view of Eqs. (117), to permit the application of Liouville's theorem to Eq. (119a). The required choice is facilitated by a knowledge of the asymptotic properties of the gamma function $\Gamma(x)$ at infinity. For one notes that, as $\zeta \to \infty$,

$$\Pi(b,\zeta) \to \prod_1^\infty \left(1 + j\frac{\zeta b}{n\pi}\right) e^{-j(\zeta b/n\pi)} = \frac{e^{-j(\zeta b/\pi)C}}{j\frac{\zeta b}{\pi}\Gamma\left(j\frac{\zeta b}{\pi}\right)},$$

where $C = 0.577$; hence from the asymptotic properties of the gamma function

$$\Pi(b,\zeta) \to \frac{1}{\sqrt{2j\zeta b}}\, e^{-j\left(C-1+\ln j\frac{\zeta b}{\pi}\right)\frac{\zeta b}{\pi}}, \qquad \zeta \to \infty. \qquad (124)$$

On using Eq. (124) in Eq. (123b), one obtains

$$\mathcal{K}_-(\zeta) = \sqrt{j\zeta b}\, e^{\chi(\zeta) - j\frac{2\zeta b}{\pi}\ln 2}, \qquad \zeta \to \infty.$$

Thus, to ensure the algebraic behavior of $\mathcal{K}_-(\zeta)$ at infinity $\chi(\zeta)$ must be chosen as

$$\chi(\zeta) = j\frac{2\zeta b}{\pi}\ln 2, \qquad (125)$$

and therefore

$$\mathcal{K}_-(\zeta) = \sqrt{j\zeta b}, \qquad \zeta \to \infty. \qquad (126)$$

From the asymptotic $\zeta \to \infty$ behavior of $\mathcal{E}(\zeta)$ and $\mathcal{K}_-(\zeta)$ given in Eqs. (117b) and (126), one deduces by reference to Eq. (119b) that the constant A is real and that $E(z) \sim z^{-\frac{1}{2}}$ as $z \to 0$. It therefore follows from Eqs. (120), (123a), and (125) that the desired circuit parameter B/Y is given by

$$\tan^{-1}\left(\frac{B}{Y}\right) = \frac{2kb}{\pi}\ln 2 + S_1\left(\frac{2kb}{\pi}\right) - 2S_1\left(\frac{kb}{\pi}\right) = \frac{2\pi d}{\lambda}, \qquad (127)$$

where the phase S_1 of the unit amplitude function $\Pi(b,k)$ is designated by

$$S_1\left(\frac{kb}{\pi}\right) = \sum_1^\infty \left(\sin^{-1}\frac{kb}{n\pi} - \frac{kb}{n\pi}\right), \qquad \frac{kb}{\pi} = \frac{2b}{\lambda}.$$

Equation (127) is the rigorous result quoted in Sec. 6·4 for the special case $b_1 = b_2 = b$ therein and if λ in Eq. (127) is replaced by λ_g (note that the b of this section is the $b/2$ of Sec. 6·4). The distance d is evidently the distance from the edge of the bifurcation to the open-circuit (zero susceptance) point in guide 1.

CHAPTER 4

TWO-TERMINAL STRUCTURES

The waveguide structures to be described in this chapter are composed of an input region that has the form of a waveguide propagating only a single mode and an output region that is either a beyond-cutoff guide or free space. In a strict sense such structures should be described by multiterminal equivalent networks. However, in many cases the behavior of the fields in the beyond-cutoff or in the free-space regions is not of primary interest. Under these circumstances the above structures may be regarded as two-terminal, i.e., one-terminal pair, networks in so far as the dominant mode in the input guide is concerned. The relative impedance at a specified terminal plane in the input guide and the wavelength of the propagating mode in the input guide suffice to determine the reflection, transmission, standing-wave, etc., characteristics of such structures. In the present chapter this information will be presented for a number of two-terminal waveguide structures encountered in practice.

LINES TERMINATING IN GUIDES BEYOND CUTOFF

4·1. Change of Cross Section, H-plane. *a. Symmetrical Case.*—An axially symmetrical junction of two rectangular guides of unequal widths but equal heights (H_{10}-mode in large rectangular guide, no propagation in small guide).

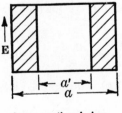
Cross sectional view

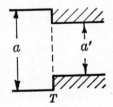
Top view

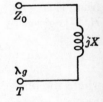
Equivalent circuit

Fig. 4·1-1.

Equivalent-circuit Parameters.—At the junction plane T,

$$\frac{X}{Z_0} = \frac{2a}{\lambda_g} \frac{X_{11}\left\{1 - \left[1 - \left(\frac{2a'}{\lambda}\right)^2\right]X_0^2\right\}}{1 - \left[1 - \left(\frac{2a'}{\lambda}\right)^2\right]X_0 X_{22} + \sqrt{1 - \left(\frac{2a'}{\lambda}\right)^2}(X_{22} - X_0)}, \quad (1a)$$

SEC. 4·1] CHANGE OF CROSS SECTION, H-PLANE

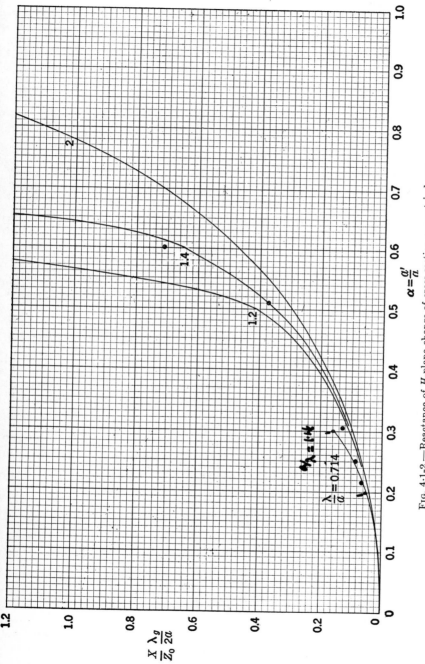

FIG. 4·1-2.—Reactance of H-plane change of cross-section *symmetrical case*. Experimental points for $\lambda/a = 1.4$.

$$\frac{X}{Z_0} \approx \frac{2a}{\lambda_g} \frac{\alpha^2}{0.429(1 - 1.56\alpha^2)(1 - 6.75\alpha^2 Q) + 0.571(1 - 0.58\alpha^2)\sqrt{1 - \left(\frac{2a'}{\lambda}\right)^2}},$$
(1b)

where X_{11}, X_{22}, and X_0 are defined in Sec. 5·24a, and

$$Q = 1 - \sqrt{1 - \left(\frac{2a}{3\lambda}\right)^2}, \quad \alpha = \frac{a'}{a}, \quad \lambda_g = \frac{\lambda}{\sqrt{1 - \left(\frac{\lambda}{2a}\right)^2}}.$$

Restrictions.—The equivalent circuit is applicable in the range $\lambda > 2a'$ provided $0.5 < a/\lambda < 1.5$. Equations (1a) and (1b) have been obtained from the data of Sec. 5·24a by terminating the circuit indicated therein with the inductive characteristic impedance of the smaller guide. This procedure is strictly valid only if the smaller guide is infinitely long; however, a length of the order of the width a' is usually sufficient. Equation (1a) is estimated to be in error by less than 1 per cent over most of the wavelength range. Equation (1b) is an approximation valid in the small-aperture range and differs from Eq. (1a) by less than 10 per cent for $\alpha < 0.4$ and for $0.5 < a/\lambda < 1$.

Numerical Results.—Figure 4·1-2 contains a plot of $X\lambda_g/Z_0 2a$ as a function of α in the wavelength range $\lambda > 2a'$, provided $0.5 < a/\lambda < 1.5$. These curves have been computed by the use of the curves in Sec. 5·24a.

Experimental Results.—A number of measured data taken at $\lambda/a = 1.4$ are indicated by the circled points in Fig. 4·1-2. These data are quite old and are not of high accuracy.

b. Asymmetrical Case.—An axially asymmetrical junction of two rectangular guides of unequal widths but equal heights (H_{10}-mode in large rectangular guide, no propagation in small guide).

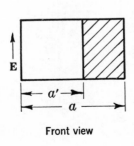
Front view

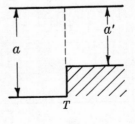
Top view

Equivalent circuit

FIG. 4·1-3.

SEC. 4·1] CHANGE OF CROSS SECTION, H-PLANE 171

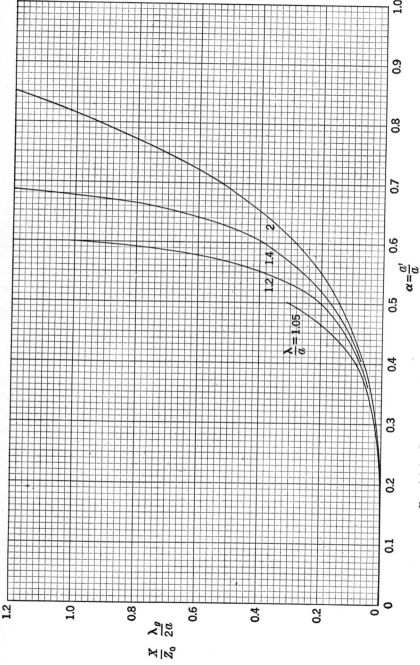

FIG. 4·1-4.—Reactance of H-plane change of cross-section *asymmetrical case*.

Equivalent-circuit Parameters:

$$\frac{X}{Z_0} = \frac{2a}{\lambda_g} \frac{X_{11}\left\{1 - \left[1 - \left(\frac{2a'}{\lambda}\right)^2\right]X_0^2\right\}}{1 - \left[1 - \left(\frac{2a'}{\lambda}\right)^2\right]X_0 X_{22} + \sqrt{1 - \left(\frac{2a'}{\lambda}\right)^2}(X_{22} - X_0)}, \quad (1a)$$

$$\frac{X}{Z_0} \approx \frac{2a}{\lambda_g} \frac{\alpha^4}{0.198(1 + 1.44\alpha^2) + 0.173(1 + 1.33\alpha^2)(1 + 41.3\alpha^4 Q)}, \quad (1b)$$

where X_{11}, X_{22}, and X_0 are defined in Sec. 5·24b, and

$$Q = 1 - \sqrt{1 - \left(\frac{a}{\lambda}\right)^2}, \quad \alpha = \frac{a'}{a}, \quad \lambda_g = \frac{\lambda}{\sqrt{1 - \left(\frac{\lambda}{2a}\right)^2}}.$$

Restrictions.—The equivalent circuit is applicable in the range $\lambda > 2a'$ provided $0.5 < a/\lambda < 1.0$. Equations (1a) and (1b) have been obtained from the data in Sec. 5·24b by terminating the circuit indicated therein with the inductive characteristic impedance of the smaller guide. Although this procedure is strictly valid only when the smaller guide is infinitely long, it usually may be employed even when the length is of the order of the width a'. Equation (1b), valid in the small-aperture range, is an approximation that differs from Eq. (1a) by less than 10 per cent when $\alpha < 0.4$ and $0.5 < a/\lambda < 1.0$.

Numerical Results.—In Fig. 4·1-4 $X\lambda_g/Z_0 2a$ is represented as a function of α in the wavelength range $\lambda > 2a'$, provided $0.5 < a/\lambda < 1.0$. These curves have been computed with the aid of the curves of Sec. 5·24b.

4·2. Bifurcation of a Rectangular Guide, H-plane.—A bifurcation of a rectangular guide by a partition of zero thickness oriented parallel to the electric field (H_{10}-mode in large rectangular guide, no propagation in smaller guides).

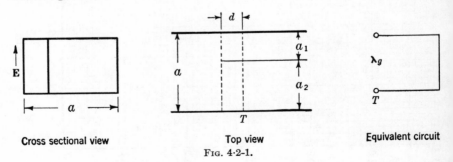

Cross sectional view   Top view   Equivalent circuit
Fig. 4·2-1.

Equivalent-circuit Parameters.—The equivalent circuit may be represented as a short circuit at the reference plane T located at a distance d

SEC. 4·2] BIFURCATION OF A RECTANGULAR GUIDE, H-PLANE 173

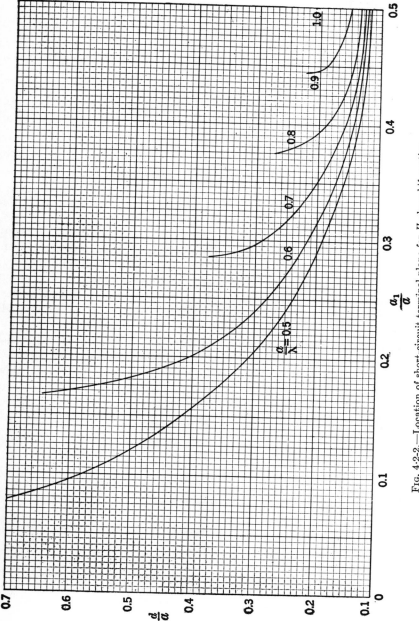

FIG. 4·2·2.—Location of short-circuit terminal plane for H-plane bifurcation.

given by

$$\frac{2\pi d}{\lambda_g} = x(1 + \alpha_1 \ln \alpha_1 + \alpha_2 \ln \alpha_2) - S_2(x;1,0) + S_1(x_1;\alpha_1,0)$$
$$+ S_1(x_2;\alpha_2,0), \quad (1a)$$

where

$$x_1 = \frac{2a_1}{\lambda_g}, \quad x_2 = \frac{2a_2}{\lambda_g}, \quad x = x_1 + x_2 = \frac{2a}{\lambda_g},$$

$$\alpha_1 = \frac{a_1}{a}, \quad \alpha_2 = \frac{a_2}{a}, \quad \lambda_g = \frac{\lambda}{\sqrt{1 - \left(\frac{\lambda}{2a}\right)^2}},$$

$$S_N(x;\alpha,0) = \sum_{n=N}^{\infty} \left(\sin^{-1} \frac{x}{\sqrt{n^2 - \alpha^2}} - \frac{x}{n} \right).$$

Restrictions.—The equivalent circuit is valid in the range $0.5 < a/\lambda < 1$ provided $\lambda \gg 2a_2$ (where $a_2 > a_1$). Equation (1a) has been obtained by the transform method and is rigorous in the above range.

Numerical Results.—In Fig. 4·2-2 there are contained curves for d/a as a function of a_1/a for various values of a/λ in the permitted range.

4·3. Coupling of a Coaxial Line to a Circular Guide.—A coaxial guide with a hollow center conductor of zero wall thickness terminating in a circular guide (principal mode in coaxial guide, no propagation in circular guide).

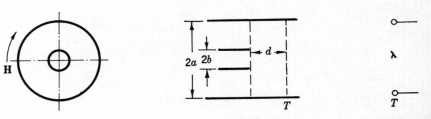

Cross sectional view   Side view   Equivalent circuit

Fig. 4·3-1.

Equivalent-circuit Parameters.—The open-circuit reference plane T is located at a distance d given by

$$\frac{2\pi d}{\lambda} = \frac{2a}{\lambda}\left(\ln \frac{1}{1-\alpha} + \alpha \ln \frac{1-\alpha}{\alpha} \right) + S_1^{J_0}(x;0) - S_1^{J_0}(\alpha x;0) - S_1^{Z_0}(x';0,c),$$
$$(1a)$$

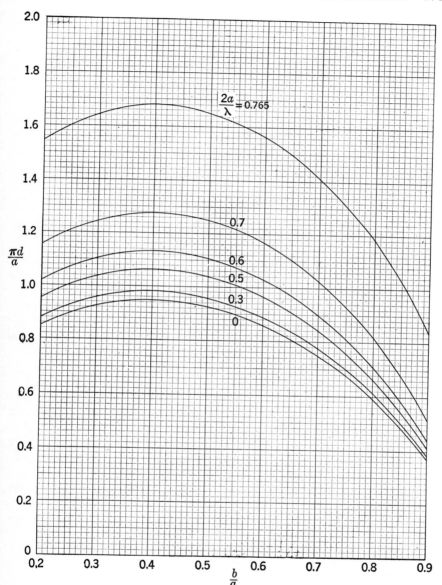

Fig. 4·3-2.—Location of open-circuit terminal plane for coaxial-circular guide junction.

$$\frac{\pi d}{a} \approx \ln \frac{1}{1-\alpha} + \alpha \ln \frac{1-\alpha}{\alpha} + (1-\alpha)\left(1.478 - \frac{1}{\gamma_1}\right), \quad (1b)$$

where

$$x = \frac{2a}{\lambda}, \qquad x' = (1-\alpha)x, \qquad c = \frac{1}{\alpha} = \frac{a}{b},$$

$$S_1^{J_0}(x;0) = \sum_{n=1}^{\infty}\left(\sin^{-1}\frac{x}{\beta_n} - \frac{x}{n}\right), \qquad J_0(\pi\beta_n) = 0,$$

$$S_1^{Z_0}(x;0,c) = \sum_{n=1}^{\infty}\left(\sin^{-1}\frac{x}{\gamma_n} - \frac{x}{n}\right), \qquad \pi\gamma_n = (c-1)\chi_{0n},$$

$$J_0(\chi_{0n})N_0(\chi_{0n}c) - N_0(\chi_{0n})J_0(\chi_{0n}c) = 0, \qquad n = 1, 2, 3 \cdots$$

Restrictions.—The equivalent circuit is valid in the wavelength range $\lambda > 2.61a$ provided the fields are rotationally symmetrical. The location of the reference plane has been determined by the transform method and is rigorous in the above range. The approximation (1b) agrees with Eq. (1a) to within 3 per cent if $(2a/\lambda < 0.3$ and $1 < a/b < 5$.

Numerical Results.—The quantity $\pi d/a$ is plotted in Fig. 4·3-2 as a function of b/a for various values of the parameter $2a/\lambda$. The summations $S_1^{J_0}$ and $S_1^{Z_0}$ are tabulated in the appendix. The roots $\pi\beta_n = \chi_{0n}$ are given in Table 2·1; the quantities $\pi\gamma_n = (c-1)\chi_{0n}$ are tabulated as a function of c in Table 2·3. For large n it is to be noted that $\beta_n \approx n - \frac{1}{4}$ and $\gamma_n \approx n$.

4·4. Rectangular to Circular Change in Cross Section.—The termination of a rectangular guide by a centered, infinitely long circular guide (H_{10}-mode in rectangular guide, no propagation in circular guide).

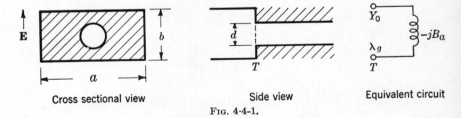

Cross sectional view   Side view   Equivalent circuit

Fig. 4·4-1.

Equivalent-circuit Parameters.—At the terminal plane T

$$\frac{B_a}{Y_0} = \frac{B}{2Y_0} + j\frac{Y_0'}{Y_0}, \qquad (1a)$$

$$\frac{B_a}{Y_0} \approx 0.685 \frac{ab\lambda_g}{d^3}, \qquad (1b)$$

SEC. 4·4] RECTANGULAR TO CIRCULAR CHANGE

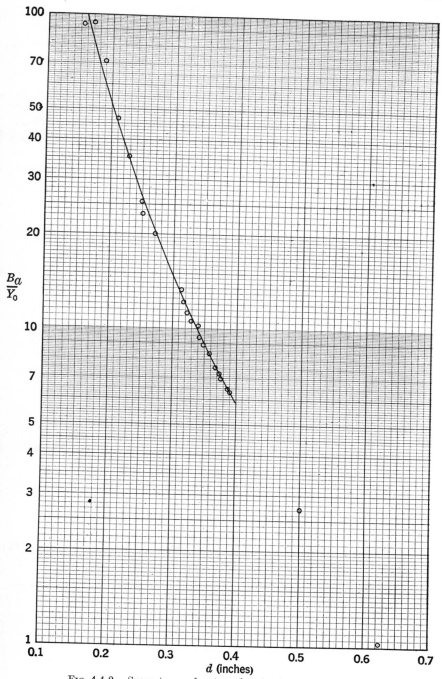

FIG. 4·4-2.—Susceptance of rectangular-circular guide junction.

where $B/2Y_0$ = one-half the quantity B/Y_0 in Eq. (1a) of Sec. 5·4a and $jY_0'/Y_0 = j$ times the quantity Y_0'/Y_0 in Eq. (1a) of Sec. 5·32.

Restrictions.—See Sec. 5·32. The approximation (1b) above agrees with Eq. (1a) in the small-aperture range $d/a \ll 1$ and $d/\lambda \ll 1$.

Numerical Results.—The parameters B/Y_0 and $jY_0'a/Y_0b$ may be obtained as functions of d/a from Figs. 5·4-2 and 5·32-2, respectively.

Experimental Results.—Data for B_a/Y_0 measured at $\lambda = 3.20$ cm for the case of a rectangular guide of dimensions $a = 0.900$ in., $b = 0.400$ in. joined to an infinitely long circular guide of variable diameter d are indicated in Fig. 4·4-2. The circled points are the measured values; the solid line is the theoretical curve computed from Eq. (1a). Since there is no justification for the extrapolation of the formulas to the case $d > b$, the theoretical curve has not been extended into this region.

4·5. Termination of a Coaxial Line by a Capacitive Gap.—A junction of a coaxial guide and a short circular guide (principal mode in coaxial guide, no propagation in circular guide).

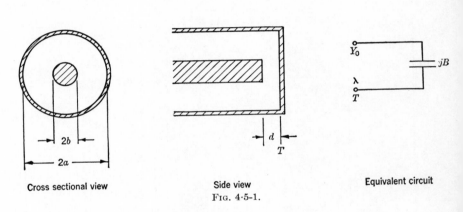

Cross sectional view   Side view   Equivalent circuit

FIG. 4·5-1.

Equivalent-circuit Parameters.—At the reference plane T

$$\frac{B}{Y_0} = \frac{4b}{\lambda} \ln \frac{a}{b}\left(\frac{\pi}{4}\frac{b}{d} + \ln \frac{a-b}{d}\right). \tag{1}$$

Restrictions.—The equivalent circuit is valid in the wavelength range $\lambda > 2(a - b)/\gamma_1$, where $\pi\gamma_1 = (a/b - 1)\chi_{01}$ may be found in Table 2·3. The susceptance has been evaluated by the small-aperture method treating the principal mode correctly and all higher modes by plane parallel approximations. Equation (1) is an approximation valid for $2\pi d/\lambda \ll 1$ and $d/(a - b) \ll 1$. For $a/b \approx 1$ the quantity $\gamma_1 \approx 1$.

LINES RADIATING INTO SPACE

4·6a. Parallel-plate Guide into Space, E-plane.—A semi-infinite parallel-plate guide of zero wall thickness radiating into free space (plane wave in parallel-plate guide incident at angle α relative to normal of terminal plane).

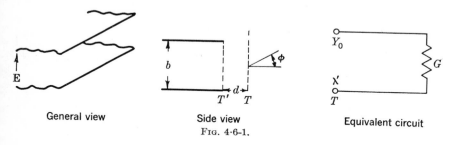

General view   Side view   Equivalent circuit
FIG. 4·6-1.

Equivalent-circuit Parameters.—At the reference plane T located at a distance d given by

$$\frac{2\pi d}{\lambda'} = x \ln \frac{2e}{\gamma x} - S_1(x;0,0), \tag{1}$$

the equivalent circuit is simply a conductance

$$\frac{G}{Y_0} = \tanh \frac{\pi x}{2}, \tag{2}$$

where

$$S_1(x;0,0) = \sum_{n=1}^{\infty} \left(\sin^{-1} \frac{x}{n} - \frac{x}{n} \right),$$

$$x = \frac{b}{\lambda'}, \quad \theta = \frac{4\pi d}{\lambda'}, \quad \lambda' = \frac{\lambda}{\cos \alpha},$$

$$e = 2.718, \quad \gamma = 1.781.$$

At the reference plane T' an alternative equivalent circuit shown in Fig. 4·6-2 is characterized by

$$\frac{G'}{Y_0} = \frac{\sinh \pi x}{\cosh \pi x + \cos \theta}, \tag{3}$$

$$\frac{G'}{Y_0} \approx \frac{\pi b}{2\lambda'}, \quad \text{for } \frac{b}{\lambda'} \ll 1, \tag{3a}$$

$$\frac{B'}{Y_0} = \frac{\sin \theta}{\cosh \pi x + \cos \theta}, \tag{4}$$

$$\frac{B'}{Y_0} \approx \frac{b}{\lambda'} \ln \frac{2e\lambda'}{\gamma b}, \quad \text{for } \frac{b}{\lambda'} \ll 1. \tag{4a}$$

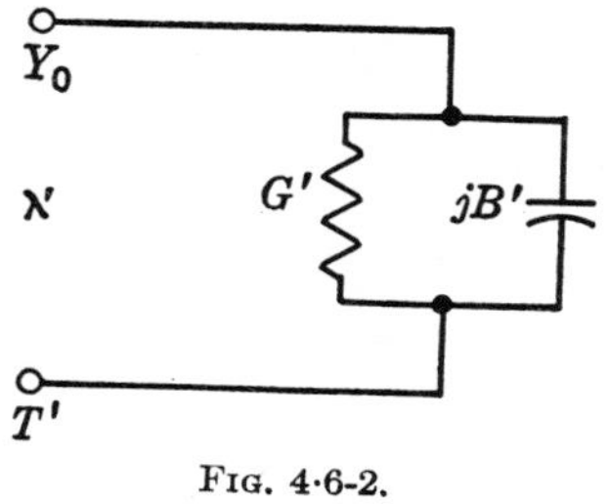

FIG. 4·6-2.

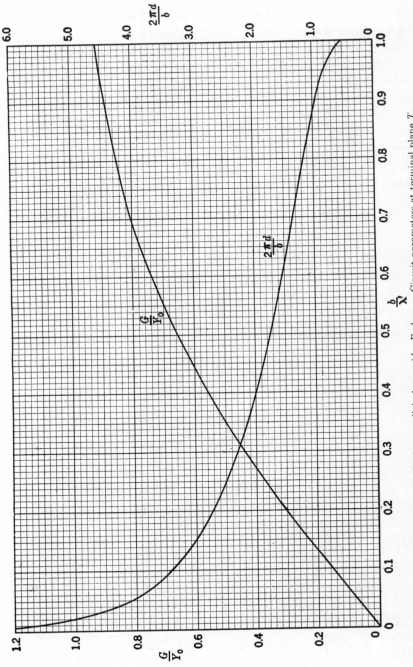

Fig. 4·6-3.—Radiating parallel plate guide, E-plane. Circuit parameters at terminal plane T.

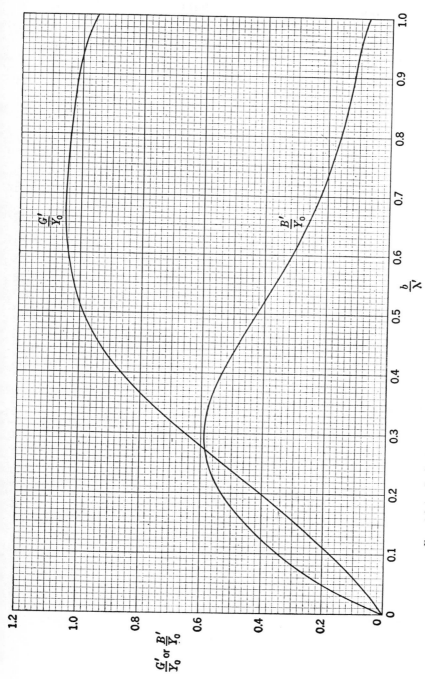

FIG. 4·6-4.—Radiating parallel plate guide, E-plane. Admittance at terminal plane T'.

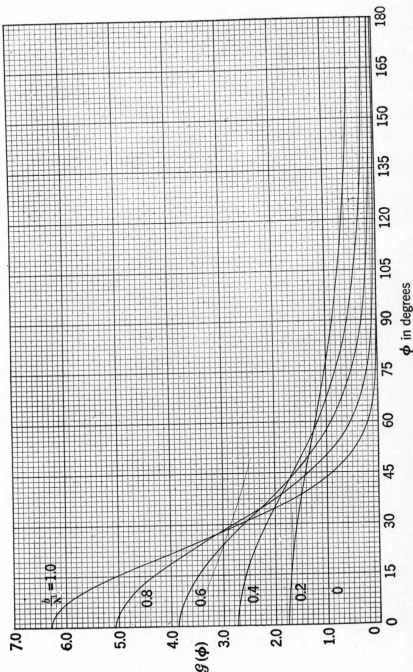

FIG. 4·6·5.—Gain pattern for radiation from parallel plate guide, E-plane.

SEC. 4·7a] *PARALLEL-PLATE GUIDE RADIATING INTO HALF SPACE* 183

The angular distribution of the emitted radiation is described by the power gain function

$$\mathcal{G}(\phi) = \mathcal{G}(0) \frac{\sin(\pi x \sin \phi)}{\pi x \sin \phi} e^{-2\pi x \sin^2 \frac{\phi}{2}}, \qquad -\pi < \phi < \pi, \qquad (5)$$

defined relative to an infinitely extended isotropic line source. The gain $\mathcal{G}(0)$ is

$$\mathcal{G}(0) = \frac{2\pi x}{1 - e^{-2\pi x}}. \qquad (5a)$$

Restrictions.—The equivalent circuit is valid in the range $b/\lambda' < 1$. The equations for the circuit parameters have been obtained by the transform method and are rigorous in the above range.

Numerical Results.—The quantities $2\pi d/b$ and G/Y_0 are plotted in Fig. 4·6-3 as a function of b/λ'. The alternative circuit parameters G'/Y_0 and B'/Y_0 are plotted as a function of b/λ' in Fig. 4·6-4. The power gain function $\mathcal{G}(\phi)$ is displayed in Fig. 4·6-5 as a function of ϕ with b/λ' as a parameter; only positive angles are indicated because the gain function is symmetrical about $\phi = 0$.

4·6b. Rectangular Guide into Bounded Space, E-plane.—A rectangular guide of zero wall thickness radiating into the space between two infinite parallel plates that form extensions of the guide sides (H_{10}-mode in rectangular guide).

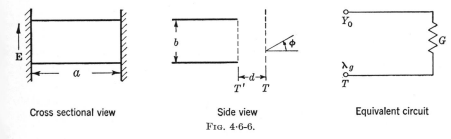

Cross sectional view   Side view   Equivalent circuit
FIG. 4·6-6.

Equivalent-circuit Parameters.—Same as in Sec. 4·6a provided λ' therein is replaced by λ_g, where

$$\lambda_g = \frac{\lambda}{\sqrt{1 - \left(\frac{\lambda}{2a}\right)^2}}.$$

Restrictions.—Same as in Sec. 4·6a with λ' replaced by λ_g.
Numerical Results.—Same as in Sec. 4·6a with λ' replaced by λ_g.

4·7a. Parallel-plate Guide Radiating into Half Space, E-plane.—A semi-infinite parallel-plate guide terminating in the plane of an infinite screen and radiating into a half space (plane wave in parallel-plate guide incident at angle α relative to normal to screen).

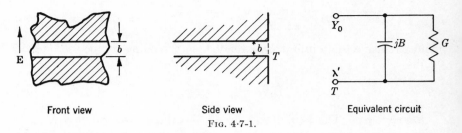

Front view   Side view   Equivalent circuit
Fig. 4·7-1.

Equivalent-circuit Parameters.—At the reference plane T

$$\frac{G}{Y_0} = \int_0^{kb} J_0(x)\,dx - J_1(kb), \tag{1a}$$

$$\frac{G}{Y_0} \approx \frac{\pi b}{\lambda'}, \tag{1b}$$

$$\frac{B}{Y_0} = \int_0^{kb} N_0(x)\,dx + N_1(kb) + \frac{2}{\pi}\frac{1}{kb}, \tag{2a}$$

$$\frac{B}{Y_0} \approx \frac{2b}{\lambda'} \ln \frac{e\lambda'}{\gamma 2b}, \tag{2b}$$

where

$$k = \frac{2\pi}{\lambda'}, \qquad \lambda' = \frac{\lambda}{\cos \alpha}, \qquad e = 2.718, \qquad \gamma = 1.781.$$

Restrictions.—The equivalent circuit is valid in the range $b/\lambda' < 1.0$. The circuit parameters have been obtained by the variational method assuming a constant electric field in the aperture at the reference plane. No estimate of accuracy is available over the entire range, but the error is no more than a few per cent for $(2b/\lambda') < 1$. Equations (1b) and (2b) are static results and agree with Eqs. (1a) and (2a), respectively, to within 5 per cent for $(2\pi b/\lambda') < 1$.

Numerical Results.—The quantities G/Y_0 and B/Y_0 are plotted in Fig. 4·7-2 as a function of b/λ'.

4·7b. Rectangular Guide Radiating into Bounded Half Space, E-plane. A rectangular guide terminating in the plane of an infinite screen and radiating into the half space bounded by two infinite parallel plates that form extensions of the guide sides (H_{10}-mode in rectangular guide).

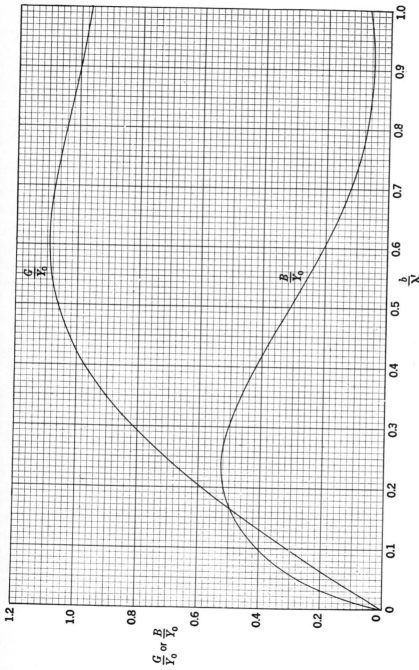

FIG. 4·7·2.—Admittance of parallel plate guide radiating into half-space, E-plane.

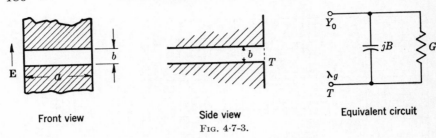

Fig. 4·7-3.

Equivalent-circuit Parameters.—Same as in Sec. 4·7a provided λ' therein is replaced by λ_g, where

$$\lambda_g = \frac{\lambda}{\sqrt{1 - \left(\dfrac{\lambda}{2a}\right)^2}}.$$

Restrictions.—Same as in Sec. 4·7a with λ' replaced by λ_g.

Numerical Results.—Same as in Sec. 4·7a with λ' replaced by λ_g.

4·8. Parallel-plate Guide into Space, H-plane.—A semi-infinite parallel-plate guide having zero thickness walls and radiating into free space (H_{10}-mode in rectangular guide of infinite height).

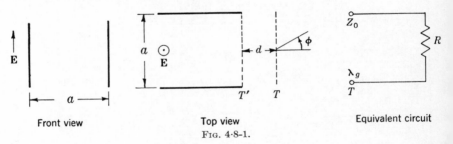

Fig. 4·8-1.

Equivalent-circuit Parameters.—At the reference plane T located at a distance d given by

$$\frac{2\pi d}{a} = \ln \frac{\lambda}{2\gamma a} + 3 - \frac{\lambda_g}{2a} \sin^{-1} \frac{\lambda}{\lambda_g}$$
$$- \frac{\lambda_g}{2a} \sum_{n=1}^{\infty} \left[\sin^{-1}\left(\frac{\dfrac{2a}{\lambda_g}}{\sqrt{(2n+1)^2 - 1}}\right) - \frac{\dfrac{2a}{\lambda_g}}{(2n+1)} \right], \quad (1)$$

the equivalent circuit is simply a resistance,

$$\frac{R}{Z_0} = \tanh \frac{\psi}{2}, \quad (2)$$

where

$$\psi = \frac{\pi a}{\lambda_g} - \frac{1}{2} \ln \frac{\lambda_g - \lambda}{\lambda_g + \lambda}$$

$$\lambda_g = \frac{\lambda}{\sqrt{1 - \left(\frac{\lambda}{2a}\right)^2}}, \quad \gamma = 1.781.$$

At the reference plane T' the equivalent circuit is that represented in Fig. 4·8-2. The corresponding circuit parameters are

$$\frac{R'}{Z_0} = \frac{\sinh \psi}{\cosh \psi + \cos \theta}, \quad \frac{X'}{Z_0} = \frac{\sin \theta}{\cosh \psi + \cos \theta} \quad (3)$$

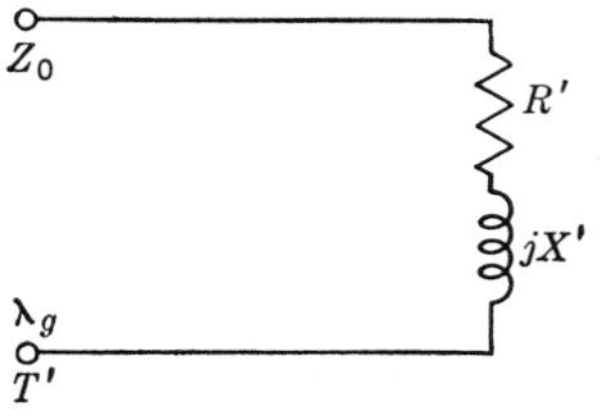

where

$$\theta = \frac{4\pi d}{\lambda_g}.$$

The angular distribution of the emitted radiation is described by the power gain function

FIG. 4·8-2.

$$\mathcal{G}(\phi) = \mathcal{G}(0) \frac{\cos\left(\frac{\pi a}{\lambda} \sin \phi\right) \cos^2 \frac{\phi}{2}}{1 - \left(\frac{2a \sin \phi}{\lambda}\right)^2} e^{-\frac{2\pi a}{\lambda} \sin^2 \frac{\phi}{2}}, \quad (4)$$

which is defined relative to an infinitely extended line source. The gain $\mathcal{G}(0)$ is

$$\mathcal{G}(0) = \frac{8a \, e^{\pi a (\lambda_g - \lambda)/\lambda \lambda_g}}{(\lambda_g + \lambda) - (\lambda_g - \lambda)e^{-2\pi a/\lambda_g}}. \quad (5)$$

Restrictions.—The equivalent circuits are valid in the wavelength range $2a/3 < \lambda < 2a$. The formulas have been obtained by the transform method and are rigorous in the above range.

Numerical Results.—In Fig. 4·8-3 $2\pi d/a$ and R/Z_0 are plotted as a function of a/λ. The alternative-circuit parameters R'/Z_0 and X'/Z_0 are shown in Fig. 4·8-4. The symmetrical gain function $\mathcal{G}(\phi)$ is displayed in Fig. 4·8-5 as a function of ϕ for several values of the parameter a/λ.

4·9. Parallel-plate Guide Radiating into Half Space, H-plane.—A semi-infinite parallel-plate guide of infinite height terminating in an infinite screen and radiating into a half space (H_{10}-mode in parallel-plate guide).

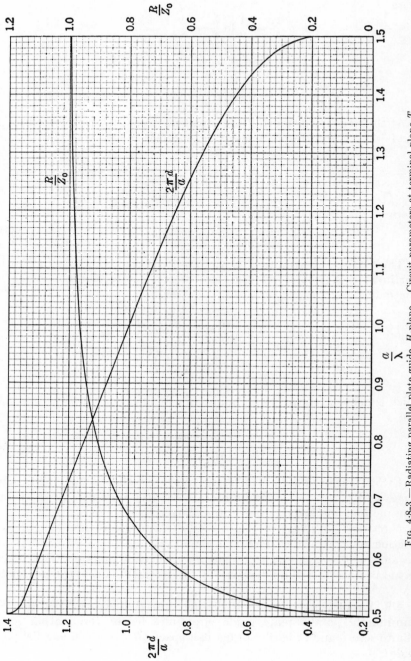

FIG. 4·8-3.—Radiating parallel plate guide, H-plane. Circuit parameters at terminal plane T.

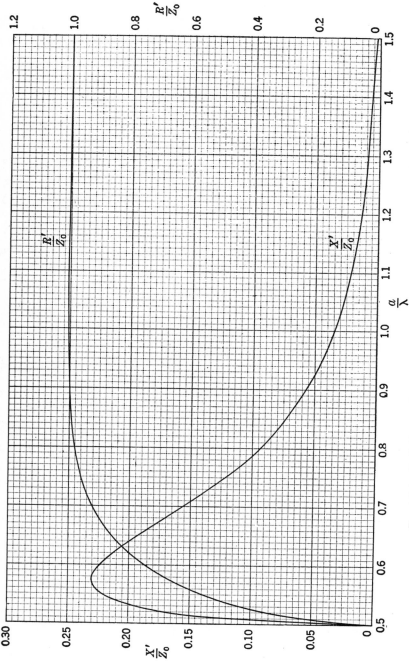

Fig. 4·8-4.—Radiating parallel plate guide, H-plane. Impedance at terminal plane T'.

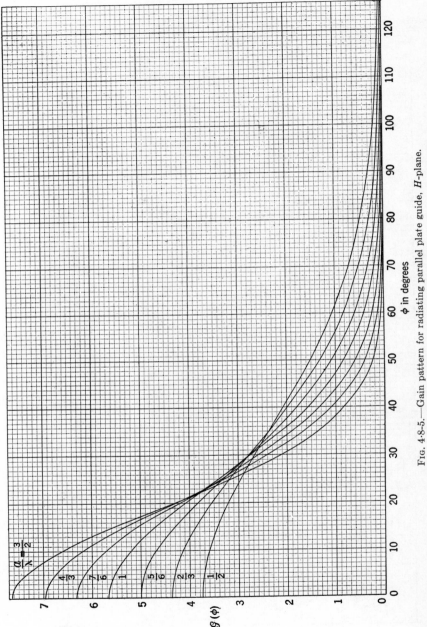

FIG. 4·8·5.—Gain pattern for radiating parallel plate guide, H-plane.

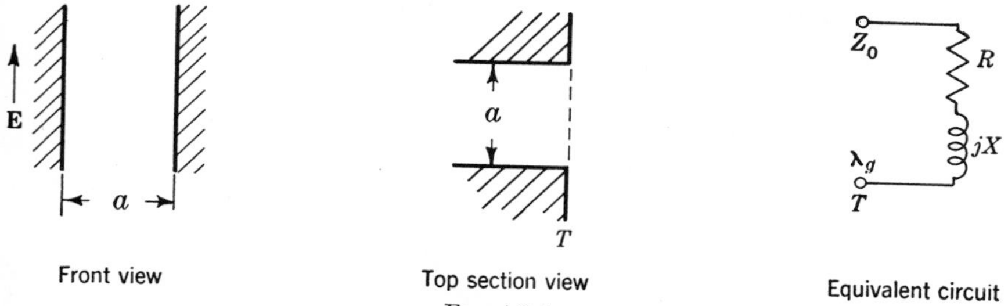

SEC. 4·9] PARALLEL-PLATE GUIDE RADIATING INTO HALF SPACE

Front view   Top section view   Equivalent circuit

FIG. 4·9-1.

Equivalent-circuit Parameters.—At the reference plane T

$$\frac{R}{Z_0} = \frac{\frac{G}{Y_0}}{\left(\frac{B}{Y_0}\right)^2 + \left(\frac{G}{Y_0}\right)^2}, \qquad \frac{X}{Z_0} = \frac{\frac{B}{Y_0}}{\left(\frac{B}{Y_0}\right)^2 + \left(\frac{G}{Y_0}\right)^2}, \qquad (1)$$

where

$$\frac{G}{Y_0} = \frac{2\pi a}{\lambda_g} \int_0^1 (1-x) \cos(\pi x) J_0(kax) \, dx$$

$$+ \frac{2a}{\lambda_g} \left[1 + 2\left(\frac{\lambda_g}{2a}\right)^2 \right] \int_0^1 \sin(\pi x) J_0(kax) \, dx, \quad (2a)$$

$$\frac{G}{Y_0} \approx 0.285 \frac{\lambda_g}{a}, \qquad \frac{a}{\lambda_g} \ll 1 \qquad (2b)$$

and

$$\frac{B}{Y_0} = \frac{2\pi a}{\lambda_g} \int_0^1 (1-x) \cos(\pi x) N_0(kax) \, dx$$

$$+ \frac{2a}{\lambda_g} \left[1 + 2\left(\frac{\lambda_g}{2a}\right)^2 \right] \int_0^1 \sin(\pi x) N_0(kax) \, dx \quad (3a)$$

$$\frac{B}{Y_0} \approx 0.156 \frac{\lambda_g}{a}, \qquad \frac{a}{\lambda_g} \ll 1 \qquad (3b)$$

$$k = \frac{2\pi}{\lambda}.$$

Restrictions.—The equivalent circuit is valid in the wavelength range $\frac{1}{2} < a/\lambda < \frac{3}{2}$. The equivalent-circuit parameters have been evaluated by the variational method assuming a lowest-mode electric-field distribution in the aperture at the reference plane. The error is estimated to be less than 10 per cent for R/Z_0 and perhaps somewhat larger for X/Z_0. Equations (2b) and (3b) are rough approximations to Eqs. (2a) and (3a), which are valid to within 10 per cent for $a/\lambda < 0.53$.

Numerical Results.—The quantities R/Z_0 and X/Z_0 of Eq. (1) are plotted in Fig. 4·9-2 as functions of a/λ for the range $0.5 < a/\lambda < 1.5$.

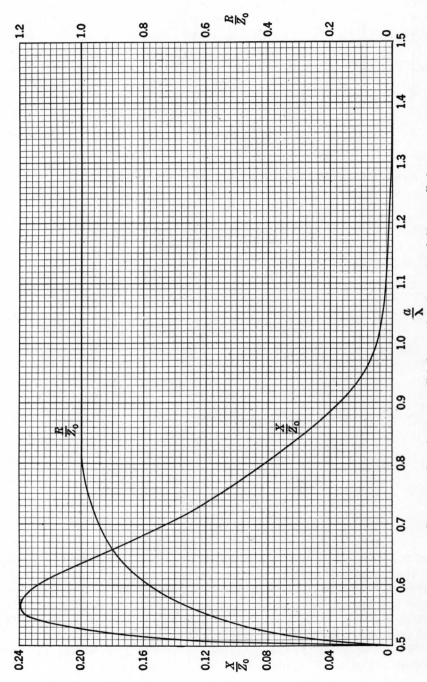

Fig. 4·9-2.—Impedance of parallel plate guide radiating into half-space, H-plane.

4·10. Apertures in Rectangular Guide. a. *Rectangular Apertures.*—

A rectangular guide terminating in a screen of finite size and radiating into space through a rectangular aperture (H_{10}-mode in rectangular guide).

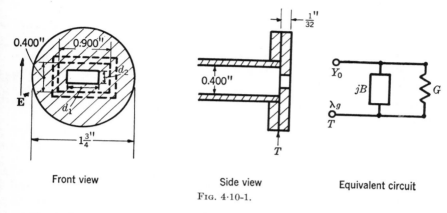

Front view   Side view   Equivalent circuit

Fig. 4·10-1.

Equivalent-circuit Parameters (Experimental).—At the reference plane T

$$\frac{G}{Y_0} + j\frac{B}{Y_0} = \frac{1-\Gamma}{1+\Gamma}. \tag{1}$$

A number of values of $\Gamma = |\Gamma|e^{j\phi} = |\Gamma|\underline{/\phi}$, measured at $\lambda = 3.20$ cm. for radiation into space through rectangular apertures in a $\frac{1}{32}$-in. thick screen, are presented in the following table:

d_1, in. \ d_2, in.	0.400 in.	0.300 in.	0.200 in.	0.100 in.	0.050 in.
0.900	0.31 /−90°	0.42 /−109°	0.54 /−123°	0.77 /−138°	0.87 /−145°
0.800	0.28 /−74°				
0.750	0.25 /−58°	0.36 /−68°	0.51 /−102°	0.68 /−120°	0.80 /−134°
0.600	0.26 /32°				
0.500	0.58 /95°	0.61 /102°	0.69 /113°	0.77 /130°	
0.400	0.88 /137°				
0.300	0.98 /161°				
0.200	1.0 /171°				

For radiation through rectangular apertures into a space bounded by "infinite" parallel plate extensions of the top and bottom sides of the rectangular guide, the values of Γ become

d_1, in. \ d_2, in.	0.400 in.	0.300 in.	0.200 in.	0.100 in.	0.050 in.
0.900	0.16|−7°	0.30|−50°	0.38|−91°	0.68|−121°	0.79|−138°
0.800	0.28|16°				
0.750	0.23|39°	0.17|21°	0.23|−39°	0.47|−95°	0.64|−115°
0.600	0.57|87°				
0.500	0.74|118°	0.74|122°	0.80|129°	0.86|140°	
0.400	0.90|144°				
0.300	0.98|157°				
0.200	1.0|168°				

For power passing through rectangular apertures into a matched guide of the same size as the input guide, the values of Γ are

d_1, in. \ d_2, in.	0.400 in.	0.300 in.	0.200 in.	0.100 in.	0.050 in.
0.750	0.10|100°	0.07|97°	0.09|−105°	0.35|−116°	0.55|−128°
0.500	0.62|123°	0.64|128°	0.69|134°	0.77|143°	

b. Circular Apertures.—A rectangular guide terminating in the plane of a screen of finite size and radiating into space through a circular aperture (H_{10}-mode in rectangular guide).

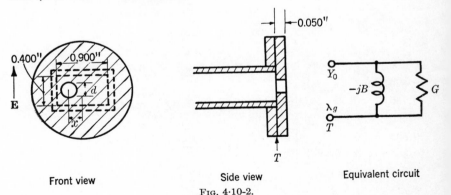

Front view   Side view   Equivalent circuit

FIG. 4·10-2.

Equivalent-circuit Parameters (Experimental).—At the reference plane T

$$\frac{G}{Y_0} - j\frac{B}{Y_0} = \frac{1-\Gamma}{1+\Gamma}. \tag{2}$$

Values of Γ measured at $\lambda = 3.20$ cm. for radiation through an off-centered 0.375-in. aperture in a screen of 0.050-in. thickness are

d, in. \ x, in.	0	0.050 in.	0.100 in.	0.150 in.	0.200 in.
0.375	0.982|159.6°	0.982|159.6°	0.983|161.6°	0.986|162.8°	0.990|165.4°

For radiation into a space bounded by "infinite" parallel-plate extensions of the top and bottom sides of the guide, the values of Γ become

d, in. \ x, in.	0	0.050 in.	0.100 in.	0.150 in.	0.200 in.
0.375	0.982|	0.982|159.9°	0.985|161.8°	0.987|163.6°	0.990|165.8°

In both cases the aperture is symmetrically located with respect to the height of the guide. The power radiated varies approximately as $\cos^2 \pi x/a$, where a is the guide width.[1]

4·11. Array of Semi-infinite Planes, H-plane.—An infinite array of semi-infinite metallic obstacles of zero thickness with edges parallel to the electric field (plane wave incident at angle θ, no propagation in parallel-plate region).

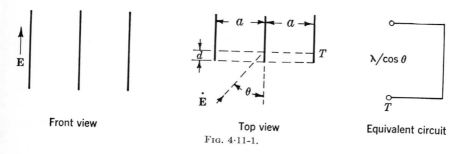

Front view   Top view   Equivalent circuit
Fig. 4·11-1.

Equivalent-circuit Parameters.—In the transmission-line representative of plane waves traveling in the directions $\pm\theta$ the equivalent circuit is simply a zero impedance at the terminal plane T. The latter is located at a distance d from the edge of the array, where

[1] *Cf.* "Representation, Measurement, and Calculation of Equivalent Circuits for Waveguide Discontinuities with Application to Rectangular Slots," Report PIB-137 (1949), Polytechnic Institute of Brooklyn, New York for more accurate data.

$$\frac{2\pi d}{\lambda/\cos\theta} = 2x \ln 2 + \sin^{-1}\frac{2x}{\sqrt{1-4y^2}} - \sin^{-1}\frac{x}{\sqrt{1-2y}} - \sin^{-1}\frac{x}{\sqrt{1+2y}}$$
$$+ S_2(2x;2y,0) - S_2(x;y,y) - S_2(x;y,-y), \quad (1a)$$
$$\frac{2\pi d}{\lambda/\cos\theta} \approx 2x \ln 2 + \sin^{-1}\frac{2x}{\sqrt{1-4y^2}} - \sin^{-1}\frac{x}{\sqrt{1-2y}} - \sin^{-1}\frac{x}{\sqrt{1+2y}}$$
$$+ \cdots \quad (1b)$$

where

$$x = \frac{a\cos\theta}{\lambda}, \qquad y = \frac{a\sin\theta}{\lambda}$$

$$S_2(x;\alpha,\beta) = \sum_{n=2}^{\infty} \left[\sin^{-1}\frac{x}{\sqrt{(n-\beta)^2 - \alpha^2}} - \frac{x}{n} \right], \quad x < \sqrt{(2-\beta)^2 - \alpha^2}.$$

Restrictions.—The equivalent circuit is valid in the wavelength range $\lambda > 2a$. Equation (1a) has been obtained by the transform method and is rigorous in the above range. Equation (1b) is an approximation that agrees with Eq. (1a) to within 3 per cent for $a/\lambda < 0.5$. The relative phase of the fields in adjacent (beyond-cutoff) guides of the parallel-plate region is $(2\pi a/\lambda)\sin\theta$.

Numerical Results.—The reference plane distance $\pi d/a$ may be obtained from Fig. 5·22-2 as a function of a/λ and θ provided the b therein is replaced everywhere by a.

4·12. Radiation from a Circular Guide, E_{01}-mode.—A semi-infinite circular guide of zero wall thickness radiating into free space (E_{01}-mode in circular guide).

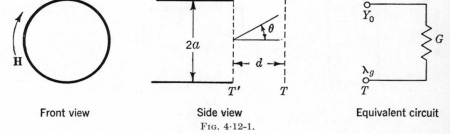

Front view   Side view   Equivalent circuit

FIG. 4·12-1.

Equivalent-circuit Parameters.—At the reference plane T located at a distance d from T', where

$$\frac{\pi d}{a} = \ln\frac{\lambda}{\gamma a} + 2 - \frac{\pi\beta_1}{\kappa a}\sin^{-1}\frac{\lambda}{\lambda_g} - \frac{1}{y}S_2^{J_0}(y;\beta_1)$$
$$+ \frac{1}{2}\int_0^{\infty} \frac{\ln\left\{e^{-2x}\sqrt{1 + \left[\frac{\pi I_0(x)}{K_0(x)}\right]^2}\right\}}{x^2 + (\pi\beta_1)^2} \frac{x\,dx}{\sqrt{x^2 + (ka)^2}}, \quad (1)$$

SEC. 4·12] RADIATION FROM A CIRCULAR GUIDE, E_{01}-MODE

$$\frac{\pi d}{a} \approx \ln\frac{\lambda}{\gamma a} + 2 - \frac{\pi\beta_1}{\kappa a}\sin^{-1}\frac{\lambda}{\lambda_g} - 0.264 - 5.91\left(\frac{\kappa a}{10}\right)^2 - 83.3\left(\frac{\kappa a}{10}\right)^4, \quad (1a)$$

the equivalent circuit is simply the conductance G defined by

$$\frac{G}{Y_0} = \tanh\frac{\psi}{2}, \quad (2)$$

where

$$\psi = \kappa a - \frac{3}{4}\ln\frac{\lambda_g + \lambda}{\lambda_g - \lambda} - \frac{\kappa a}{\pi}\int_0^\infty \frac{\tan^{-1}\left[\frac{K_0(x)}{\pi I_0(x)}\right]}{x^2 + (\pi\beta_1)^2} \frac{x\,dx}{\sqrt{x^2 + (ka)^2}}, \quad (3)$$

$$\psi \approx \kappa a - \frac{3}{4}\ln\frac{\lambda_g + \lambda}{\lambda_g - \lambda} \quad (3a)$$

and

$$k = \frac{2\pi}{\lambda}, \quad \kappa = \frac{2\pi}{\lambda_g}, \quad \lambda_g = \frac{\lambda}{\sqrt{1 - (\lambda/2.61a)^2}}, \quad y = \frac{\kappa a}{\pi}.$$

$$S_2^{J_0}(y;\beta_1) = \sum_{n=2}^\infty \left(\sin^{-1}\frac{y}{\sqrt{\beta_n^2 - \beta_1^2}} - \frac{y}{n}\right), \quad J_0(\pi\beta_n) = 0.$$

An alternative equivalent circuit at the reference plane T' is shown in Fig. 4·12-2. The corresponding circuit parameters are

$$\left.\begin{array}{l}\dfrac{G'}{Y_0} = \dfrac{\sinh\psi}{\cosh\psi + \cos 2\kappa d'} \\[2mm] \dfrac{B'}{Y_0} = \dfrac{\sin 2\kappa d}{\cosh\psi + \cos 2\kappa d'}\end{array}\right\} \quad (4)$$

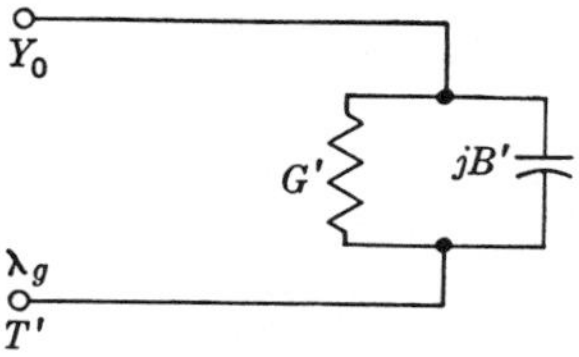

Fig. 4·12-2.

The angular distribution of the emitted radiation is symmetrical about the guide axis. The power gain function defined relative to an isotropic point source is

$$\mathcal{G}(\theta) = 2\kappa a\,\frac{J_1(\pi\beta_1)}{\sin\theta}\,\frac{J_0(ka\sin\theta)}{1 - \left(\dfrac{ka}{\pi\beta_1}\sin\theta\right)^2}\,\frac{E(ka,ka\cos\theta)}{E(ka,\kappa a) - E(ka,-\kappa a)}, \quad (5)$$

$$\mathcal{G}(\theta) \approx \theta^2 \qquad\text{for } \theta \ll 1, \quad (5a)$$

$$\mathcal{G}(\theta) \approx \frac{1}{\theta'^2}\,\frac{1}{\dfrac{4}{\pi^2}\left[\ln\left(\gamma\dfrac{ka}{2}\theta'\right)\right]^2 + 1} \qquad\text{for } \theta' = (\pi - \theta) \ll 1, \quad (5b)$$

where

$$E(x,y) = (x - y)^{3/2}\,e^{y + F(x,y)}, \quad (6)$$

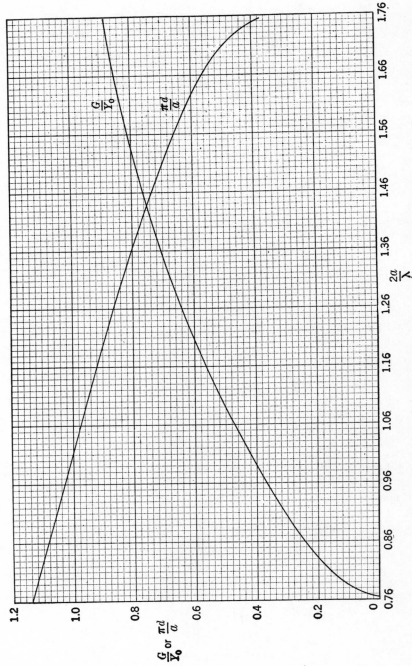

FIG. 4·12-3.—Radiating circular guide, E_{01}-mode. Circuit parameters at terminal plane T.

SEC. 4·12] RADIATION FROM A CIRCULAR GUIDE, E_{01}-MODE

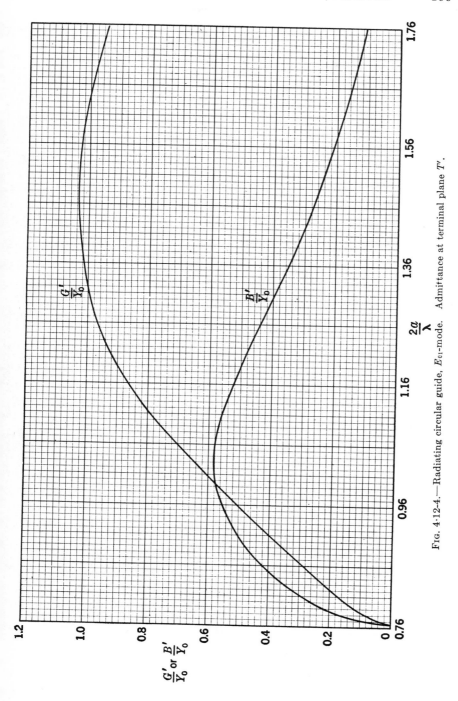

FIG. 4·12-4.—Radiating circular guide, E_{01}-mode. Admittance at terminal plane T'.

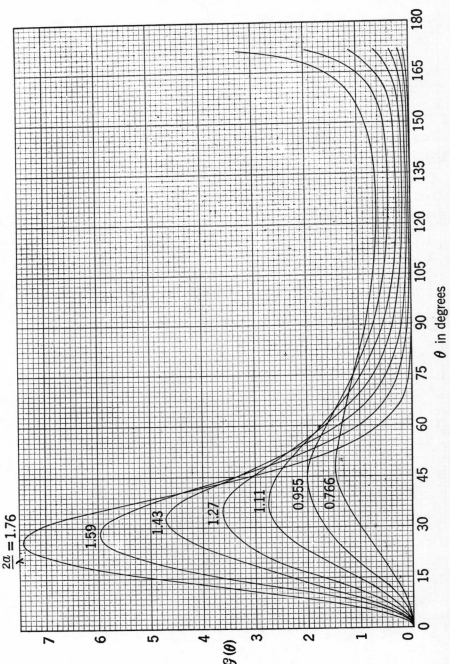

FIG. 4.12-5.—Gain pattern for radiating circular guide, E_{01}-mode.

$$E(x,y) \approx (x - y)^{3/2} e^y, \tag{6a}$$

$$F(x,y) = \frac{1}{\pi} \int_0^\infty \frac{\tan^{-1}\left[\frac{K_0(t)}{\pi I_0(t)}\right]}{\sqrt{t^2 + x^2} + y} \frac{t\, dt}{\sqrt{t^2 + x^2}}.$$

Restrictions.—The above formulas have been obtained by the transform method and, for the indicated field symmetry, are rigorous in the wavelength range $1.14a < \lambda < 2.61a$. For $\kappa a < 1$ the approximate Eq. (1a) is correct to within 2 per cent and Eq. (3a) to within 1 per cent. Equation (6a) is correct to within 1 per cent for $y > 0$, i.e., $\theta > 90°$.

Numerical Results.—Curves of $\pi d/a$ and G/Y_0 as a function of $2a/\lambda$ are presented in Fig. 4·12-3. In Fig. 4·12-4 G'/Y_0 and B'/Y_0 are plotted as a function of $2a/\lambda$. The graph of the power gain function $\mathcal{G}(\theta)$ as a function of θ with $2a/\lambda$ as a parameter is given in Fig. 4·12-5.

4·13. Radiation from a Circular Guide, H_{01}-mode.—A semi-infinite circular guide of zero wall thickness radiating into free space (H_{01}-mode in circular guide).

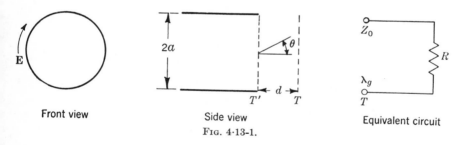

Front view Side view Equivalent circuit
Fig. 4·13-1.

Equivalent-circuit Parameters.—At the reference plane T located at a distance d defined by

$$\frac{\pi d}{a} = \ln\frac{\lambda}{\gamma a} + 2 - \frac{\pi \beta_1'}{\kappa a}\sin^{-1}\frac{\lambda}{\lambda_g} - \frac{1}{y} S_2^{J'_0}(y;\beta_1')$$

$$+ \frac{1}{2}\int_0^\infty \frac{\ln\left\{e^{-2x}\sqrt{1 + \left[\frac{\pi I_1(x)}{K_1(x)}\right]^2}\right\}}{x^2 + (\pi\beta_1')^2} \frac{x\, dx}{\sqrt{x^2 + (ka)^2}}, \tag{1}$$

$$\frac{\pi d}{a} \approx \ln\frac{\lambda}{\gamma a} + 2 - \frac{\pi\beta_1'}{\kappa a}\sin^{-1}\frac{\lambda}{\lambda_g} - 0.383\left(\frac{\kappa a}{10}\right)^2 - 0.379\left(\frac{\kappa a}{10}\right)^4, \tag{1a}$$

the equivalent circuit is simply the resistance R defined by

$$\frac{R}{Z_0} = \tanh\frac{\psi}{2}, \tag{2}$$

where

$$\psi = \kappa a - \frac{1}{4} \ln \frac{\lambda_g + \lambda}{\lambda_g - \lambda} + \frac{\kappa a}{\pi} \int_0^\infty \frac{\tan^{-1}\left[\frac{K_1(x)}{\pi I_1(x)}\right]}{x^2 + (\pi\beta_1')^2} \frac{x \, dx}{\sqrt{x^2 + (ka)^2}}, \quad (3)$$

$$\psi \cong \kappa a - \frac{1}{4} \ln \frac{\lambda_g + \lambda}{\lambda_g - \lambda}, \quad (3a)$$

and

$$k = \frac{2\pi}{\lambda}, \quad \kappa = \frac{2\pi}{\lambda_g}, \quad \lambda_g = \frac{\lambda}{\sqrt{1 - \left(\frac{\lambda}{1.64a}\right)^2}}, \quad y = \frac{\kappa a}{\pi},$$

$$S_2^{J_0'}(y;\beta_1') = \sum_{n=2}^\infty \sin^{-1}\left(\frac{y}{\sqrt{\beta_n'^2 - \beta_1'^2}} - \frac{y}{n}\right), \quad J_0'(\pi\beta_n') = 0.$$

An alternative equivalent circuit at the reference plane T' is indicated in Fig. 4·13-2. The corresponding circuit parameters are

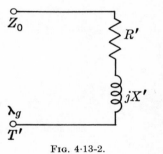

Fig. 4·13-2.

$$\left.\begin{array}{l} \dfrac{R'}{Z_0} = \dfrac{\sinh \psi}{\cosh \psi + \cos(2\kappa d)}, \\[6pt] \dfrac{X'}{Z_0} = \dfrac{\sin 2\kappa d}{\cosh \psi + \cos 2\kappa d}. \end{array}\right\} \quad (4)$$

The angular distribution of the emitted radiation is symmetrical about the guide axis. The power gain function defined relative to an isotropic point source is

$$\mathcal{G}(\theta) = \frac{2k\kappa a^2}{\pi\beta_1'} \frac{J_0(\pi\beta_1') J_1(ka \sin\theta)}{1 - \left(\dfrac{ka}{\pi\beta_1'} \sin\theta\right)^2} \frac{H(ka, ka\cos\theta)}{H(ka, -\kappa a), -H(ka, \kappa a)}, \quad (5)$$

$$\begin{array}{ll} \mathcal{G}(\theta) \approx \theta^2 & \text{for } \theta \ll 1, \\ \mathcal{G}(\theta) \approx (\pi - \theta)^2 & \text{for } (\pi - \theta) \ll 1, \end{array}$$

where

$$H(x,y) = \sqrt{x - y} \, e^{y - F(x,y)}, \quad (6)$$

$$H(x,y) \approx \sqrt{x - y} \, e^y, \quad (6a)$$

$$F(x,y) = \frac{1}{\pi} \int_0^\infty \frac{\tan^{-1}\left[\dfrac{K_1(t)}{\pi I_1(t)}\right]}{\sqrt{t^2 + x^2} + y} \frac{t \, dt}{\sqrt{t^2 + x^2}}.$$

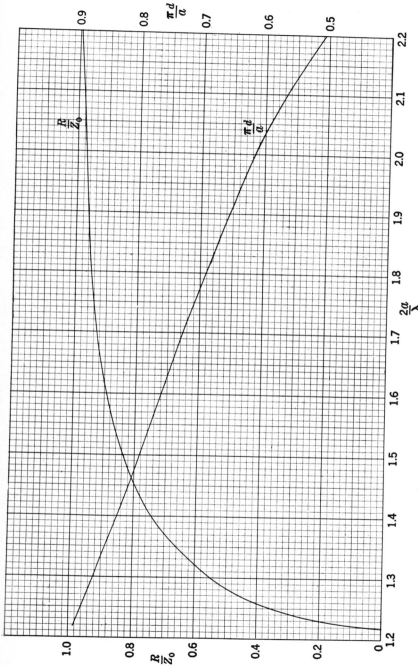

FIG. 4·13-3.—Radiating circular guide, H_{01}-mode. Circuit parameters at terminal plane T.

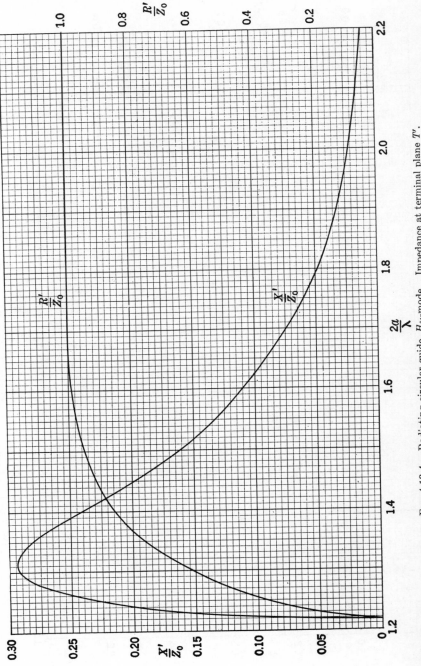

FIG. 4·13-4.—Radiating circular guide, H_{01}-mode. Impedance at terminal plane T'.

SEC. 4·13] RADIATION FROM A CIRCULAR GUIDE, H_{01}-MODE

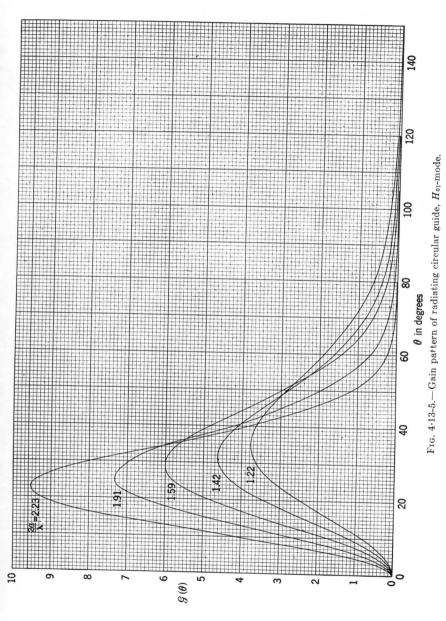

FIG. 4·13-5.—Gain pattern of radiating circular guide, H_{01}-mode.

Restrictions.—The above formulas have been obtained by the transform method and, for the indicated field symmetry, are rigorous in the wavelength range $0.896a < \lambda < 1.64a$. The approximation (1a) is correct to within 5 per cent and (3a) to within 1 per cent. Equation (6a) is correct to within 2 per cent for $y > 0$; i.e., $\theta < 90°$.

Numerical Results.—Curves of $\pi d/a$ and R/Z_0 as a function of $2a/\lambda$ are presented in Fig. 4·13-3. In Fig. 4·13-4 R'/Z_0 and X'/Z_0 are plotted as a function of $2a/\lambda$. The graph of the power gain function $\mathcal{G}(\theta)$ vs. θ with $2\pi a/\lambda$ as a parameter is given in Fig. 4·13-5; the monotonic decrease to zero of the curves in the range $150° < \theta < 180°$ is omitted.

4·14. Radiation from a Circular Guide, H_{11}-mode.—A semi-infinite circular guide of zero wall thickness radiating into free space (H_{11}-mode in circular guide).

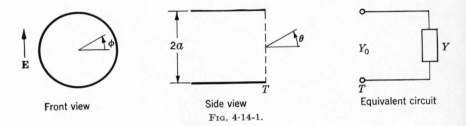

Front view   Side view   Equivalent circuit

Fig. 4·14-1.

Equivalent-circuit Parameters.—At the reference plane T the voltage reflection coefficient is

$$\Gamma = -L \frac{A_1 + jA_2}{\dfrac{1}{A_1} + jA_2} = \frac{1 - \dfrac{Y}{Y_0}}{1 + \dfrac{Y}{Y_0}}, \qquad (1)$$

where

$$L = \frac{\xi_0 + \kappa}{\xi_0 - \kappa}\left(\frac{\lambda_g - \lambda}{\lambda_g + \lambda}\right) e^c e^{j2\Phi},$$

$$A_1 = \frac{\lambda_g + \lambda}{\lambda_g - \lambda},$$

$$A_2 = \frac{\pi a}{\lambda}\left(1 + \frac{\xi_0}{k}\right)^2 e^{c_1} e^{j2\Phi_1},$$

and

$$\Phi = \frac{2a}{\lambda_g}\left(\ln\frac{\lambda}{\gamma a} + 2\right) - \beta_1' \sin^{-1}\frac{\lambda}{\lambda_g} - S_2^{J'_1}\left(\frac{2a}{\lambda_g};\beta_1'\right)$$
$$+ \frac{a}{\lambda_g}\int_0^\infty \frac{x\,dx}{x^2 + (\pi\beta_1')^2}\frac{1}{\sqrt{x^2 + (ka)^2}}\ln\left\{\left(1 + \left[\frac{\pi I_1'(x)}{K_1'(x)}\right]^2\right)e^{-4x}\right\},$$

SEC. 4·14] RADIATION FROM A CIRCULAR GUIDE, H_{11}-MODE

$$\Phi_1 = \frac{2a}{\lambda} + S_1^{J_1}\left(\frac{2a}{\lambda};0\right) - S_2^{J'_1}\left(\frac{2a}{\lambda};0\right)$$

$$- \frac{1}{4\pi}\int_0^\infty \left[1 - \frac{ka}{\sqrt{x^2+(ka)^2}}\right] \ln\left\{\frac{1 + \left[\frac{\pi I'_1(x)}{K'_1(x)}\right]^2}{1 + \left[\frac{\pi I_1(x)}{K_1(x)}\right]^2}\right\} \frac{dx}{x},$$

$$C = -\frac{2\pi a}{\lambda_g} + \frac{2a}{\lambda_g}\int \frac{x\,dx}{x^2+(\pi\beta'_1)^2}\frac{1}{\sqrt{x^2+(ka)^2}} \tan^{-1}\left[-\frac{K'_1(x)}{\pi I'_1(x)}\right],$$

$$C_1 = -\frac{1}{\pi}\int \left[1 - \frac{ka}{\sqrt{x^2+(ka)^2}}\right]\left[\tan^{-1}\frac{K_1(x)}{\pi I_1(x)} + \tan^{-1}\frac{-K'_1(x)}{\pi I'_1(x)}\right]\frac{dx}{x},$$

and

$$k = \frac{2\pi}{\lambda}, \quad \kappa = \frac{2\pi}{\lambda_g}, \quad \lambda_g = \frac{\lambda}{\sqrt{1 - \left(\frac{\lambda}{3.41a}\right)^2}}, \quad \gamma = 1.781,$$

$$(\xi_0 a)^2 = (ka)^2 - (x_0 - jy_0)^2, \quad x_0 = 0.5012, \quad y_0 = 0.6435,$$

$$S_1^{J_1}(x;0) = \sum_{n=1}^\infty \left(\sin^{-1}\frac{x}{\beta_n} - \frac{x}{n}\right), \quad J_1(\pi\beta_n) = 0,$$

$$S_2^{J'_1}(x;y) = \sum_{n=2}^\infty \left(\sin^{-1}\frac{x}{\sqrt{\beta_n'^2 - y^2}} - \frac{x}{n}\right), \quad J'_1(\pi\beta'_n) = 0.$$

The angular distribution of the emitted radiation is described by the power gain function $G(\theta,\phi)$, where θ and ϕ are the polar and azimuthal angles. This function, defined relative to an isotropic point source (i.e., normalized so that its integral over all angles is 4π) is given by

$$G(\theta,\phi) = \frac{4k\kappa a^2}{\pi\beta'_1} \cdot \frac{J_1(\pi\beta'_1)}{H(ka,\kappa) - H(ka,-\kappa)} \cdot \left[\cos^2\phi \frac{J'_1(ka\sin\theta)}{1-\left(\frac{ka\sin\theta}{\pi\beta'_1}\right)^2} H(ka,ka\cos\theta) \right.$$

$$\left. + 2\sin^2\phi \frac{J_1(ka\sin\theta)}{\sin\theta} E(ka,ka\cos\theta)\right],$$

where

$$H(x,y) = \frac{(x+y)^{\frac{1}{2}} e^y}{|x + \xi_0 a|^2 |y + \xi_0 a|^2} \left|\frac{x-y}{x+y} + jA_2\right|^2$$

$$\cdot e^{\frac{1}{\pi}\int_0^\infty \frac{t\,dt}{\sqrt{t^2+x^2}}\left(\frac{1}{\sqrt{t^2+x^2+y}} + \frac{1}{\sqrt{t^2+x^2+x}}\right)\tan^{-1}\frac{-K'_1(t)}{\pi I'_1(t)}}$$

$$E(x,y) = \frac{e^y}{\sqrt{x+y}} e^{-\frac{1}{\pi}\int_0^\infty \frac{t\,dt}{\sqrt{t^2+x^2}}\left(\frac{1}{\sqrt{t^2+x^2+y}} + \frac{1}{\sqrt{t^2+x^2+x}}\right)\tan^{-1}\frac{K_1(t)}{\pi I_1(t)}}.$$

The choice of coordinate system is such that $\phi = 0$ corresponds to the magnetic plane, $\phi = \pi/2$ to the electric plane (i.e., $H_z \sim \cos \phi$ and $H_\phi \sim \sin \phi$).

Restrictions.—The above equations have been obtained by the transform method and are rigorous in the wavelength range $2.61a < \lambda < 3.41a$.

Numerical Results.—Graphical plots of the above equations are not available. Tabulations of the summation functions S^{J_1} and $S^{J'_1}$ are presented in the Appendix. Several of the Bessel function roots $\chi_{1n} = \pi\beta_n$ and $\chi'_{1n} = \pi\beta'_n$ are given in Tables 2·1 and 2·2.

4·15. Coaxial Line with Infinite Center Conductor.—A coaxial line with a semi-infinite outer conductor of zero thickness and an infinite center conductor (principal mode in coaxial guide).

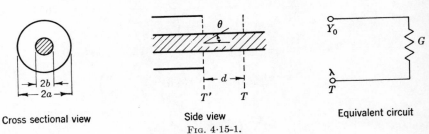

Cross sectional view   Side view   Equivalent circuit

Fig. 4·15-1.

Equivalent-circuit Parameters.—At the reference plane T located at a distance d from the reference plane T', where

$$\frac{\pi d}{a-b} = \ln \frac{e\lambda}{\gamma(a-b)} - \frac{\pi}{k(a-b)} S_1^{Z_0}\left(y;0,\frac{a}{b}\right)$$
$$+ \frac{1}{2}\frac{1}{1-\frac{b}{a}} \int_0^\infty \frac{dx}{x} \frac{1}{\sqrt{x^2+(ka)^2}} \ln\left(\sqrt{1+\left(\frac{\pi I_0(x)}{K_0(x)}\right)^2} e^{-2x}\right)$$
$$- \frac{1}{2}\frac{\frac{b}{a}}{1-\frac{b}{a}} \int_0^\infty \frac{dx}{x} \frac{1}{\sqrt{x^2+(kb)^2}} \ln\left(\sqrt{1+\left(\frac{\pi I_0(x)}{K_0(x)}\right)^2} e^{-2x}\right), \quad (1)$$

$$\frac{\pi d}{a-b} \approx \ln \frac{e\lambda}{\gamma(a-b)} - \sum_1^\infty \left[\frac{1}{n}\left(\frac{1}{\gamma_n}-1\right) + \frac{2}{3}\left(\frac{a-b}{\lambda}\right)^2 \left(\frac{1}{n\gamma_n}\right)^3 + \cdots\right],$$
$$a \approx b \gg \lambda. \quad (1a)$$

the equivalent circuit is simply the conductance G defined by

$$\frac{G}{Y_0} = \tanh \frac{\psi}{2}, \quad (2)$$

SEC. 4·15] COAXIAL LINE WITH INFINITE CENTER CONDUCTOR 209

where

$$\psi = k(a - b) - \frac{1}{\pi} \int_0^\infty \frac{dx}{x} \tan^{-1} \frac{K_0(x)}{\pi I_0(x)} \left[\frac{1}{\sqrt{1 + \left(\frac{x}{ka}\right)^2}} - \frac{1}{\sqrt{1 + \left(\frac{x}{kb}\right)^2}} \right] \quad (3)$$

$$\psi \cong k(a - b), \quad a \approx b \gg \lambda, \quad (3a)$$

and

$$(c - 1)\chi_{0n} = \pi\gamma_n, \quad J_0(\chi_{0n})N_0(c\chi_{0n}) - J_0(c\chi_{0n})N_0(\chi_{0n}) = 0,$$
$$n = 1, 2, \cdots,$$

$$S_1^{z_0}\left(y;0,\frac{a}{b}\right) = \sum_1^\infty \left[\sin^{-1}\left(\frac{y}{\gamma_n}\right) - \frac{y}{n} \right], \quad k = \frac{2\pi}{\lambda}, \quad y = \frac{k(a-b)}{\pi}, \quad c = \frac{a}{b}.$$

At the reference plane T' the parameters of the alternative equivalent circuit indicated in the Fig. 4·15-2 are

$$\left. \begin{array}{l} \dfrac{G'}{Y_0} = \dfrac{\sinh \psi}{\cosh \psi + \cos 2kd}, \\[2mm] \dfrac{B'}{Y_0} = \dfrac{\sin 2kd}{\cosh \psi + \cos 2kd}. \end{array} \right\} \quad (4)$$

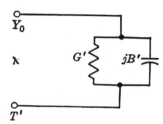

Fig. 4·15-2.

The angular distribution of the emitted radiation is described by the power gain function (defined relative to an isotropically radiating point source),

$$\mathcal{G}(\theta) = \frac{4}{\pi} \frac{J_0(kb \sin \theta)N_0(ka \sin \theta) - J_0(ka \sin \theta)N_0(kb \sin \theta)}{\sin^2\theta \, [J_0^2(kb \sin \theta) + N_0^2(kb \sin \theta)]} \cdot \frac{\frac{F_a(k \cos \theta)}{F_b(k \cos \theta)}}{\frac{F_a(k)}{F_b(k)} - \frac{F_a(-k)}{F_b(-k)}}, \quad (5)$$

$$\mathcal{G}(\theta) \approx \frac{1}{\theta^2} \frac{1}{\frac{4}{\pi^2}\left[\ln\left(\frac{\gamma kb\theta}{2}\right)\right]^2 + 1}, \quad \theta \ll 1, \quad \gamma = 1.781, \quad (5a)$$

$$\mathcal{G}(\theta) \approx \frac{1}{(\pi - \theta)^2} \frac{1}{\frac{4}{\pi^2}\left\{\ln\left[\frac{\gamma kb}{2}(\pi - \theta)\right]\right\}^2 + 1}, \quad \pi - \theta \ll 1 \quad (5b)$$

where

$$F_a(\xi) = e^{\xi a} e^{\frac{1}{\pi} \int_0^\infty \frac{x \, dx}{\sqrt{x^2 + (ka)^2}} \frac{\tan^{-1}\left[\frac{K_0(x)}{\pi I_0(x)}\right]}{\sqrt{x^2 + (ka)^2 + \xi a}}} \quad (6)$$

$$\approx e^{\xi a}, \quad \xi > 0 \quad (6a)$$

and $F_b(\xi)$ is obtained by simply replacing a in $F_a(\xi)$ with b.

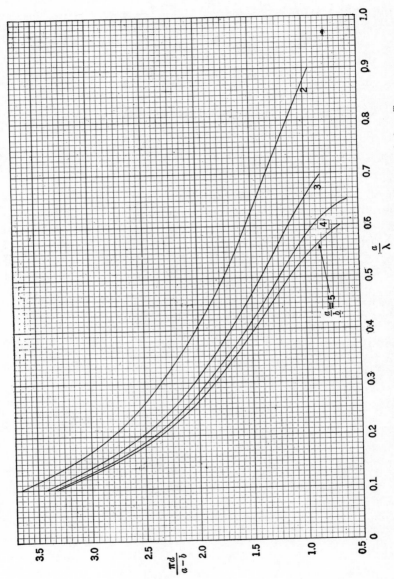

FIG. 4·15-3.—Coaxial guide radiating into space. Location of terminal plane T.

SEC. 4·15] COAXIAL LINE WITH INFINITE CENTER CONDUCTOR

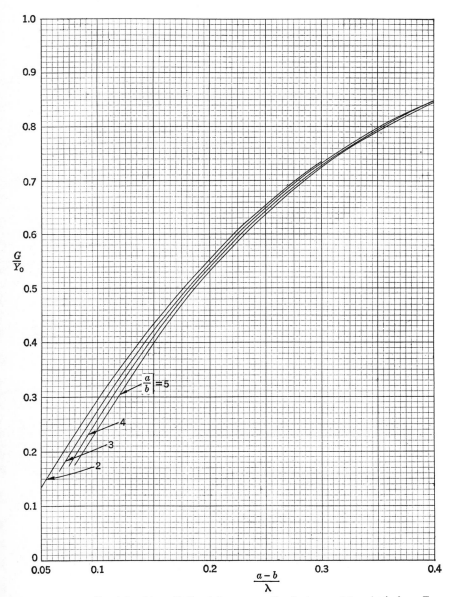

FIG. 4·15-4.—Coaxial guide radiating into space. Conductance at terminal plane T.

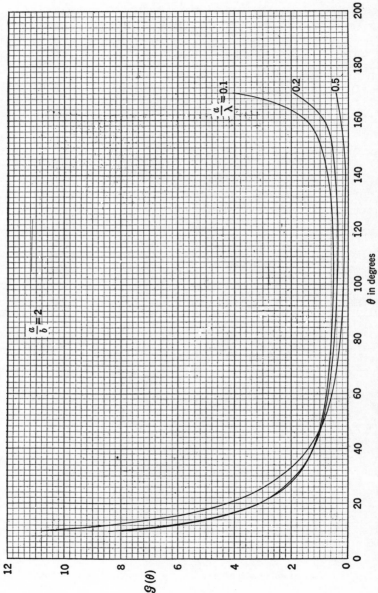

FIG. 4·15·5.—Gain pattern for coaxial guide radiating into space.

Restrictions.—The problem is treated by the transform method. The equivalent circuits and the above formulas are rigorous in the wavelength range

$$2\frac{a-b}{\gamma_1\left(\frac{a}{b}\right)} < \lambda < \infty.$$

The quantity $\gamma_1(a/b)$, as well as the quantities $\gamma_n(a/b)$, may be obtained from Table 2·3. When $a \approx b \gg \lambda$, $\gamma_n(a/b) \approx n$ and the formulas go over into those for the radiation from parallel plates, E-plane, whose separation is $2(a-b)$ (see Sec. 4·6a). Equation (1a) is accurate to within 12 per cent for $b/a > 0.2$, $a/\lambda > 0.1$. Equation (3a) is accurate to within 15 per cent for $b/a > 0.5$, $a/\lambda > 0.1$. Equation (3a) is in error by 42 per cent for $b/a = 0.2$, $a/\lambda = 0.1$. Equation (6a) is valid to within a few per cent for $\xi > 0$, i.e., $\theta < 90°$.

Numerical Results.—Figure 4·15-3 contains a plot of $\pi d/(a-b)$ as a function of a/λ for various values of a/b. Figure 4·15-4 shows the variation of G/Y_0 with $(a-b)/\lambda$ for a few values of a/b. The gain function $\mathcal{G}(\theta)$ is plotted in Fig. 4·15-5 as a function of θ for $a/b = 2$ and a few values of a/λ. The summation $S_1^{z_0}(y;0,c)$ is tabulated in the Appendix.

4·16. Coaxial Line Radiating into Semi-infinite Space.—A semi-infinite coaxial line terminating in the plane of an infinite metallic screen and radiating into free space (principal mode in coaxial guide).

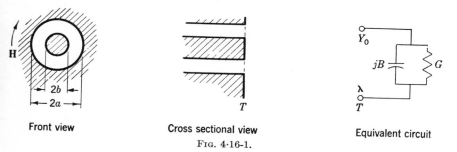

Front view   Cross sectional view   Equivalent circuit
Fig. 4·16-1.

Equivalent-circuit Parameters.—At the terminal plane T

$$\frac{G}{Y_0} = \frac{1}{\ln \frac{a}{b}} \int_0^{\pi/2} \frac{d\theta}{\sin \theta} [J_0(ka \sin \theta) - J_0(kb \sin \theta)]^2, \quad (1a)$$

$$\frac{G}{Y_0} \approx \frac{2}{3} \frac{1}{\ln \frac{a}{b}} \left[\frac{\pi^2(b^2 - a^2)}{\lambda^2}\right]^2, \quad \frac{a}{\lambda}, \frac{b}{\lambda} \ll 1. \quad (1b)$$

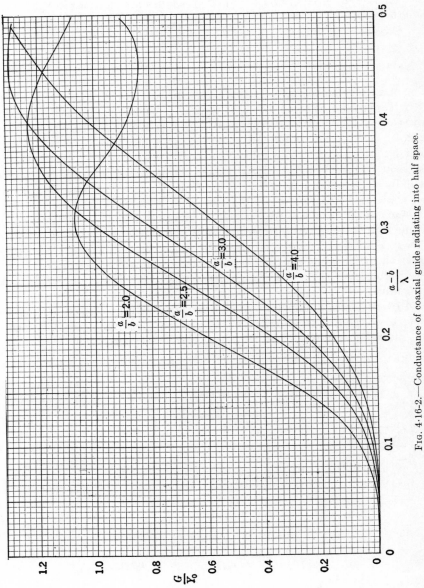

Fig. 4·16-2.—Conductance of coaxial guide radiating into half space.

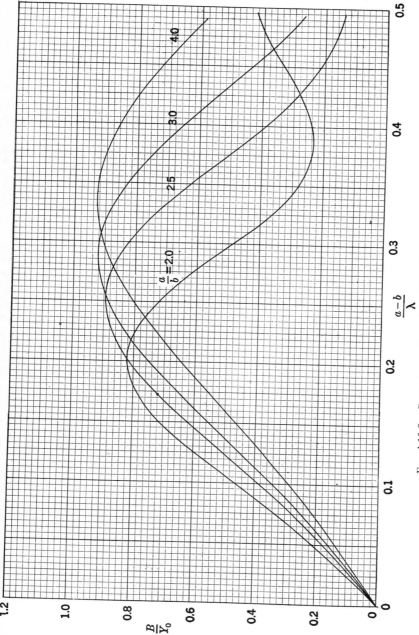

FIG. 4·16-3.—Susceptance of coaxial guide radiating into half space.

$$\frac{B}{Y_0} = \frac{1}{\pi \ln \frac{a}{b}} \int_0^\pi \left[2Si(k\sqrt{a^2+b^2-2ab\cos\phi}) - Si\left(2ka\sin\frac{\phi}{2}\right) - Si\left(2kb\sin\frac{\phi}{2}\right) \right] d\phi, \quad (2a)$$

$$\frac{B}{Y_0} \approx \frac{8(a+b)}{\lambda \ln \frac{a}{b}} \left[E\left(\frac{2\sqrt{ab}}{a+b}\right) - 1 \right], \quad \frac{a}{\lambda}, \frac{b}{\lambda} \ll 1, \quad (2b)$$

where $Si(x)$ is the sine-integral function, $E(x)$ is the complete elliptic integral of the second kind, and $k = 2\pi/\lambda$.

Restrictions.—The equivalent circuit is valid in the wavelength range $\lambda > 2(a-b)/\gamma_1$, where γ_1 is determined from the first root $\chi = \chi_{01}$ of

$$J_0(\chi)N_0\left(\chi \frac{a}{b}\right) - J_0\left(\chi \frac{a}{b}\right)N_0(\chi) = 0, \quad \chi = \frac{\pi}{\frac{a}{b}-1}\gamma_1,$$

and may be found from Table 2·3. The circuit parameters have been obtained by the variational method assuming a principal mode electric aperture field and are presumed to be in error by less than 10 per cent over most of the range of validity. The approximate Eq. (1b) agrees with Eq. (1a) to within 15 per cent for $(a-b)/\lambda < 0.10$ $a/b \geqq 3$; similarly Eq. (2b) agrees with Eq. (2a) to within 15 per cent for $(a-b)/\lambda < 0.10$ and $a/b \geqq 2$. The accuracy is improved in both cases for larger values of a/b.

Numerical Results.—In Figs. 4·16-2, -3 the quantities G/Y_0 and B/Y_0 are plotted as functions of $(a-b)/\lambda$, for several values of a/b.

CHAPTER 5

FOUR-TERMINAL STRUCTURES

A structure that contains a geometrical discontinuity is designated as a four-terminal, or two-terminal-pair, waveguide structure if it comprises an input and an output region each in the form of a waveguide propagating only a single mode. The over-all description of the propagating modes is effected by representation of the input and output waveguides as transmission lines and by representation of the discontinuity as a four-terminal lumped-constant circuit. The transmission lines together with the lumped-constant circuit form a four-terminal network that determines the reflection, transmission, standing-wave, etc., properties of the over-all structure. The quantitative description of the transmission lines requires the indication of their characteristic impedance and propagation wavelength; the description of the four-terminal circuit requires, in general, the specification of three circuit parameters and the locations of the input and output terminal planes.

In the various sections of this chapter a number of basic four-terminal waveguide structures will be represented at specified terminal planes by a four-terminal electrical network. The circuit elements of this network are specified by their reactance or susceptance values. The latter do not, in general, correspond to constant, i.e., frequency-independent, inductances and capacitances, but this does not impair their usefulness. The choice of terminal planes as well as of the form of the equivalent network is not unique; other equivalent forms, which are desirable in particular applications, may be readily obtained by the methods outlined in Sec. 3·3. As stated above, five impedances and two propagation wavelengths are employed for the description of the general four-terminal structure. In the presentation of this information it is most convenient to specify all impedances relative to the characteristic impedance of the input transmission line, although any other impedance can be employed as a norm. The propagation wavelengths of the input and output transmission lines will be indicated explicitly in the equivalent-circuit representation of the given structure. When both the input and output guides are identical, this explicit indication will sometimes be omitted if no confusion is likely.

A number of free-space structures are included in the present chapter. Under appropriate conditions of excitation, the scattering of plane waves by gratings and arrays in free space can be treated (*cf.* Sec. 2·6) in the same manner as scattering by a discontinuity in a waveguide. In both cases the scattering is described by a four-terminal network of the type described in the preceding paragraphs. The applicability of such a description is restricted to the wavelength range in which the higher diffraction orders, i.e., higher modes, cannot be propagated.

STRUCTURES WITH ZERO THICKNESS

5·1. Capacitive Obstacles and Windows in Rectangular Guide.

a. Window Formed by Two Obstacles.—Window formed by zero thickness obstacles with edges perpendicular to the electric field (H_{10}-mode in rectangular guide).

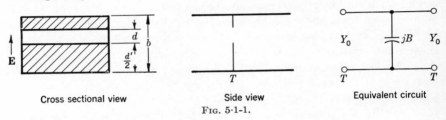

Cross sectional view   Side view   Equivalent circuit

FIG. 5·1-1.

Equivalent-circuit Parameters.—At the terminal plane T for the unsymmetrical case $d' \neq b - d$:

$$\frac{B}{Y_0} = \frac{4b}{\lambda_g} \left\{ \ln\left[\csc\frac{\pi d}{2b} \csc\frac{\pi}{2b}(d'+d)\right] + \frac{2Q_1 \cos^2\frac{\pi d}{2b} \cos^2\frac{\pi}{2b}(d'+d)}{1 + Q_1 \sin^2\frac{\pi d}{2b} \sin^2\frac{\pi}{2b}(d'+d)} \right.$$

$$\left. + Q_2 \left[3 \cos^2\frac{\pi d}{2b} \cos^2\frac{\pi}{2b}(d'+d) - \cos^2\frac{\pi d}{2b} - \cos^2\frac{\pi}{2b}(d'+d)\right]^2 \right\}, \quad (1a)$$

where

$$Q_n = \frac{1}{\sqrt{1 - \left(\frac{2b}{n\lambda_g}\right)^2}} - 1.$$

For the symmetrical case $d' = b - d$:

$$\frac{B}{Y_0} = \frac{4b}{\lambda_g}\left[\ln\left(\csc\frac{\pi d}{2b}\right) + \frac{Q_2 \cos^4\frac{\pi d}{2b}}{1 + Q_2 \sin^4\frac{\pi d}{2b}}\right.$$

$$\left. + \frac{1}{16}\left(\frac{b}{\lambda_g}\right)^2 \left(1 - 3\sin^2\frac{\pi d}{2b}\right)^2 \cos^4\frac{\pi d}{2b}\right], \quad (2a)$$

SEC. 5·1] CAPACITIVE OBSTACLES 219

$$\frac{B}{Y_0} \cong \frac{4b}{\lambda_g} \left\{ \ln\left(\frac{2b}{\pi d}\right) + \frac{1}{6}\left(\frac{\pi d}{2b}\right)^2 + \frac{1}{2}\left(\frac{b}{\lambda_g}\right)^2 \left[1 - \frac{1}{2}\left(\frac{\pi d}{2b}\right)^2\right]^4 \right\}, \qquad \frac{d}{b} \ll 1, \quad (2b)$$

$$\frac{B}{Y_0} \cong \frac{2b}{\lambda_g} \left[\left(\frac{\pi d'}{2b}\right)^2 + \frac{1}{6}\left(\frac{\pi d'}{2b}\right)^4 + \frac{3}{2}\left(\frac{b}{\lambda_g}\right)^2 \left(\frac{\pi d'}{2b}\right)^4 \right], \qquad \frac{d'}{b} \ll 1. \quad (2c)$$

Restrictions.—The equivalent circuit is valid in the range $b/\lambda_g < \frac{1}{2}$ for the unsymmetrical case and $b/\lambda_g < 1$ for the symmetrical case. Equations (1a) and (2a) have been obtained by the equivalent static method employing a static field in the aperture due to the incidence of the two lowest modes; the higher-mode attenuation constants have been approximated by $n\pi/b$ for $n \geq 3$. Equation (1a) is applicable in the range $2b/\lambda_g < 1$ with an estimated error that rises to less than 5 per cent at the lowest wavelength range. Equation (2a) is applicable in the range $b/\lambda_g < 1$ with an error of less than about 5 per cent and in the range $2b/\lambda_g < 1$ to within 1 per cent. Equation (2b) is a small-aperture approximation that agrees with Eq. (2a) to within 5 per cent in the range $d/b < 0.5$ and $b/\lambda_g < 0.5$. The small-obstacle approximation (2c) agrees with Eq. (2a) to within 5 per cent in the range $d/b > 0.5$ and $b/\lambda_g < 0.4$.

Numerical Results.—The quantities $B\lambda_g/Y_0 b$ and $Y_0 b/B\lambda_g$, as obtained from Eq. (2a), are plotted in Fig. 5·1-4 as a function of d/b with b/λ_g as a parameter.

b. Window Formed by One Obstacle.—Window formed by a zero thickness obstacle with its edge perpendicular to the electric field (H_{10}-mode in rectangular guide).

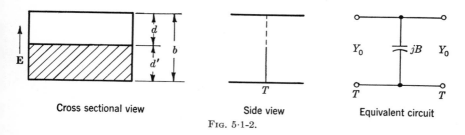

Cross sectional view   Side view   Equivalent circuit

FIG. 5·1-2.

Equivalent-circuit Parameters.—Same as Eqs. (2a) to (2c) except that λ_g is replaced by $\lambda_g/2$.

Restrictions.—Same as for Eqs. (2a) to (2c) except that λ_g is replaced by $\lambda_g/2$.

Numerical Results.—If the λ_g in Fig. 5·1-4 is replaced by $\lambda_g/2$, one obtains a plot of $B\lambda_g/2Y_0 b$ as a function of d/b with $2b/\lambda_g$ as a parameter, where B/Y_0 is now the relative susceptance of a window formed by one obstacle.

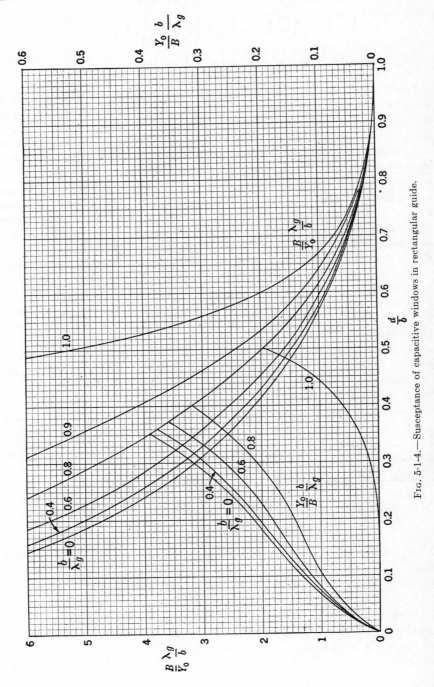

Fig. 5·1-4.—Susceptance of capacitive windows in rectangular guide.

c. Symmetrical Obstacle.

A symmetrical obstacle of zero thickness with its edges perpendicular to the electric field (H_{10}-mode in rectangular guide).

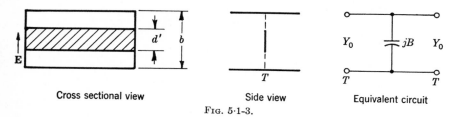

Cross sectional view Side view Equivalent circuit
Fig. 5·1-3.

Equivalent-circuit Parameters.—Same as for Eqs. (2a) to (2c).
Restrictions.—Same as for Eqs. (2a) to (2c).
Numerical Results.—Same as for Eq. (2a) and plotted in Fig. 5·1-4.

5·2. Inductive Obstacles and Windows in Rectangular Guide.

a. Symmetrical Window.—Symmetrical window formed by zero thickness obstacles with edges parallel to the electric field (H_{10}-mode in rectangular guide).

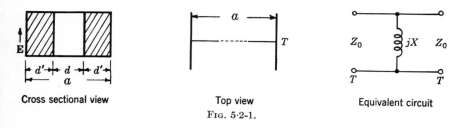

Cross sectional view Top view Equivalent circuit
Fig. 5·2-1.

Equivalent-circuit Parameters.—At the terminal plane T

$$\frac{X}{Z_0} = \frac{a}{\lambda_g} \tan^2 \frac{\pi d}{2a} \left\{ 1 + \frac{3}{4} \left[\frac{1}{\sqrt{1 - \left(\frac{2a}{3\lambda}\right)^2}} - 1 \right] \sin^2 \frac{\pi d}{a} \right.$$

$$\left. + 2\left(\frac{a}{\lambda}\right)^2 \left[1 - \frac{4}{\pi} \frac{E(\alpha) - \beta^2 F(\alpha)}{\alpha^2} \cdot \frac{E(\beta) - \alpha^2 F(\beta)}{\beta^2} - \frac{1}{12} \sin^2 \frac{\pi d}{a} \right] \right\}, \quad (1a)$$

$$\frac{X}{Z_0} \approx \frac{a}{\lambda_g} \tan^2 \frac{\pi d}{2a} \left[1 + \frac{1}{6}\left(\frac{\pi d}{\lambda}\right)^2 \right], \qquad \frac{d}{a} \ll 1, \quad (1b)$$

$$\frac{X}{Z_0} \approx \frac{a}{\lambda_g} \cot^2 \frac{\pi d'}{a} \left[1 + \frac{2}{3}\left(\frac{\pi d'}{\lambda}\right)^2 \right], \qquad \frac{d'}{a} \ll 1, \quad (1c)$$

where

$$\alpha = \sin \frac{\pi d}{2a}, \qquad \beta = \cos \frac{\pi d}{2a}$$

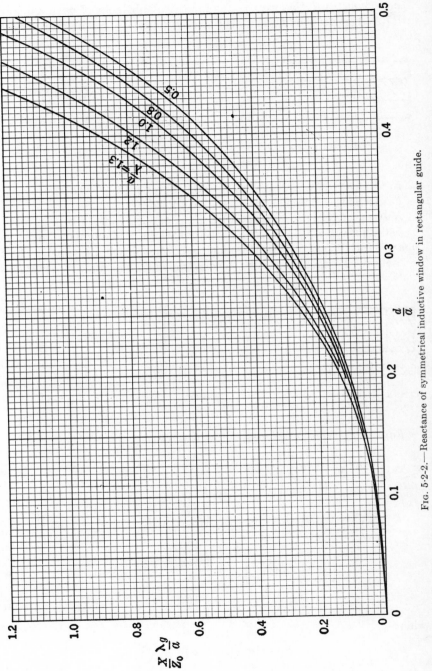

Fig. 5·2-2.—Reactance of symmetrical inductive window in rectangular guide.

SEC. 5·2] INDUCTIVE OBSTACLES

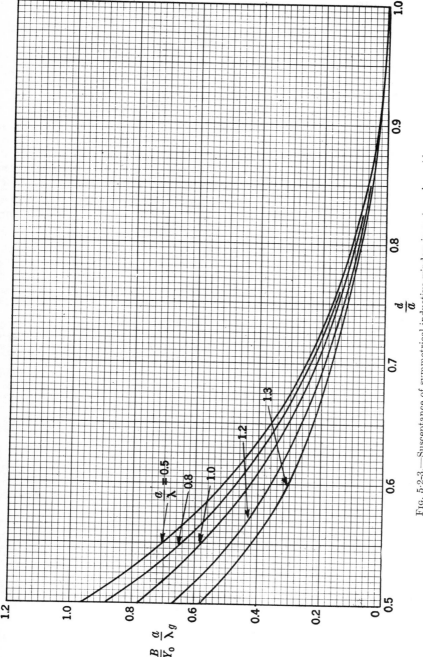

FIG. 5·2-3.—Susceptance of symmetrical inductive window in rectangular guide.

and $F(\alpha)$, $E(\alpha)$ are complete elliptic integrals of the first and second kinds, respectively.

Restrictions.—The equivalent circuit is applicable in the wavelength range $\frac{2}{3}a < \lambda < 2a$. Equation (1a) has been derived by the equivalent static method employing the static aperture field set up by an incident lowest mode and, in addition, the higher-mode attenuation constant approximations

$$\sqrt{\left(\frac{n\pi}{a}\right)^2 - \left(\frac{2\pi}{\lambda}\right)^2} \simeq \frac{n\pi}{a}\left[1 - \frac{1}{2}\left(\frac{2a}{n\lambda}\right)^2\right]$$

for $n \geq 5$. In the range $a < \lambda < 2a$ Eq. (1a) is estimated to be in error by less than 1 per cent; for $\frac{2}{3}a < \lambda < a$ the error is larger, but no estimate is available. The term in $(a/\lambda)^2$ of Eq. (1a) accounts for not more than 5 per cent of X/Z_0. The approximate form (1b), valid in the small-aperture range, agrees with Eq. (1a) to within 4 per cent for $d < 0.5a$ and $a < 0.9\lambda$. Equation (1c) is an approximate form valid in the small-obstacle range; for $d' \leq 0.2a$ and $a < 0.9\lambda$ it agrees with Eq. (1a) to within 5 per cent.

Numerical Results.—As obtained from Eq. (1a), $X\lambda_g/Z_0 a$ is plotted in Fig. 5·2-2 as a function of d/a for the range 0 to 0.5 and for various values of a/λ. In Fig. 5·2-3 the inverse quantity $Ba/Y_0\lambda_g = Z_0 a/X\lambda_g$ is given as a function of d/a in the range 0.5 to 1.

b. *Asymmetrical Window.*—Asymmetrical window formed by a zero thickness obstacle with its edges parallel to the electric field (H_{10}-mode in rectangular guide).

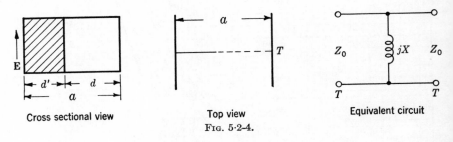

Cross sectional view Top view Equivalent circuit
FIG. 5·2-4.

Equivalent-circuit Parameters.—At the terminal plane T

$$\frac{X}{Z_0} = \frac{a}{\lambda_g} \frac{\tan^2 \frac{\pi d}{2a}}{1 + \csc^2 \frac{\pi d}{2a}} \left\{ 1 + \frac{8\alpha^4 \beta^2 Q}{1 + \alpha^2 + \beta^6(\beta^4 + 6\alpha^2)Q} \right.$$
$$\left. + 2\left(\frac{a}{\lambda}\right)^2 \left[1 - 2\frac{\alpha^2 + 2\beta^2 \ln \beta}{\alpha^4(1 + \alpha^2)} - \frac{2\alpha^4 \beta^2}{1 + \alpha^2}\right] \right\}, \quad (1a)$$

SEC. 5·2]   INDUCTIVE OBSTACLES   225

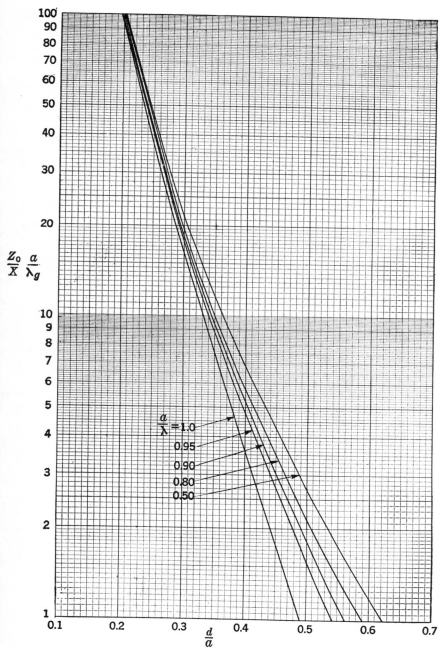

FIG. 5·2-5.—Susceptance of asymmetrical inductive window in rectangular guide.

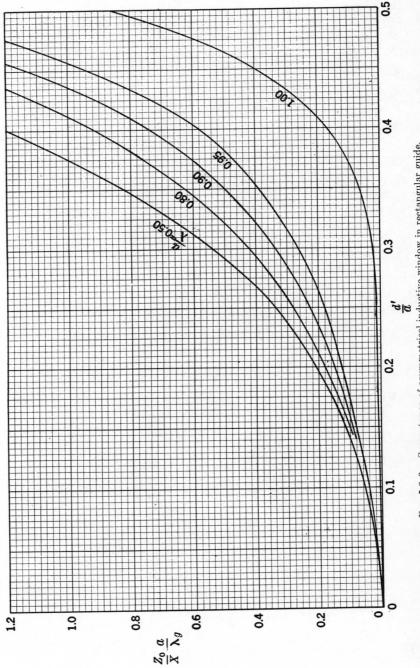

FIG. 5·2-6.—Susceptance of asymmetrical inductive window in rectangular guide.

SEC. 5·2]   INDUCTIVE OBSTACLES

$$\frac{X}{Z_0} \approx \frac{a}{\lambda_g}\left(\frac{\pi d}{2a}\right)^4 \left[1 - \frac{2}{3}\left(\frac{\pi d}{2a}\right)^2\right]\left[1 + \frac{4}{3}\left(\frac{a}{\lambda}\right)^2\left(\frac{\pi d}{2a}\right)^2\right], \quad \frac{d}{a} \ll 1, \quad (1b)$$

$$\frac{Z_0}{X} \approx \frac{2\lambda_g}{a}\left(\frac{\pi d'}{2a}\right)^2 \left[1 + \frac{3}{2}\left(\frac{\pi d'}{2a}\right)^2\right]\left[1 + 4\left(\frac{a}{\lambda}\right)^2\left(\frac{\pi d'}{2a}\right)^2 \ln \frac{\pi d'}{2a}\right], \quad \frac{d'}{a} \ll 1, \quad (1c)$$

where

$$\alpha = \sin\frac{\pi d}{2a}, \quad \beta = \cos\frac{\pi d}{2a}, \quad Q = \frac{1}{\sqrt{1 - \left(\frac{a}{\lambda}\right)^2}} - 1.$$

Restrictions.—The equivalent circuit is applicable in the wavelength range $a < \lambda < 2a$. Equation (1a) has been derived by the equivalent static method employing the static aperture field for two lowest modes incident and, in addition, higher-mode approximations similar to those indicated in Sec. 5·2a for $n \geq 4$. Equation (1a) is applicable in the wavelength range $a < \lambda < 2a$ with an estimated error of about 1 per cent. The asymptotic form Eq. (1b) valid for the small-aperture range agrees with Eq. (1a) to within 5 per cent if $d/a < 0.3$ and $a/\lambda < 0.8$. Equation (1c) is valid in the small-obstacle range; for $d'/a < 0.2$ and $a/\lambda < 0.8$ it agrees with Eq. (1a) to within 10 per cent.

Numerical Results.—As obtained from Eq. (1a), $Z_0a/X\lambda_g$ is plotted in Fig. 5·2-5 as a function of d/a in the range 0.1 to 0.7. Similarly in Fig. 5·2-6 there is a plot of $Z_0a/X\lambda_g$ as a function of d'/a in the range 0 to 0.5. In both figures a/λ is employed as a parameter.

c. Symmetrical Obstacle.—A centered symmetrical obstacle of zero thickness with its edges parallel to the electric field (H_{10}-mode in rectangular guide).

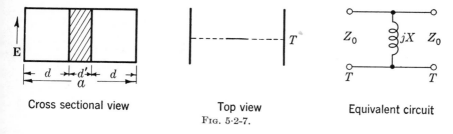

Cross sectional view   Top view   Equivalent circuit
Fig. 5·2-7.

Equivalent-circuit Parameters:

$$\frac{X}{Z_0} = \frac{a}{\lambda_g}\left\{\frac{(1+\alpha^2)F(\beta) - 2E(\beta)}{2E(\beta) - \alpha^2 F(\beta)} + \frac{2}{27}\left(\frac{a}{\lambda}\right)^2\left[\frac{2(2\alpha^2 - 1)E(\beta) - \alpha^2(3\alpha^2 - 1)F(\beta)}{2E(\beta) - \alpha^2 F(\beta)}\right]^2\right\}, \quad (1a)$$

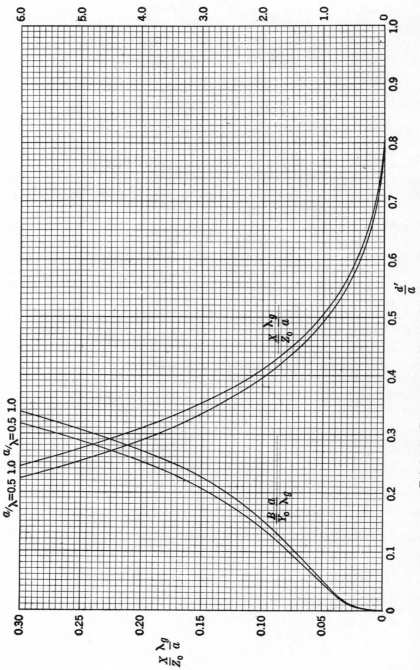

Fig. 5·2-8.—Reactance of inductive obstacle in rectangular guide.

$$\frac{X}{Z_0} \approx \frac{a}{2\lambda_g} \left[\ln\left(\frac{8}{\pi e^2}\frac{a}{d'}\right) + \frac{4}{27}\left(\frac{a}{\lambda}\right)^2 \right], \qquad \frac{d'}{a} \ll 1, \qquad (1b)$$

$$\frac{X}{Z_0} \approx \frac{2a}{\lambda_g} \left(\frac{\pi d}{2a}\right)^4 \left[1 + 12\left(\frac{a}{\lambda}\right)^2 \left(\frac{\pi d}{2a}\right)^4 \right], \qquad \frac{d}{a} \ll 1, \qquad (1c)$$

where

$$\alpha = \sin\frac{\pi d'}{2a}, \qquad \beta = \cos\frac{\pi d'}{2a}.$$

Restrictions.—The equivalent circuit is applicable in the wavelength range $2a/3 < \lambda < 2a$. Equation (1a) has been derived by the equivalent static method employing the static obstacle current set up by an incident lowest mode and using higher-mode approximations similar to those indicated in Sec. 5·2a for $n \geqq 3$. Equation (1a) is applicable in the wavelength range $2a/3 < \lambda < 2a$ with an estimated error of a few per cent for $a < \lambda < 2a$. The asymptotic form Eq. (1b), valid in the small-obstacle range, agrees with Eq. (1a) to within 10 per cent provided $d'/a \leqq 0.15$ and $a/\lambda \leqq 1$. Similarly Eq. (1c), valid in the small-aperture range, agrees with Eq. (1a) to within 10 per cent for $d/a < 0.25$ and $a/\lambda < 1$.

Numerical Results.—As obtained from Eq. (1a), $Ba/Y_0\lambda_g = Z_0a/X\lambda_g$ is plotted in Fig. 5·2-8 as a function of d'/a with a/λ as a parameter.

5·3. Capacitive Windows in Coaxial Guide. *a. Disk on Inner Conductor.*—A window formed by a circular metallic disk of zero thickness on the inner conductor of a coaxial line (principal mode in coaxial guide).

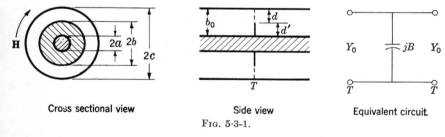

Cross sectional view   Side view   Equivalent circuit

Fig. 5·3-1.

Equivalent-circuit Parameters.—At the terminal plane T

$$\frac{B}{Y_0} = \frac{2b_0}{\lambda} A_1 \left[4\ln\left(\csc\frac{\pi d}{2b_0}\right) + \frac{4A\cos^4\frac{\pi d}{2b_0}}{1 + A\sin^4\frac{\pi d}{2b_0}} \right.$$
$$\left. + \left(\frac{b_0}{\lambda}\right)^2 \left(1 - 3\sin^2\frac{\pi d}{2b_0}\right)^2 \cos^4\frac{\pi d}{2b_0} + A_2 \right], \qquad (1a)$$

$$\frac{B}{Y_0} \approx \frac{2b_0 A_1}{\lambda} \left\{ 4 \ln\left(\frac{2b_0}{\pi d}\right) + \frac{2}{3}\left(\frac{\pi d}{2b_0}\right)^2 + 8\left(\frac{b_0}{\lambda}\right)^2 \left[1 - \frac{1}{2}\left(\frac{\pi d}{2b_0}\right)^2\right]^4 + A_2 \right\},$$
$$\frac{d}{b_0} \ll 1, \quad (1b)$$

$$\frac{B}{Y_0} \approx \frac{2b_0 A_1}{\lambda} \left[2\left(\frac{\pi d'}{2b_0}\right)^2 + \frac{1}{3}\left(\frac{\pi d'}{2b_0}\right)^4 + 12\left(\frac{b_0}{\lambda}\right)^2 \left(\frac{\pi d'}{2b_0}\right)^4 + A_2 \right],$$
$$\frac{d'}{b_0} \ll 1, \quad (1c)$$

where $d = c - b$, $\quad d' = b - a$, $\quad b_0 = c - a$,

$$A = \frac{1}{\sqrt{1 - \left(\frac{2b_0}{\lambda}\right)^2}} - 1,$$

$$A_1 = \frac{b}{a} \frac{\ln \frac{c}{a}}{\frac{c}{a} - 1} \left(\frac{\frac{c}{b} - 1}{\ln \frac{c}{b}}\right)^2,$$

$$A_2 = \frac{\pi^2 \frac{a}{b}}{\gamma_1 \sqrt{1 - \left(\frac{2b_0}{\gamma_1 \lambda}\right)^2}} \frac{\frac{c}{a} - 1}{\frac{J_0^2(\chi)}{J_0^2(\chi c/a)} - 1} \left[\frac{J_0(\chi) N_0\left(\frac{\chi b}{a}\right) - N_0(\chi) J_0\left(\frac{\chi b}{a}\right)}{\frac{c}{b} - 1}\right]^2$$
$$- \frac{1}{\sqrt{1 - \left(\frac{2b_0}{\lambda}\right)^2}} \left(\frac{2}{\pi} \frac{b_0}{d} \sin \frac{\pi d}{b_0}\right)^2,$$

and $\chi = \dfrac{\pi \gamma_1}{c/a - 1} = \chi_{01}$ is the first nonvanishing root of

$$J_0(\chi) N_0\left(\frac{\chi c}{a}\right) - N_0(\chi) J_0\left(\frac{\chi c}{a}\right) = 0.$$

Restrictions.—The equivalent circuit is valid in the wavelength range $\lambda > 2(c - a)/\gamma_1$ provided the fields are rotationally symmetrical. The susceptance has been evaluated by means of a variational method treating the first higher E-mode correctly and all higher modes by plane parallel approximations. Equation (1a) is estimated to be correct to within a few per cent for $c/a < 5$ and for wavelengths not too close to cutoff of the first higher mode. Equation (1b) is a small-aperture approximation that agrees with Eq. (1a) to within 5 per cent in the range $d/b_0 < 0.5$ and $2b_0/\lambda < 0.5$. Equation (1c) is a small-obstacle approximation and agrees with Eq. (1a) to within 5 per cent in the range $d/b_0 > 0.5$, $2b_0/\lambda < 0.4$.

Numerical Results.—For $c/a = 1$ the graph of $B\lambda/Y_0 2b_0$ as a function of d/b_0 with $2b_0/\lambda$ as a parameter may be obtained from Fig. 5·1-4 if the

SEC. 5·3]   CAPACITIVE WINDOWS IN COAXIAL GUIDE   231

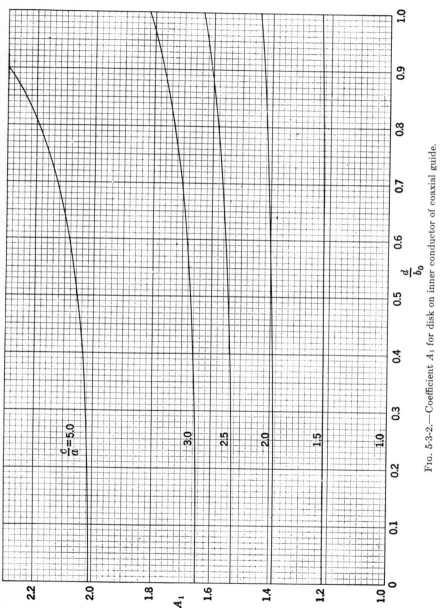

FIG. 5·3-2.—Coefficient A_1 for disk on inner conductor of coaxial guide.

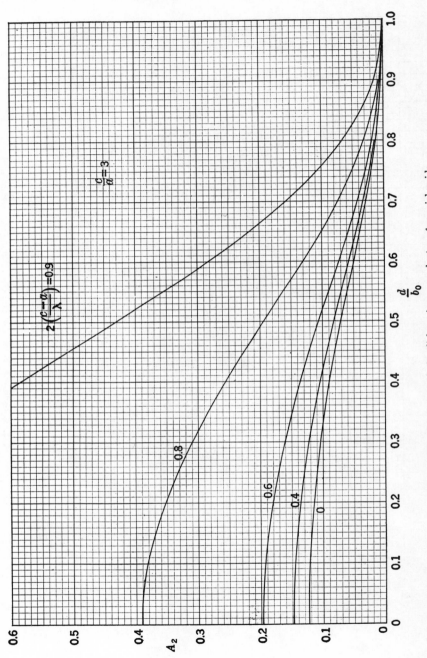

Fig. 5·3-3.—Coefficient A_2 for disk on inner conductor of coaxial guide.

SEC. 5·3] CAPACITIVE WINDOWS IN COAXIAL GUIDE

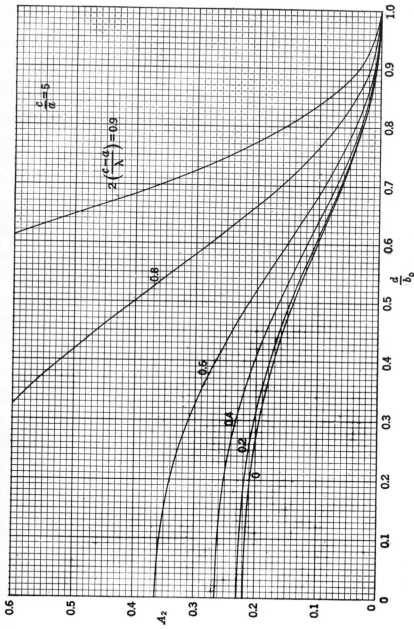

FIG. 5·3-4.—Coefficient A_2 for disk on inner conductor of coaxial guide.

λ_g and b therein are replaced by the $\lambda/2$ and b_0 of this section. For $c/a > 1$, $B\lambda/Y_0 2b_0$ may be obtained from its value for $c/a = 1$ by addition of the term A_2 and multiplication of the resulting sum by A_1. The wavelength-independent term A_1 is plotted in Fig. 5·3-2 as a function of d/b_0 for several values of c/a. In Figs. 5·3-3, 5·3-4 graphs of A_2 as a function of d/b_0 with $2b_0/\lambda$ as a parameter are shown for $c/a = 3$ and 5; $A_2 = 0$ for the case $c/a = 1$. The root $\chi(c/a)$ may be obtained from Table 2·3 (note that the c therein is the c/a of this section).

b. Disk on Outer Conductor.—A window formed by a metallic disk of zero thickness on the outer conductor of a coaxial line (principal mode in coaxial guide).

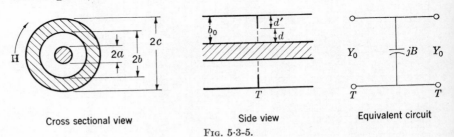

Cross sectional view · Side view · Equivalent circuit

FIG. 5·3-5.

Equivalent-circuit Parameters.—The equivalent-circuit parameters are the same as in Eqs. (1a) to (1c) of Sec. 5·3a except that now

$$d = b - a, \qquad d' = c - b, \qquad b_0 = c - a,$$

$$A_1 = \frac{a}{b} \frac{\ln \frac{c}{a}}{\frac{c}{a} - 1} \left(\frac{\frac{b}{a} - 1}{\ln \frac{b}{a}} \right)^2,$$

$$A_2 = \frac{\pi^2 \frac{c}{b}}{\gamma_1 \sqrt{1 - \left(\frac{2b_0}{\gamma_1 \lambda}\right)^2}} \frac{1 - \frac{a}{c}}{1 - \frac{J_0^2(\chi)}{J_0^2(\chi a/c)}} \left[\frac{J_0(\chi) N_0\left(\frac{\chi b}{c}\right) - N_0(\chi) J_0\left(\frac{\chi b}{c}\right)}{1 - \frac{a}{b}} \right]^2 - \frac{1}{\sqrt{1 - \left(\frac{2b_0}{\lambda}\right)^2}} \left(\frac{2}{\pi} \frac{b_0}{d} \sin \frac{\pi d}{b_0} \right)^2$$

and the root $\chi = \dfrac{\pi \gamma_1}{1 - a/c} = \dfrac{c}{a} \chi_{01}$ is defined by

$$J_0(\chi) N_0 \left(\frac{\chi a}{c}\right) - N_0(\chi) J_0 \left(\frac{\chi a}{c}\right) = 0.$$

Restrictions.—Same as in Sec. 5·3a.

SEC. 5·3] CAPACITIVE WINDOWS IN COAXIAL GUIDE

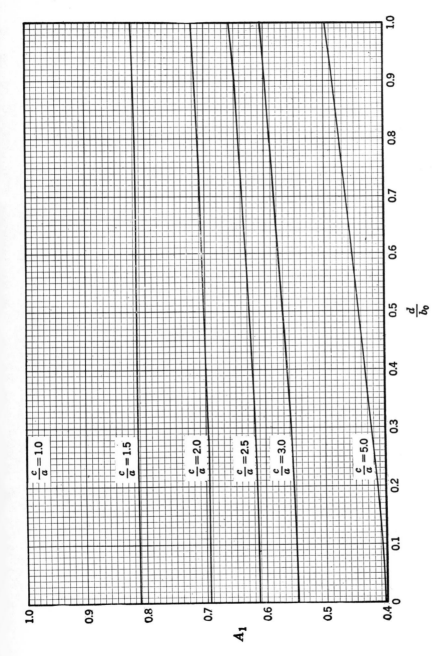

FIG. 5·3·6.—Coefficient A_1 for disk on outer conductor of coaxial guide.

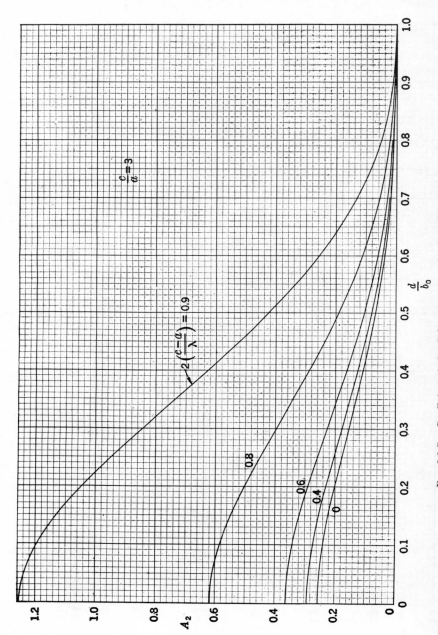

FIG. 5·3-7.—Coefficient A_2 for disk on outer conductor of coaxial guide.

SEC. 5·3] CAPACITIVE WINDOWS IN COAXIAL GUIDE   237

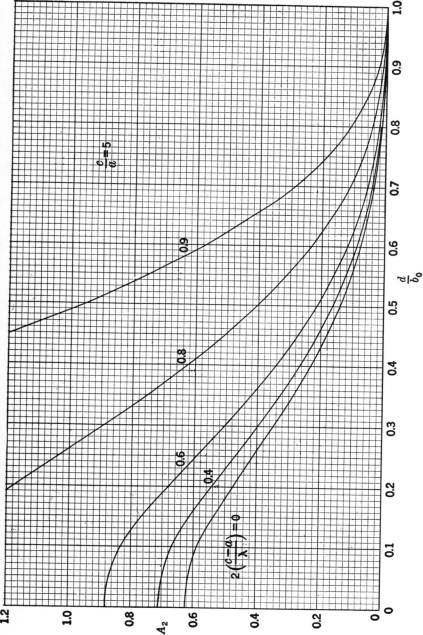

FIG. 5·3-8.—Coefficient A_2 for disk on outer conductor of coaxial guide.

Numerical Results.—For $c/a = 1$ the graph of $B\lambda/Y_0 2b_0$ as a function of d/b_0 with $2b_0/\lambda$ as a parameter may be obtained from Fig. 5·1-4 if λ_g and b therein are replaced by the $\lambda/2$ and b_0 of this section. For $c/a > 1$, $B\lambda/Y_0 2b_0$ may be obtained from its value for $c/a = 1$ by addition of the term A_2 and multiplication of the resulting sum by A_1. The wavelength-independent term A_1 is plotted in Fig. 5·3-6 as a function of d/b_0 for several values of c/a. In Figs. 5·3-7, 5·3-8 graphs of A_2 as a function of d/b_0 with $2b_0/\lambda$ as a parameter are shown for $c/a = 3$ and 5; $A_2 = 0$ for the case $c/a = 1$. The root χ_{01} may be obtained from Table 2·3 (note that the c therein is the c/a of this section).

5·4. Circular and Elliptical Apertures in Rectangular Guide. *a. Centered Circular Aperture.*—A centered circular aperture in a transverse metallic plate of zero thickness (H_{10}-mode in rectangular guide).

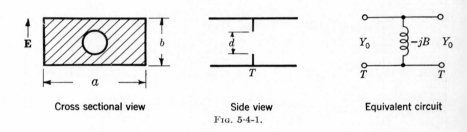

Cross sectional view   Side view   Equivalent circuit

Fig. 5·4-1.

Equivalent-circuit Parameters.—At the terminal plane T

$$\frac{B}{Y_0} = \frac{\lambda_g}{a}\left[\frac{\pi b}{24d\, j_1^2(x)} - 1 + A_1 - \left(\frac{a}{\lambda}\right)^2 A_2\right], \tag{1a}$$

$$\frac{B}{Y_0} \approx \frac{\lambda_g}{a}\left[\frac{\pi b}{24d\, j_1^2(x)} - 1\right] \approx \frac{3}{2\pi}\frac{ab\lambda_g}{d^3}, \qquad d \ll b, \tag{1b}$$

where

$$A_1 = \frac{b}{4d\, j_1^2(x)} \sum{}' (-1)^n \left[\frac{3\theta - 6\tan\frac{\theta}{2} + \tan^3\frac{\theta}{2}}{9}\right.$$

$$\left. - \left(\frac{\beta}{\alpha}\right)^2 \frac{1 - 2\cos\theta \sin^2\frac{\theta}{2} - \left(\tan\frac{\theta}{2} - 1\right)^2}{15\tan\frac{\theta}{2}}\right],$$

$$A_1 \approx -\frac{1}{96}\frac{bd}{a^3 j_1^2(x)} \sum{}' \frac{(-1)^n}{\alpha^3}\left[1 - \left(\frac{\beta}{\alpha}\right)^2\right], \qquad d \ll b,$$

CIRCULAR AND ELLIPTICAL APERTURES

$$A_2 = \frac{4b^3}{a^3 j_1^2(x)} \sum{}'' \left[n^2 + \frac{\left(\frac{mb}{a}\right)^2}{1 + \sqrt{1 - \left(\frac{2b}{\lambda\alpha}\right)^2}} \right] \frac{j_1^2\left(\frac{\alpha x a}{b}\right)}{\alpha^5},$$

$$A_2 \approx \frac{2b}{a j_1^2(x)} \left[\frac{\pi^3 d}{120 a}\left(1 - \frac{10}{3\pi}\frac{d}{b}\right) + \left(\frac{b}{a}\right)^4 \sum{}'' \frac{m^2 j_1^2\left(\frac{\alpha x a}{b}\right)}{\alpha^5} \right], \quad \begin{array}{l} d \ll b \\ b \ll \lambda \end{array}$$

and

$$x = \frac{\pi d}{2a}, \quad \alpha = \sqrt{n^2 + \left(\frac{mb}{a}\right)^2},$$

$$\sin\theta = \frac{d}{\alpha a}, \quad \beta = \sqrt{n^2 - \left(\frac{mb}{a}\right)^2},$$

$$j_1(x) = \frac{1}{x}\left(\frac{\sin x}{x} - \cos x\right)$$

$$\approx \frac{x}{3}, \quad x \ll 1,$$

$$\sum = \sum_{n=-\infty}^{+\infty} \sum_{m=-\infty}^{+\infty} \quad \text{(omit } n = 0, m = 0 \text{ term)},$$

$$\sum{}'' = \sum_{n=0,\pm 2,\pm 4,\cdots}^{\pm\infty} \sum_{m=1,3,5,\cdots}^{\infty} \quad \text{(omit } n = 0, m = 1 \text{ term)}.$$

Restrictions.—The equivalent circuit is valid in the wavelength range $2a > \lambda > 2a/3$ provided the H_{11}-mode is not propagating. The circuit parameter B/Y_0 has been computed by a variational method employing as an aperture field the static small-hole electric-field distribution. The error is estimated to lie within 10 per cent for $d < 0.9b$, provided λ is not too close to cutoff of the next propagating mode. The expressions (1b) are approximations to Eq. (1a) in the small-aperture range $d \ll b$. The restriction to zero thickness should be emphasized; thickness effects are considered in Sec. 8·10.

Numerical Results.—The quantity B/Y_0 of Eq. (1a) has been plotted as a function of d/a in Fig. 5·4-2 for various values of λ/a and a/b.

b. *Small Elliptical or Circular Aperture.*—A small elliptical or circular aperture in a transverse metallic plate of zero thickness (H_{10}-mode in rectangular guide).

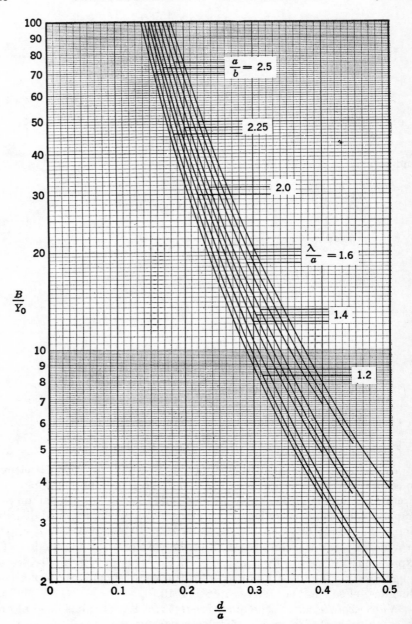

FIG. 5·4-2.—Relative susceptance of centered circular aperture.

CIRCULAR AND ELLIPTICAL APERTURES

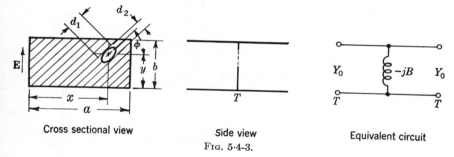

Fig. 5·4-3.

Equivalent-circuit Parameters.—For an elliptical aperture

$$\frac{B}{Y_0} = \frac{\lambda_g}{a}\left(\frac{a^2 b}{4\pi M} - 1\right), \tag{2}$$

where

$$M = (M_1 \cos^2 \phi + M_2 \sin^2 \phi) \sin^2 \frac{\pi x}{a} \tag{3}$$

and

$$M_1 = \frac{d_1^3}{6}\frac{\pi}{4}\frac{\epsilon^2}{F(\epsilon) - E(\epsilon)}, \tag{4a}$$

$$M_1 \approx \frac{d_1^3}{6}\frac{\pi}{4}\frac{1}{\ln\frac{4d_1}{d_2} - 1}, \quad \epsilon \ll 1, \tag{4b}$$

$$M_2 = \frac{d_1 d_2^2}{6}\frac{\pi}{4}\frac{\epsilon^2}{E(\epsilon) - (1 - \epsilon^2)F(\epsilon)}, \tag{5a}$$

$$M_2 \approx \frac{d_1 d_2^2}{6}\frac{\pi}{4}, \quad \epsilon \ll 1, \tag{5b}$$

$$\epsilon = \sqrt{1 - \left(\frac{d_2}{d_1}\right)^2},$$

and $F(\epsilon)$ and $E(\epsilon)$ are complete elliptic integrals of the first and second kinds, respectively.

For the special case of a circular aperture $d_1 = d_2 = d$ and

$$M = \frac{1}{6}d^3 \sin^2 \frac{\pi x}{a}. \tag{6}$$

Restrictions.—The equivalent circuit is valid in the wavelength range $2a > \lambda > a$. The parameter B/Y_0 has been obtained by an integral equation method employing the small-aperture assumptions: the largest dimension of the aperture is small compared with λ/π, the aperture is relatively far removed from the guide walls, and the wavelength is not too close to cutoff of the second mode. In this approxi-

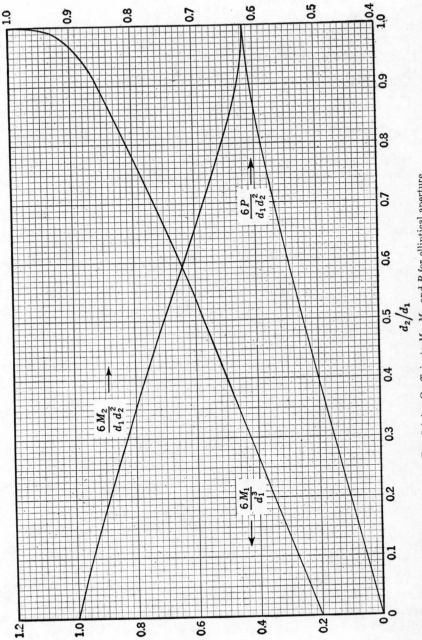

Fig. 5·4·4.—Coefficients M_1, M_2, and P for elliptical aperture.

mation B/Y_0 is independent of position along the direction of electric field. The approximations (4b) and (5b) are asymptotic forms of Eqs. (4a) and (5a) and are valid for small eccentricities ϵ.

Numerical Results.—The coefficients M_1 and M_2, plotted as a function of d_2/d_1, may be obtained from Fig. 5·4-4.

5·5. Elliptical and Circular Apertures in Circular Guide.—A small elliptical or circular aperture in a transverse metallic plate of zero thickness (H_{11}-mode in circular guide).

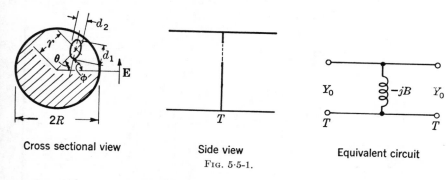

Cross sectional view   Side view   Equivalent circuit
FIG. 5·5-1.

Equivalent-circuit Parameters.—At the reference plane T for an elliptical aperture of orientation $\phi = 0$ or $\pi/2$ and angular positions $\theta = 0$, $\pi/2$, π, or $3\pi/2$

$$\frac{B}{Y_0} = \frac{\lambda_g}{4R}\left[\frac{(2R)^3}{8.40M} - 2.344\right], \tag{1}$$

where

$$M = M_1\left[2J_1'(\alpha r)\cos\theta\,\cos(\phi - \theta) - 2\frac{J_1(\alpha r)}{\alpha r}\sin\theta\,\sin(\phi - \theta)\right]^2$$
$$+ M_2\left[2J_1'(\alpha r)\cos\theta\,\sin(\phi - \theta) + 2\frac{J_1(\alpha r)}{\alpha r}\sin\theta\,\cos(\phi - \theta)\right]^2, \tag{2}$$

$$\alpha R = 1.841, \qquad \lambda_g = \frac{\lambda}{\sqrt{1 - \left(\dfrac{\lambda}{3.412R}\right)^2}},$$

and the coefficients M_1 and M_2 are defined in Eqs. (4) and (5) of Sec. 5·4b.

For a centered elliptical aperture, $r = 0$ and Eq. (2) reduces to

$$M = M_1\cos^2\phi + M_2\sin^2\phi. \tag{3}$$

For a circular aperture, $d_1 = d_2 = d$ and Eq. (2) reduces to

$$M = \frac{d^3}{6}\left\{[2J_1'(\alpha r)\cos\theta]^2 + \left[\frac{2J_1(\alpha r)}{\alpha r}\sin\theta\right]^2\right\}. \tag{4}$$

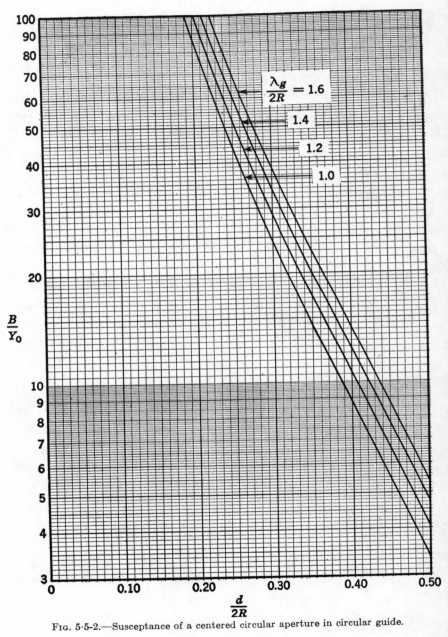

Fig. 5·5-2.—Susceptance of a centered circular aperture in circular guide.

SEC. 5·5] ELLIPTICAL AND CIRCULAR APERTURES 245

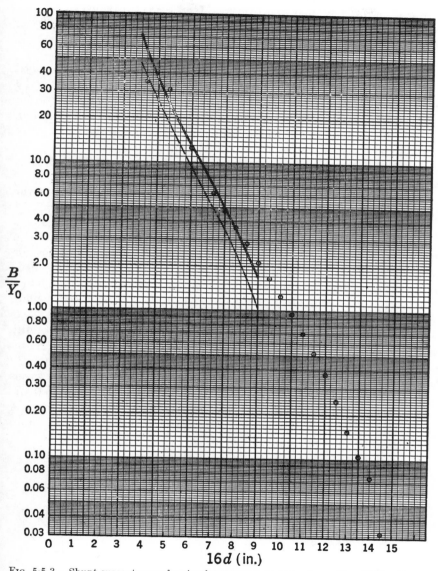

FIG. 5·5-3.—Shunt susceptance of a circular aperture in a circular guide of $\frac{15}{16}$ in. diameter. (Points, experimental; curves, theoretical.)

For a centered circular aperture, $r = 0$ and Eq. (2) reduces to

$$M = \frac{d^3}{6}. \tag{5}$$

Restrictions.—The equivalent circuit is applicable in the wavelength range $2.61R < \lambda < 3.41R$. The above-mentioned restrictions on ϕ and θ serve to ensure that only a single H_{11}-mode of the indicated polarization can be propagated. The comments mentioned under *Restrictions* in Sec. 5·4b also apply to this case except that B/Y_0 is not independent of position along the direction of electric-field intensity.

Numerical Results.—The coefficients M_1 and M_2 are plotted as a function of d_2/d_1 in Fig. 5·4-4. The relative susceptance B/Y_0, as obtained from Eqs. (1) and (5), is plotted vs. $d/2R$ in Fig. 5·5-2 for the special case of a centered circular aperture in the small aperture range.

Experimental Results.—Measurements of B/Y_0 taken at $\lambda = 3.20$ cm in a circular guide of $\frac{15}{16}$ in. diameter are shown by the circled points in Fig. 5·5-3. These rough data apply to centered apertures in a transverse metallic plate of $\frac{1}{32}$ in. thickness. For comparison the dotted curve shows values of B/Y_0 obtained from Eq. (1) with the aperture diameter d replaced by $(d - \frac{1}{32}$ in.) to account approximately for the effect of plate thickness. The solid curve is a corresponding plot of B/Y_0 vs. d as obtained from Eq. (1) for the case of zero thickness. For plates of finite thickness (see Sec. 5·16) the equivalent circuit should be represented by a tee rather than by a simple shunt circuit at the terminal plane T; for the $\frac{1}{32}$ in. thick plate the relative series reactance of the tee varies approximately linearly from 0 to about 0.05 as d varies from $2R$ to $0.4R$.

5·6. Small Elliptical and Circular Apertures in Coaxial Guide.—A small elliptical or circular aperture in a transverse metallic plate of zero thickness (principal mode in coaxial guide).

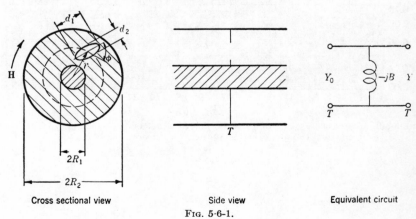

Cross sectional view   Side view   Equivalent circuit

FIG. 5·6-1.

SEC. 5·7] ANNULAR WINDOW IN CIRCULAR GUIDE

Equivalent-circuit Parameters.—At the reference plane T

$$\frac{B}{Y_0} = \frac{\lambda r^2 \ln \frac{R_2}{R_1}}{M}, \tag{1}$$

where

$$M = M_1 \cos^2 \phi + M_2 \sin^2 \phi. \tag{2}$$

The coefficients M_1 and M_2 are given as a function of d_2/d_1 in Eqs. (4) and (5) of Sec. 5·4b. For the special case of a circular aperture, $d_1 = d_2 = d$ and Eq. (2) becomes

$$M = \frac{d^3}{6}. \tag{3}$$

Restrictions.—The equivalent circuit is applicable as long as only the principal mode can be propagated. Otherwise, comments are the same as in Sec. 5·4b.

Numerical Results.—The coefficients M_1 and M_2 are plotted as a function of d_2/d_1 in Fig. 5·4-4.

5·7. Annular Window in Circular Guide.—An annular window in a metallic plate of zero thickness (H_{01}-mode in circular guide).

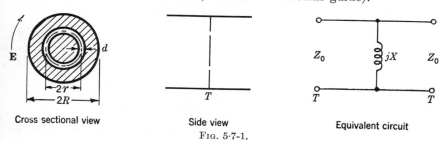

Cross sectional view   Side view   Equivalent circuit
FIG. 5·7-1.

Equivalent-circuit Parameters.—At the terminal plane T

$$\frac{X}{Z_0} = \frac{r}{\lambda_g}\left(\frac{\pi d}{2R}\right)^2 \frac{J_1^2\left(\frac{3.83r}{R}\right)}{0.162}, \tag{1}$$

where

$$\lambda_g = \frac{\lambda}{\sqrt{1 - \left(\frac{\lambda}{1.64R}\right)^2}}.$$

Restrictions.—The equivalent circuit is valid in the wavelength range $0.896R < \lambda < 1.64R$. It is to be noted that modes other than the H_{01}-mode can be propagated in this range. The equivalent circuit describes the junction effect for the H_{01}-mode only. Equation (1) is obtained by an integral equation method employing the small-aperture approximations and is applicable in the above wavelength range with an estimated error of less than 10 per cent if $d \leq 0.1R$ and $0.3R \leq r \leq 0.8R$. The error, which is estimated by comparison of Eq. (1) with a more accurate numerical variational expression, becomes smaller as r approaches $0.55R$.

Numerical Results.—In Fig. 5·7-2 the quantity $X\lambda_g/Z_0R$ is plotted as a function of r/R with d/R as a parameter.

5·8. Annular Obstacles in Circular Guide.—An annular metallic strip of zero thickness (E_{01}-mode in circular guide).

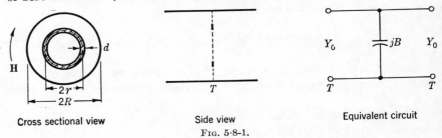

Cross sectional view   Side view   Equivalent circuit
Fig. 5·8-1.

Equivalent-circuit Parameters.—At the terminal plane T

$$\frac{B}{Y_0} = \frac{r}{\lambda_g} \left(\frac{\pi d}{R}\right)^2 \frac{J_1^2\left(2.405\, \frac{r}{R}\right)}{0.269}, \tag{1}$$

where

$$\lambda_g = \frac{\lambda}{\sqrt{1 - \left(\dfrac{\lambda}{2.61R}\right)^2}}.$$

Restrictions.—The equivalent circuit is valid in the wavelength range $1.14R < \lambda < 2.61R$. Equation (1) has been obtained by an integral equation method employing small-obstacle approximations. It is to be noted that modes other than the E_{01}-mode can be propagated in this range. The equivalent circuit describes the junction effect for the E_{01}-mode only. Equation (1) is valid for strips of small width; the error is estimated to be within 10 per cent if $d \leq 0.10R$ and $0.3R \leq r \leq 0.7R$.

Numerical Results.—In Fig. 5·8-2 the quantity $B\lambda_g/Y_0R$ is plotted as a function of r/R with d/R as a parameter.

STRUCTURES WITH FINITE THICKNESS

5·9. Capacitive Obstacles of Finite Thickness. *a. Window Formed by Two Obstacles.*—Window formed by obstacles of small but finite thickness with edges perpendicular to the electric field (H_{10}-mode in rectangular guide).

SEC. 5·7]   ANNULAR WINDOW IN CIRCULAR GUIDE   249

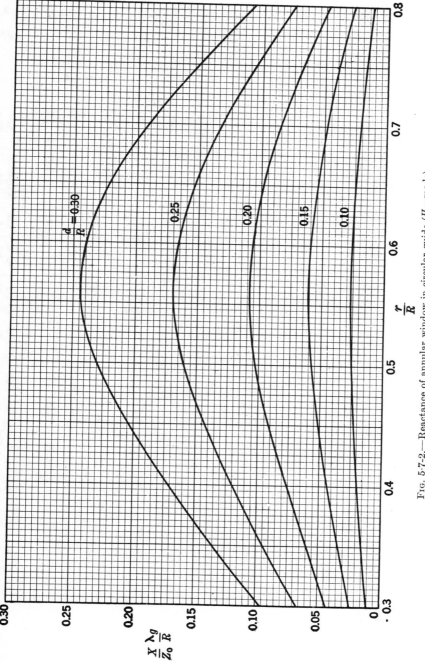

FIG. 5·7-2.—Reactance of annular window in circular guide (H_{01}-mode).

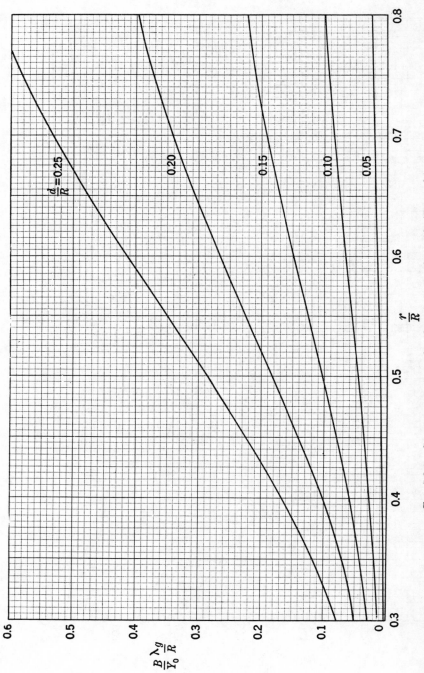

Fig. 5·8-2.—Susceptance of annular obstacle in circular guide (E_{01}-mode).

CAPACITIVE OBSTACLES OF FINITE THICKNESS

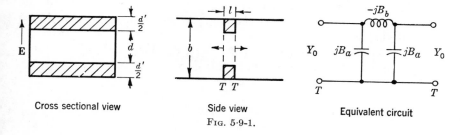

Fig. 5·9-1.

Equivalent-circuit Parameters.—At the reference planes T

$$\frac{B_a}{Y_0} = \frac{B_1}{Y_0} + \frac{\dot{b}}{d} \tan \frac{\pi l}{\lambda_g}, \qquad (1)$$

$$\frac{B_b}{Y_0} = \frac{b}{d} \csc \frac{2\pi l}{\lambda_g}, \qquad (2)$$

where

$$\frac{B_1}{Y_0} = \frac{2b}{\lambda_g} \left[\ln \sec\left(\frac{\pi}{2} \frac{d'}{b} g\right) - \frac{\pi}{2} \frac{d'l}{bd} + \frac{A \sin^4 \frac{\pi d'}{2b}}{1 + A \cos^4 \frac{\pi d'}{2b}} \right.$$

$$\left. + \frac{1}{16} \left(\frac{b}{\lambda_g}\right)^2 \left(1 - 3 \cos^2 \frac{\pi d'}{2b}\right)^2 \sin^4 \frac{\pi d'}{2b} \right], \quad (3a)$$

$$\frac{B_1}{Y_0} \approx \frac{b}{\lambda_g} \left[\left(\frac{\pi d'}{2b} g\right)^2 + \frac{1}{6}\left(\frac{\pi d'}{2b} g\right)^4 - \frac{\pi}{2} \frac{d'}{b} \frac{l}{d} + \frac{3}{2}\left(\frac{b}{\lambda_g}\right)^2 \left(\frac{\pi d'}{2b}\right)^4 \right], \quad (3b)$$

and

$$g = \frac{\alpha'}{E(\alpha') - \alpha^2 F(\alpha')} \approx 1 + \frac{l}{\pi d'} \ln\left(\frac{4\pi}{e} \cdot \frac{d'}{l}\right) \text{ for } \frac{l}{d'} \ll 1,$$

$$\frac{l}{d'} = \frac{E(\alpha) - \alpha'^2 F(\alpha)}{E(\alpha') - \alpha^2 F(\alpha')},$$

$$A = \frac{1}{\sqrt{1 - \left(\dfrac{b}{\lambda_g}\right)^2}} - 1, \qquad \alpha^2 = 1 - \alpha'^2.$$

Restrictions.—The equivalent circuit is valid in the wavelength range $b/\lambda_g < 1$. The above equations have been obtained by the equivalent static method in which the two lowest modes have been treated correctly. Equation (2) is in error by less than 2 per cent. Equation (3a) is in error by less than 5 per cent when $d'/b < 0.5$ and

$l/d < 0.5$; for $d'/b > 0.5$ or $l/d > 0.5$ the results of Sec. 8·8 should be employed. The approximation (3b) agrees with Eq. (3a) to within 5 per cent when $d'/b < 0.5$ and $b/\lambda_g < 0.4$.

Numerical Results.—The quantity $B_1\lambda_g/Y_0 b$ may be obtained as a function of $d/b = 1 - d'/b$ by the addition of

$$2 \ln \frac{\sec\left(\frac{\pi d'}{2b} g\right)}{\sec \frac{\pi}{2} \frac{d'}{b}} - \pi \frac{d'l}{bd}$$

to one-half the quantity $B\lambda_g/Y_0 b$ plotted in Fig. 5·1-4. The quantity g plotted as a function of l/d' may be obtained from Figs. 5·9-4 and 5·9-5.

b. *Window Formed by One Obstacle.*—Window formed by an obstacle of small but finite thickness with edges perpendicular to the electric field (H_{10}-mode in rectangular guide).

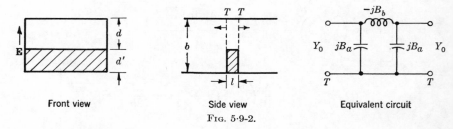

Fig. 5·9-2.

Equivalent-circuit Parameters.—Same as in Sec. 5·9a except that λ_g is replaced by $\lambda_g/2$.

Restrictions.—Same as in Sec. 5·9a except that λ_g is replaced by $\lambda_g/2$.

Numerical Results.—If λ_g in Fig. 5·1-4 is everywhere replaced by $\lambda_g/2$, there is obtained a plot of $B\lambda_g/Y_0 2b$ as a function of d/b with $2b/\lambda_g$ as a parameter. The addition of

$$2 \ln \frac{\sec\left(\frac{\pi}{2} \frac{d'}{b} g\right)}{\sec \frac{\pi}{2} \frac{d'}{b}} - \frac{\pi d' l}{bd}$$

to one-half $B\lambda_g/Y_0 2b$ yields the quantity $B_1\lambda_g/Y_0 2b$. In Figs. 5·9-4 and 5·9-5, g is plotted as a function of l/d'.

c. *Symmetrical Obstacle.*—A symmetrical obstacle of small but finite thickness with edges perpendicular to the electric field (H_{10}-mode in rectangular guide).

SEC. 5·9] CAPACITIVE OBSTACLES OF FINITE THICKNESS 253

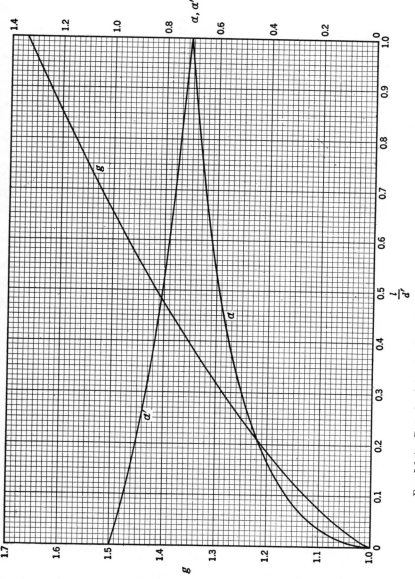

FIG. 5·9·4.—Data for determination of equivalent circular diameters of rectangular posts.

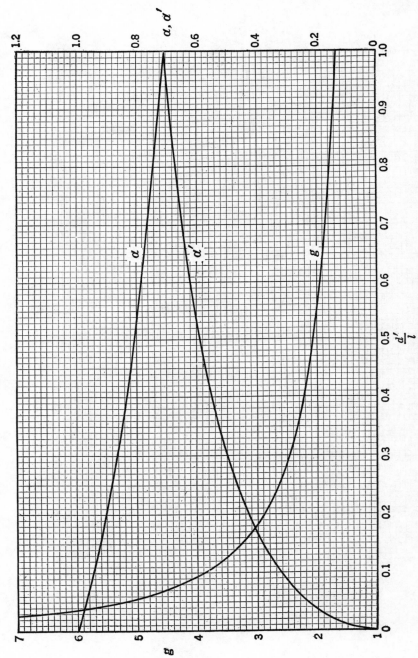

Fig. 5·9-5.—Data for determination of equivalent circular diameters of rectangular posts.

SEC. 5·10] INDUCTIVE OBSTACLES OF FINITE THICKNESS

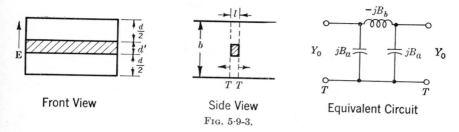

Front View   Side View   Equivalent Circuit

FIG. 5·9-3.

Equivalent-circuit Parameters.—Same as in Sec. 5·9a.
Restrictions.—Same as in Sec. 5·9a.
Numerical Results.—Same as in Sec. 5·9a.

5·10. Inductive Obstacles of Finite Thickness. (a) *Symmetrical Window.*—A symmetrical window formed by obstacles of elliptical or rectangular cross section with edges parallel to the electric field (H_{10}-mode in rectangular guide).

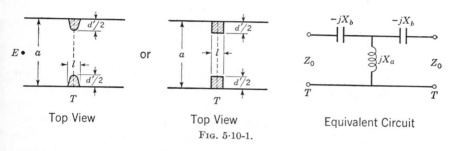

Top View   Top View   Equivalent Circuit

FIG. 5·10-1.

Equivalent-circuit Parameters.—At the terminal plane T

$$\frac{X_a}{Z_0} = \frac{2a}{\lambda_g}\left(\frac{a}{\pi D'}\right)^2, \qquad \frac{\pi D'}{\lambda} \ll 1, \tag{1}$$

$$\frac{X_b}{Z_0} = \frac{a}{8\lambda_g}\left(\frac{\pi D_1}{a}\right)^4, \qquad \frac{\pi D_1}{\lambda} \ll 1. \tag{2}$$

For a window formed by obstacles of elliptical cross section the equivalent circular diameters D' and D_1 are

$$D' = \sqrt{\frac{d'(l+d')}{2}}, \tag{3}$$

$$D_1 = \sqrt[4]{\frac{ld'(l+d')^2}{2}}. \tag{4}$$

For a window formed by obstacles of rectangular cross section the equivalent circular diameter D' is

$$D' = \frac{d'}{\sqrt{2}} \frac{\alpha'}{E(\alpha') - \alpha^2 F(\alpha')}, \tag{5a}$$

$$D' \approx \frac{d'}{\sqrt{2}} \left(1 + \frac{l}{\pi d'} \ln \frac{4\pi d'}{el}\right), \qquad \frac{l}{d'} \ll 1, \tag{5b}$$

where α is determined by

$$\frac{l}{d'} = \frac{E(\alpha) - \alpha'^2 F(\alpha)}{E(\alpha') - \alpha^2 F(\alpha')}, \qquad \alpha = \sqrt{1 - \alpha'^2}, \qquad e = 2.718;$$

the equivalent circular diameter D_1 is

$$D_1 = \sqrt[4]{\frac{\alpha^2 \alpha'^2}{3}} \frac{l}{E(\alpha') - \alpha^2 F(\alpha')} = \sqrt[4]{\frac{\alpha^2 \alpha'^2}{3}} \frac{d'}{E(\alpha) - \alpha'^2 F(\alpha)}, \tag{6a}$$

$$D_1 \approx \sqrt[4]{\frac{4}{3\pi} l d'^3}, \qquad \frac{l}{d'} \ll 1, \tag{6b}$$

where α is determined by

$$\frac{l}{d'} = \frac{E(\alpha') - \alpha^2 F(\alpha')}{E(\alpha) - \alpha'^2 F(\alpha)}, \qquad \alpha = \sqrt{1 - \alpha'^2}.$$

The functions $F(\alpha)$ and $E(\alpha)$ are the complete elliptic integrals of the first and second kind, respectively.

Restrictions.—The equivalent circuit is valid in the wavelength range $2a > \lambda > 2a/3$. The circuit parameters have been evaluated by an integral-equation method employing small-obstacle approximations. These evaluations have been presented only in the static approximation $\pi D'/\lambda$ and $\pi D_1/\lambda \ll 1$; the wavelength correction to Eq. (1) is of the same order as that in Eq. (1a) of Sec. 5·2a (note that $d'/2$ of this section is the d' of Sec. 5·2a). The equivalent circular diameters D' and D_1 have been evaluated by conformal mapping methods. In the range D', $D_1 < 0.2a$ and $a < \lambda$ Eqs. (1) and (2) are estimated to be in error by less than 10 per cent.

Numerical Results.—The equivalent circular diameters for obstacles of rectangular cross section may be obtained from the curves of Figs. 5·9-4 and 5·9-5. The relation between $\sqrt{2}\, D'/d'$ and l/d' given in Eq. (5a) is identical with the g vs. l/d' relation plotted in Fig. 5·9-4. The D_1/l vs. d'/l relation given in Eq. (6a) may be obtained from the α, α', g vs. l/d' curves of Figs. 5·9-4 and 5·9-5 provided the l/d' therein is replaced by d'/l.

b. *Asymmetrical Window.*—An asymmetrical window formed by an obstacle of elliptical or rectangular cross section with edges parallel to the electric field (H_{10}-mode in rectangular guide).

SEC. 5·11] SOLID INDUCTIVE POST IN RECTANGULAR GUIDE

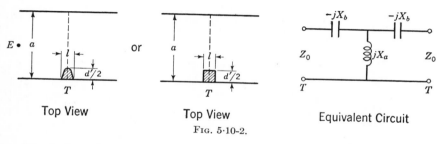

Fig. 5·10-2.

Equivalent-circuit Parameters.—At the terminal plane T

$$\frac{X_a}{Z_0} = \frac{4a}{\lambda_g}\left(\frac{a}{\pi D'}\right)^2, \qquad \frac{\pi D'}{\lambda} \ll 1, \tag{7}$$

$$\frac{X_b}{Z_0} = \frac{a}{16\lambda_g}\left(\frac{\pi D_1}{a}\right)^2, \qquad \frac{\pi D_1}{\lambda} \ll 1. \tag{8}$$

For obstacles of elliptical or rectangular cross section the equivalent circular diameters D' and D_1 are given in Eqs. (3) to (6).

Restrictions.—The equivalent circuit is valid in the wavelength range $a < \lambda < 2a$. The circuit parameters have been evaluated by an integral-equation method subject to small-obstacle approximations. These evaluations are given in Eqs. (7) and (8) only in the static approximation $\pi D'/\lambda$, $\pi D_1/\lambda \ll 1$; the wavelength correction to Eq. (7) is of the same order as that in Eq. (1a) of Sec. 5·2b (note that the $d'/2$ of this section is the d' of Sec. 5·2b). For D', $D_1 < 0.2a$ and $a < \lambda$ Eqs. (7) and (8) are estimated to be in error by less than 10 per cent.

Numerical Results.—Same as in Sec. 5·10a.

5·11. Solid Inductive Post in Rectangular Guide. *a. Off-centered Post.*—A solid metallic obstacle of circular cross section with axis parallel to the electric field (H_{10}-mode in rectangular guide).

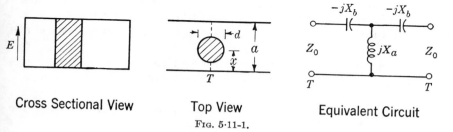

Fig. 5·11-1.

Equivalent-circuit Parameters.—At the reference plane T

$$\frac{X_a}{Z_0} - \frac{X_b}{2Z_0} = \frac{a}{2\lambda_g}\csc^2\frac{\pi x}{a}\left[S_0 - \left(\frac{\pi d}{2\lambda}\right)^2 - \left(\frac{\pi d}{2a}\right)^2\left(S_0\cot\frac{\pi x}{a} - S_1\right)^2\right], \tag{1}$$

$$\frac{X_b}{Z_0} = \frac{a}{\lambda_g}\left(\frac{\pi d}{a}\right)^2 \sin^2\frac{\pi x}{a}, \tag{2}$$

where

$$S_0 = \ln\left(\frac{4a}{\pi d}\sin\frac{\pi x}{a}\right) - 2\sin^2\frac{\pi x}{a} + 2\sum_{n=2}^{\infty}\sin^2\frac{n\pi x}{a}\left[\frac{1}{\sqrt{n^2-\left(\frac{2a}{\lambda}\right)^2}} - \frac{1}{n}\right],$$

$$S_1 = \frac{1}{2}\cot\frac{\pi x}{a} - \sin\frac{2\pi x}{a} + \sum_{n=2}^{\infty}\sin\frac{2n\pi x}{a}\left[\frac{n}{\sqrt{n^2-\left(\frac{2a}{\lambda}\right)^2}} - 1\right].$$

Restrictions.—The equivalent circuit is applicable in the wavelength range $a < \lambda < 2a$. The circuit parameters indicated in Eqs. (1) and (2) have been obtained by a variational method employing a constant, a cosine, and a sine term in the expression for the obstacle current. These equations represent the first two terms of an expansion in powers of $(\pi d/a)^2$. Although it is difficult to state the limits of accuracy, the results are believed to be reliable to within a few per cent for $d/a < 0.10$ and $0.8 > x/a > 0.2$. A more precise value for the case $x/a = 0.5$ is given in Sec. 5·11b.

Numerical Results.—Numerical values of X_a/Z_0, as obtained from Eqs. (1) and (2), are given as a function of d/a with x/a as a parameter in the graphs of Figs. 5·11-2, 5·11-3, and 5·11-4. These curves refer to the wavelengths $\lambda/a = 1.4$, 1.2, and 1.1, respectively. Values of the parameter $X_b\lambda_g/2Z_0 a$ may be obtained as a function of d/a from the curve in Fig. 5·11-5 on multiplication of the ordinates in the latter by $\sin^2(\pi x/a)$.

Experimental Results.—Measurements of the circuit parameters were performed at $\lambda = 3.20$ cm in a rectangular guide of dimensions $a = 0.90$ in. and $b = 0.40$ in. on posts for which $d/a \approx 0.04$. These data agree to within an experimental accuracy of a few per cent with the above theory in the range $0.2 < x/a < 0.8$.

b. *Centered Post.*—A symmetrically located post of circular cross section aligned parallel to the electric field (H_{10}-mode in rectangular guide).

Equivalent-circuit Parameters.—At the reference plane T and for $x = a/2$ (cf. Fig. 5·11-1)

$$\frac{X_a}{Z_0} - \frac{X_b}{2Z_0} = \frac{a}{2\lambda_g}\left[S_0 - \left(\frac{\pi d}{2\lambda}\right)^2 - \frac{5}{8}\left(\frac{\pi d}{2\lambda}\right)^4 - 2\left(\frac{\pi d}{2\lambda}\right)^4\left(S_2 - 2S_0\frac{\lambda^2}{\lambda_g^2}\right)^2\right], \tag{3}$$

SEC. 5·11] SOLID INDUCTIVE POST IN RECTANGULAR GUIDE

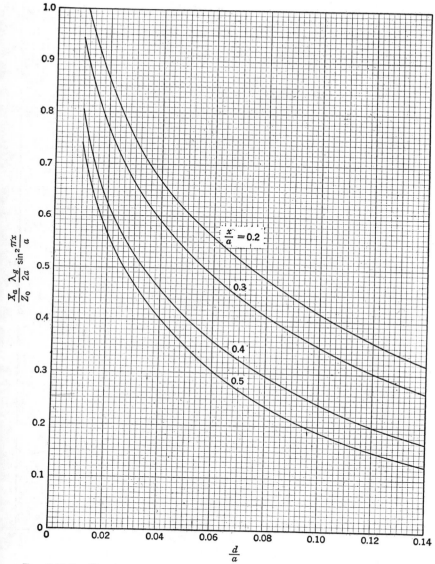

FIG. 5·11-2.—Shunt reactance of inductive post in rectangular guide ($\lambda/a = 1.4$).

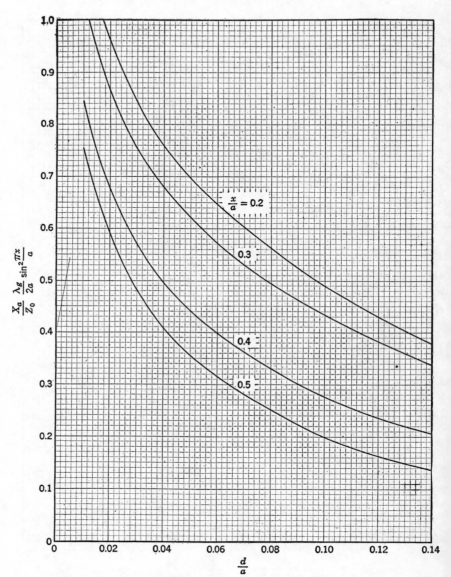

Fig. 5·11-3.—Shunt reactance of inductive post in rectangular guide ($\lambda/a = 1.2$).

SEC. 5·11] SOLID INDUCTIVE POST IN RECTANGULAR GUIDE

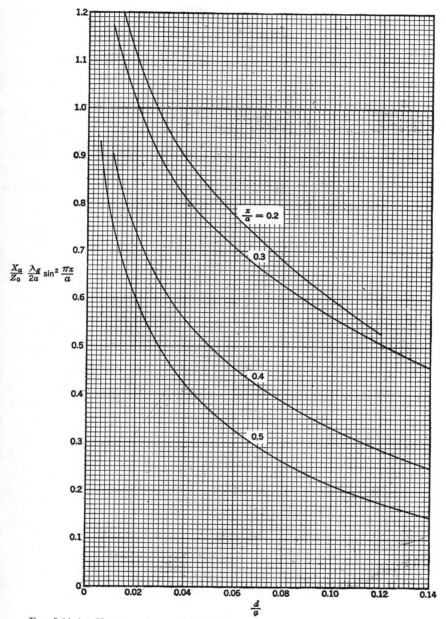

FIG. 5·11-4.—Shunt reactance of inductive post in rectangular guide ($\lambda/a = 1.1$).

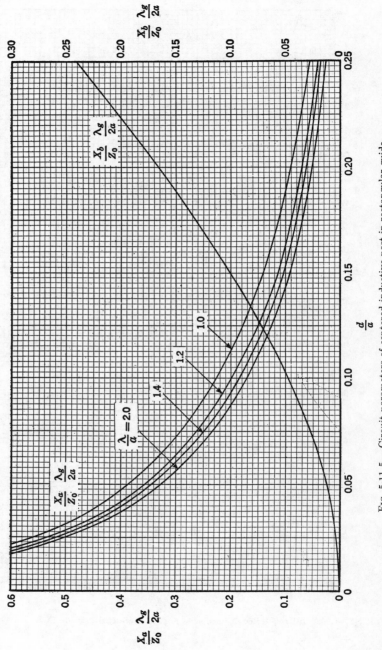

FIG. 5·11-5.—Circuit parameters of centered inductive post in rectangular guide.

SEC. 5·11] SOLID INDUCTIVE POST IN RECTANGULAR GUIDE

$$\frac{X_b}{Z_0} = \frac{a}{\lambda_g} \frac{\left(\frac{\pi d}{a}\right)^2}{1 + \frac{1}{2}\left(\frac{\pi d}{\lambda}\right)^2 \left(S_2 + \frac{3}{4}\right)}, \quad (4a)$$

$$\frac{X_b}{Z_0} \approx \frac{a}{\lambda_g} \frac{\left(\frac{\pi d}{a}\right)^2}{1 + \frac{11}{24}\left(\frac{\pi d}{a}\right)^2}, \quad (4b)$$

where

$$S_0 = \ln \frac{4a}{\pi d} - 2 + 2 \sum_{n=3,5,\cdots}^{\infty} \left[\frac{1}{\sqrt{n^2 - \left(\frac{2a}{\lambda}\right)^2}} - \frac{1}{n}\right],$$

$$S_2 = \ln \frac{4a}{\pi d} - \frac{5}{2} + \frac{11}{3}\left(\frac{\lambda}{2a}\right)^2 - \left(\frac{\lambda}{a}\right)^2 \sum_{n=3,5,\cdots}^{\infty} \left[\sqrt{n^2 - \left(\frac{2a}{\lambda}\right)^2} - n + \frac{2}{n}\left(\frac{a}{\lambda}\right)^2\right].$$

Restrictions.—The equivalent circuit is valid in the wavelength range $2a > \lambda > 2a/3$. Equations (3) and (4) have been evaluated by a variational method employing one stage of approximation beyond the results presented in Sec. 5·11a. These equations are accurate to within a few per cent for $d/a < 0.20$. Equation (4b) differs from Eq. (4a) by less than 3 per cent in the wavelength range $2a > \lambda > a$ and for $d/a < 0.2$.

Numerical Results.—Curves of $X_a \lambda_g / Z_0 2a$ vs. d/a are plotted in Fig. 5·11-5 for several values of λ/a. A graph of $X_b \lambda_g / Z_0 2a$ vs. d/a, as obtained from Eq. (4b), is also plotted in Fig. 5·11-5.

Experimental Results.—Measurements of the circuit parameters were carried out at guide wavelengths of 1.76 and 2.00 in. in a rectangular guide of dimensions $a = 0.90$ in. and $b = 0.40$ in. These data agree with the theoretical values for the range $d/a < 0.15$ to within an experimental accuracy of a few per cent.

c. Noncircular Posts.—A solid post of elliptical or rectangular cross section with axis parallel to the electric field (H_{10}-mode in rectangular guide).

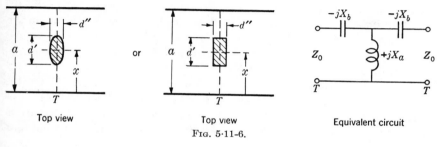

Top view   Top view   Equivalent circuit

FIG. 5·11-6.

Equivalent-circuit Parameters.—At the reference plane T

$$\frac{X_a}{Z_0} - \frac{X_b}{2Z_0} = \frac{a}{2\lambda_g} \csc^2 \frac{\pi x}{a} \left\{ \ln\left(\frac{4a}{\pi d_0} \sin \frac{\pi x}{a}\right) - 2 \sin^2 \frac{\pi x}{a} \right.$$

$$\left. + \sum_{n=2}^{\infty} 2 \sin^2 \frac{n\pi x}{a} \left[\frac{1}{\sqrt{n^2 - \left(\frac{2a}{\lambda}\right)^2}} - \frac{1}{n} \right] \right\}, \quad (5)$$

$$\frac{X_b}{Z_0} = \frac{a}{\lambda_g} \left(\frac{\pi d_1}{a}\right)^2 \sin^2 \frac{\pi x}{a}. \quad (6)$$

For a post of elliptical cross section

$$d_0 = \frac{d' + d''}{2}, \quad d_1 = \sqrt{d'' d_0}. \quad (7)$$

For a post of rectangular cross section

$$d_0 = \frac{d'}{2} \frac{1}{E(\alpha) - \alpha'^2 F(\alpha)} = \frac{d''}{2} \frac{1}{E(\alpha') - \alpha^2 F(\alpha')}, \quad (8a)$$

$$d_0 \approx \frac{d'}{2} \left[1 + \frac{d''}{\pi d'} \ln\left(4\pi e \frac{d'}{d''}\right) \right], \quad \frac{d''}{d'} \ll 1, \quad (8b)$$

$$d_1 = \frac{d''}{\sqrt{2}} \frac{\alpha'}{E(\alpha') - \alpha^2 F(\alpha')}, \quad (9a)$$

$$d_1 \approx \frac{d''}{\sqrt{2}} \left[1 + \frac{d'}{\pi d''} \ln\left(\frac{4\pi}{e} \frac{d''}{d'}\right) \right], \quad \frac{d'}{d''} \ll 1, \quad (9b)$$

where α is given by

$$\frac{d''}{d'} = \frac{E(\alpha') - \alpha^2 F(\alpha')}{E(\alpha) - \alpha'^2 F(\alpha)}, \quad \alpha' = \sqrt{1 - \alpha^2}, \quad e = 2.718,$$

and $F(\alpha)$ and $E(\alpha)$ are complete elliptic integrals of the first and second kinds, respectively.

Restrictions.—The equivalent circuit is valid in the wavelength range $2a > \lambda > a$. The circuit parameters X_a and X_b have been evaluated by a variational method employing the small-obstacle approximations $d_0 \ll \lambda$, as is evident by comparison with the results of Sec. 5·11a. The "equivalent circular diameters" d_0 and d_1 have been computed in the static approximation by conformal mapping methods. The above results are estimated to be correct to within a few per cent in the range $d_0, d_1 < 0.10a$ and $0.7 > x/a > 0.3$.

Numerical Results.—In the range of applicability of Eqs. (5) to (9), numerical values of X_a/Z_0 and X_b/Z_0 may be obtained from the curves of Figs. 5·11-2 to 5·11-5 provided the equivalent diameter d_0 or d_1 is employed in place of the circular diameter d. The equivalent circular

SEC. 5·11] SOLID INDUCTIVE POST IN RECTANGULAR GUIDE   265

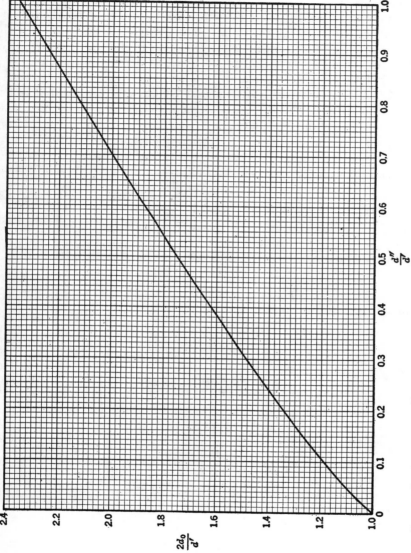

FIG. 5·11-7.—Equivalent circular diameter d_0 of a rectangular post ($2d_0/d'$ vs. d''/d' same as $2d_0/d''$ vs. d'/d'').

diameter d_0 of Eq. (8a) for a rectangular post is plotted in Fig. 5·11-? as a function of the ratio d''/d'. Since the variation of $2d_0/d'$ vs. d''/d' is identical with that of $2d_0/d''$ vs. d'/d'', a plot of the former in the range $0 < d''/d' < 1$ yields d_0 for the entire range of d''/d'. The curve of $\sqrt{2}\, d_1/d''$ vs. d'/d'', as given by Eq. (9a), is identically the curve of ϵ vs. l/d' plotted in Figs. 5·9-4 and 5·9-5.

5·12. Dielectric Posts in Rectangular Guide.—A cylindrical dielectric obstacle of circular cross section aligned parallel to the electric field (H_{10}-mode in rectangular guide).

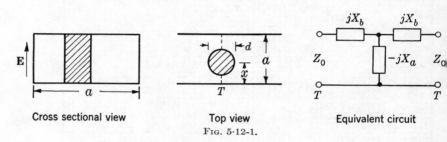

Cross sectional view   Top view   Equivalent circuit
Fig. 5·12-1.

Equivalent-circuit Parameters.—For an obstacle with a real dielectric constant $\epsilon' = \epsilon/\epsilon_0$

$$\frac{X_a}{Z_0} - \frac{X_b}{2Z_0} = \frac{a}{2\lambda_g} \csc^2 \frac{\pi x}{a} \left[\frac{J_0(\beta)}{J_0(\alpha)} \frac{1}{\beta J_0(\alpha) J_1(\beta) - \alpha J_0(\beta) J_1(\alpha)} - S_0 + \frac{\alpha^2}{4} \right], \quad (1a)$$

$$\frac{X_a}{Z_0} - \frac{X_b}{2Z_0} \approx \frac{a}{2\lambda_g} \csc^2 \frac{\pi x}{a} \left[\frac{2}{(\epsilon'-1)\alpha^2} - S_0 - \frac{1}{4}\frac{\epsilon'-3}{\epsilon'-1} \right], \quad (1b)$$

$$\frac{X_b}{Z_0} = \frac{\dfrac{2a}{\lambda_g}\left(\dfrac{\pi d}{a}\right)^2 \sin^2 \dfrac{\pi x}{a}}{\dfrac{\alpha^2 J_1(\beta)}{J_1(\alpha)} \dfrac{1}{\alpha J_0(\alpha)J_1(\beta) - \beta J_0(\beta) J_1(\alpha)} - 2}, \quad (2a)$$

$$\frac{X_b}{Z_0} \approx \frac{a}{8\lambda_g}\left(\frac{a}{\lambda}\right)^2 (\epsilon'-1)\left(\frac{\pi d}{a}\right)^4 \sin^2 \frac{\pi x}{a}\left(1 + \frac{\epsilon'-2}{6}\alpha^2\right), \quad (2b)$$

where

$$\alpha = \frac{\pi d}{\lambda}, \qquad \beta = \sqrt{\epsilon'}\,\frac{\pi d}{\lambda},$$

$$S_0 = \ln\left(\frac{4a}{\pi d}\sin\frac{\pi x}{a}\right) - 2\sin^2\frac{\pi x}{a} + 2\sum_{n=2}^{\infty}\sin^2\frac{n\pi x}{a}\left[\frac{1}{\sqrt{n^2 - \left(\frac{2a}{\lambda}\right)^2}} - \frac{1}{n}\right].$$

For an obstacle with a complex dielectric constant $\epsilon' - j\epsilon'' = \epsilon/\epsilon_0$, the above formulas are still valid provided ϵ', X_a/Z_0, and X_b/Z_0 are replaced by $\epsilon' - j\epsilon''$, $j(Z_a/Z_0)$, and $-j(Z_b/Z_0)$, respectively. The complex impedances Z_a and Z_b represent the values of the shunt and series arms of the equivalent circuit of Fig. 5·12-1.

It is to be noted that resonant effects occur for large values of ϵ' with attendant changes in sign of the circuit elements. In this connection the representation of Fig. 3·14 is advantageous for investigating conditions under which complete transmission occurs.

Restrictions.—The equivalent circuit is applicable in the wavelength range $2a > \lambda > a$, and for the centered cylinder ($x = a/2$) in the wider range $2a > \lambda > 2a/3$. The approximations and method employed in the derivations of Eqs. (1) and (2) are similar to those involved in Sec. 5·11a, though not so accurate for the off-centered posts. Equations (1a) and (2a) are estimated to be in error by only a few per cent in the range $d/a < 0.15$ and $0.2 < x/a < 0.8$ provided also that neither X_a/Z_0 nor X_b/Z_0 are too close to resonance. In the range $\sqrt{\epsilon'} < 4$ Eqs. (1b) and (2b) agree with Eqs. (1a) and (2a), respectively, to within a few per cent. For relatively small diameters, say $d/a < 0.1$, or values of ϵ' close to unity ($\epsilon'' \ll 1$), the equivalent circuit reduces to a simple shunt element.

Numerical Results.—In view of the large number of parameters $(d/a, x/a, \lambda/a, \epsilon', \epsilon'')$ involved in the general result, numerical computations have been made only for a centered post ($x/a = 0.5$) with real dielectric constant ($\epsilon'' = 0$) and for a wavelength $\lambda/a = 1.4$. The former is not a serious restriction, for the essential dependence upon x/a is contained in the multiplicative factors in Eqs. (1) and (2). Tables of X_a/Z_0 and X_b/Z_0 vs. $\sqrt{\epsilon'}$ are presented below for several values of d/a.

$\sqrt{\epsilon'}$	$d/a = 0.05$		$d/a = 0.10$		$d/a = 0.15$	
	$\dfrac{X_a}{Z_0}$	$\dfrac{X_b}{Z_0}$	$\dfrac{X_a}{Z_0}$	$\dfrac{X_b}{Z_0}$	$\dfrac{X_a}{Z_0}$	$\dfrac{X_b}{Z_0}$
2	12.9	0.000059	3.12	0.00093	1.39	0.0049
3	4.57	0.000158	1.03	0.00262	0.443	0.0145
4	2.25	0.00030	0.438	0.0053	0.184	0.0323
5	1.25	0.00049	0.187	0.0092	0.085	0.068
6	0.731	0.00073	0.057	0.0153	0.067*	0.16*
7	0.422	0.00104	−0.028	0.0249	0.5*	1.0*
8	0.224	0.00141	−0.063	0.042	−0.3*	−0.41*
9	0.089	0.00186	−0.083	0.078	−0.24	−0.21
10	−0.0028	0.00242	−0.05*	0.21*	−0.28	−0.15
11	−0.078	0.00312	−0.6*	−0.87*	−0.61	−0.119
12	−0.132	0.0040	−0.3*	−0.17*	+0.230	−0.100
13	−0.174	0.0051	−0.28*	−0.11*	+0.006	−0.082
14	−0.207	0.0065	−0.29	−0.079	−0.040	−0.061
15	−0.234	0.0083	−0.34	−0.066	−0.048	−0.021

The starred values of X_a/Z_0, as computed from Eqs. (1a) and (2a), correspond to values of X_b/Z_0 near antiresonance and are of questionable accuracy. The first antiresonance in the series reactance X_b occurs for a dielectric constant $\epsilon' \approx (3\lambda/4d)^2$, whereas that for the shunt reactance occurs at $\epsilon' \approx (5\lambda/4d)^2$.

5·13. Capacitive Post in Rectangular Guide.—A metallic obstacle of circular cross section with axis perpendicular to the electric field (H_{10}-mode in rectangular guide).

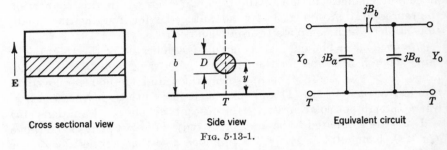

Cross sectional view   Side view   Equivalent circuit

Fig. 5·13-1.

Equivalent-circuit Parameters.—At the reference plane T

$$\frac{B_a}{Y_0} = \frac{2b}{\lambda_g}\left(\frac{\pi D}{2b}\right)^2 \frac{1}{A_2}, \tag{1a}$$

$$\frac{B_a}{Y_0} \approx \frac{2b}{\lambda_g}\left(\frac{\pi D}{2b}\right)^2, \qquad \frac{D}{b} \ll 1, \tag{1b}$$

$$\frac{B_b}{Y_0} = \frac{\lambda_g}{2b}\left(\frac{2b}{\pi D}\right)^2 A_1 - \frac{b}{\lambda_g}\left(\frac{\pi D}{2b}\right)^2 \frac{1}{A_2}, \tag{2a}$$

$$\frac{B_b}{Y_0} \cong \frac{\lambda_g}{2b}\left(\frac{2b}{\pi D}\right)^2, \qquad \frac{D}{b} \ll 1, \tag{2b}$$

where

$$A_1 = 1 + \frac{1}{2}\left(\frac{\pi D}{\lambda_g}\right)^2\left(\ln\frac{b\,\csc\frac{\pi y}{b}}{\pi D} + \frac{3}{4}\right)$$

$$+ \left(\frac{\pi D}{\lambda_g}\right)^2 \sum_{m=1}^{\infty} \cos^2\frac{m\pi y}{b}\left[\frac{1}{\sqrt{m^2 - \left(\frac{2b}{\lambda_g}\right)^2}} - \frac{1}{m}\right],$$

$$A_2 = 1 + \frac{1}{2}\left(\frac{\pi D}{\lambda_g}\right)^2\left[\frac{11}{4} - \ln\left(\frac{b}{\pi D}\csc\frac{\pi y}{b}\right)\right] + \frac{1}{4}\left(\frac{\pi D}{2b}\right)^2\left(\csc^2\frac{\pi y}{b} - \frac{1}{3}\right)$$

$$- 2\left(\frac{\pi D}{2b}\right)^2 \sum_{m=1}^{\infty} \cos^2\frac{m\pi y}{b}\left[m - \frac{1}{2m}\left(\frac{2b}{\lambda_g}\right)^2 - \sqrt{m^2 - \left(\frac{2b}{\lambda_g}\right)^2}\right].$$

Restrictions.—The equivalent circuit is valid in the wavelength range $2b/\lambda_g < 1$. Equations (1a) and (2a) have been derived by the variational method; the angular current distribution on the obstacle was assumed to be the sum of an even constant function and an odd sine function. These equations are estimated to be in error by a few per cent in the

SEC. 5·13]   CAPACITIVE POST IN RECTANGULAR GUIDE   269

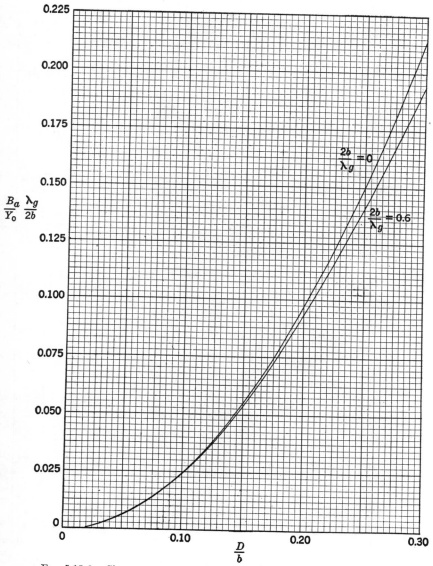

FIG. 5·13-2.—Shunt susceptance of capacitive post in rectangular guide.

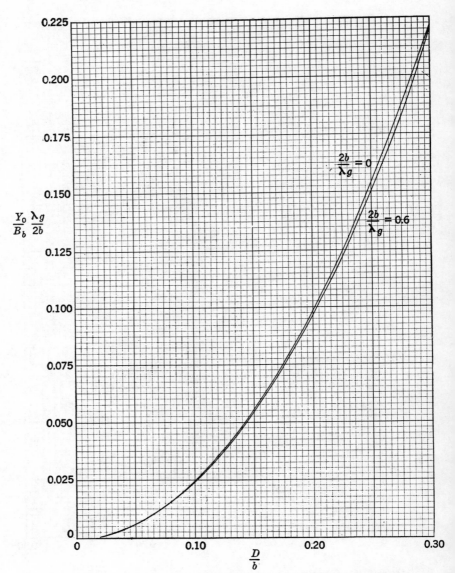

Fig. 5·13-3.—Series reactance of capacitive post in rectangular guide.

range $D/b < 0.1$ and $0.2 < y/b < 0.8$. For $y/b = 0.5$ approximations (1b) and (2b) agree with Eqs. (1a) and (2a) to within 10 per cent for $D/b < 0.3$ and $2b/\lambda_g < 0.4$. The variation of the circuit parameters with y/b is of the order of a few per cent.

Numerical Results.—Figure 5·13-2 contains a plot of $B_a\lambda_g/Y_02b$ of Eq. (1a) as a function of D/b with $2b/\lambda_g$ as a parameter. In Fig. 5·13-3 $Y_0\lambda_g/B_b 2b$ of Eq. (2a) is plotted vs. D/b with $2b/\lambda_g$ as a parameter. Both curves apply to the symmetrical case $y/b = 0.5$.

5·14. Post of Variable Height in Rectangular Guide.—A centered metallic cylindrical post of variable height with axis parallel to the dominant-mode electric field (H_{10}-mode in rectangular guide).

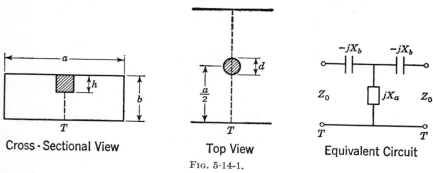

Fig. 5·14-1.

Equivalent-circuit Parameters. Experimental.—The equivalent-circuit parameters at the terminal plane T have been measured in rectangular guide of dimensions $a = 0.90$ in. and $b = 0.40$ in. The measured data are tabulated below as a function of h/b for a number of post diameters d and wavelengths λ. For posts with a flat base:

$d = \frac{1}{16}$ in., $\lambda = 3.4$ cm, $\lambda_g = 2.000$ in.

h/b (in.)	0.249	0.497	0.746	0.871	0.921	0.934	0.993	1.000
X_b/Z_0	0.005	0.010	0.014	0.017	0.018	0.018	0.020	0.020
X_a/Z_0	-6.481	-1.015	-0.894	-0.035	+0.016	0.031	0.151	0.241

$d = \frac{1}{16}$ in., $\lambda = 3.2$ cm, $\lambda_g = 1.763$ in.

h/b (in.)	0.254	0.505	0.756	0.829	0.943	0.961	1.000
X_b/Z_0	0.006	0.011	0.017	0.019	0.021	0.022	0.023
X_a/Z_0	-6.204	-0.906	-0.122	-0.028	+0.083	0.112	0.277

$d = \frac{1}{16}$ in., $\lambda = 3.0$ cm, $\lambda_g = 1.561$ in.

h/b (in.)	0.246	0.504	0.629	0.755	0.784	0.845	0.898	1.000
X_b/Z_0	0.005	0.013	0.016	0.019	0.019	0.021	0.022	0.025
X_a/Z_0	-6.384	-0.763	-0.277	-0.053	-0.017	+0.047	0.088	0.341

$d = \frac{1}{8}$ in., $\lambda = 3.4$ cm, $\lambda_g = 2.001$ in.

h/b (in.)	0.258	0.507	0.758	0.882	0.970	1.000
X_b/Z_0	0.016	0.035	0.054	0.065	0.073	0.076
X_a/Z_0	−3.179	−0.606	−0.147	−0.052	+0.028	0.107

$d = \frac{1}{8}$ in., $\lambda = 3.2$ cm, $\lambda_g = 1.764$ in.

h/b (in.)	0.251	0.501	0.759	0.834	0.882	0.965	1.000
X_b/Z_0	0.017	0.038	0.061	0.068	0.073	0.081	0.085
X_a/Z_0	−3.37	−0.591	−0.129	−0.058	−0.020	+0.040	0.126

$d = \frac{1}{8}$ in., $\lambda = 3.0$ cm, $\lambda_g = 1.561$ in.

h/b (in.)	0.240	0.488	0.745	0.818	0.923	1.000
X_b/Z_0	0.019	0.044	0.069	0.077	0.086	0.098
X_a/Z_0	−3.333	−0.596	−0.109	−0.050	+0.027	0.147

$d = \frac{1}{4}$ in., $\lambda = 3.4$ cm, $\lambda_g = 2.000$ in.

h/b (in.)	0.252	0.499	0.760	0.925	1.000
X_b/Z_0	0.047	0.101	0.174	0.227	0.256
X_a/Z_0	−1.775	−0.468	−0.166	−0.053	+0.026

$d = \frac{1}{4}$ in., $\lambda = 3.2$ cm, $\lambda_g = 1.761$ in.

h/b (in.)	0.262	0.505	0.755	0.880	0.924	1.000
X_b/Z_0	0.052	0.111	0.191	0.240	0.267	0.291
X_a/Z_0	−1.717	−0.477	−0.182	−0.088	−0.038	+0.033

$d = \frac{1}{4}$ in., $\lambda = 3.0$ cm, $\lambda_g = 1.561$ in.

h/b (in.)	0.250	0.502	0.750	0.880	0.940	1.000
X_b/Z_0	0.056	0.121	0.211	0.270	0.300	0.335
X_a/Z_0	−1.859	−0.494	−0.179	−0.085	−0.040	+0.023

For a post with a hemispherical base of diameter d:

$d = \frac{1}{4}$ in., $\lambda = 3.4$ cm, $\lambda_g = 1.999$ in.

h/b (in.)	0.248	0.477	0.751	0.950	0.988	1.000
X_b/Z_0	0.033	0.083	0.151	0.214	0.304	0.409
X_a/Z_0	−3.373	−0.778	−0.241	−0.079	+0.037	0.118

$d = \frac{1}{4}$ in., $\lambda = 3.2$ cm, $\lambda_g = 1.762$ in.

h/b (in.)	0.252	0.540	0.735	0.925	1.000
X_b/Z_0	0.036	0.105	0.162	0.228	0.262
X_a/Z_0	−3.268	−0.575	−0.263	−0.098	+0.037

SEC. 5·16] CIRCULAR OBSTACLE IN CIRCULAR GUIDE

$d = \frac{1}{4}$ in., $\lambda = 3.0$ cm, $\lambda_g = 1.564$ in.

h/b (in.)	0.250	0.504	0.743	0.874	1.000
X_b/Z_0	0.039	0.107	0.187	0.240	0.304
X_a/Z_0	−3.208	−0.726	−0.272	−0.146	+0.045

The resonant effects exhibited by the shunt reactance X_a are to be noted.

5·15. Spherical Dent in Rectangular Guide.—A centered spherical dent in the broad face of a rectangular guide (H_{10}-mode in rectangular guide).

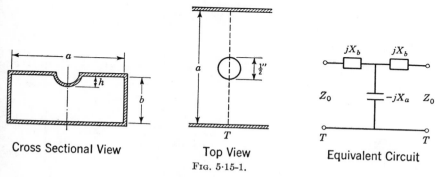

Cross Sectional View   Top View   Equivalent Circuit

FIG. 5·15-1.

Equivalent-circuit Parameters. Experimental.—A spherical dent of variable height h was formed by pressure of a $\frac{1}{2}$-in. steel ball on the broad face of a rectangular guide with dimensions $a = 0.90$ in. and $b = 0.40$ in. (0.050-in. wall). The measured circuit parameters X_a/Z_0 and X_b/Z_0 at the terminal plane T are shown in Fig. 5·15-2 as a function of h/b at the wavelengths $\lambda = 3.4$ cm ($\lambda_g = 2.006$ in.), $\lambda = 3.2$ cm ($\lambda_g = 1.769$ in.), and $\lambda = 3.0$ cm ($\lambda_g = 1.564$ in.). These data are questionable at the larger values of h/b because of the deformation of the guide walls.

5·16. Circular Obstacle of Finite Thickness in Circular Guide.—A centered circular aperture in a plate of finite thickness transverse to the axis of a circular guide (H_{11}-mode in circular guide).

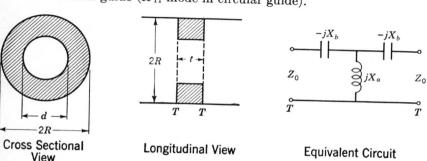

Cross Sectional View   Longitudinal View   Equivalent Circuit

FIG. 5·16-1.

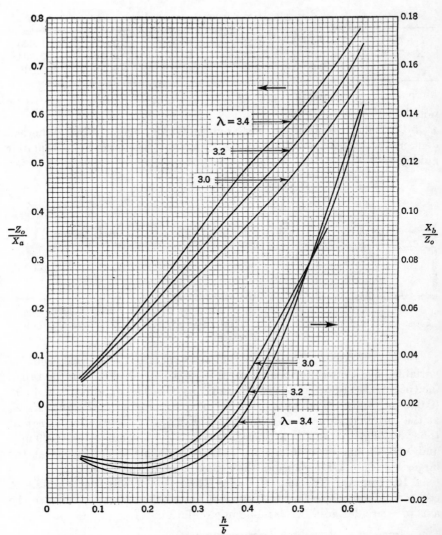

FIG. 5·15-2.—Measured shunt and series reactance of a spherical dent in the broad face of a rectangular guide.

Equivalent Circuit Parameters. Experimental.—The circuit parameters Z_0/X_a and X_b/Z_0 measured at $\lambda = 3.20$ cm ($\lambda_g = 2.026$ in.) as a function of plate thickness for a $\frac{9}{16}$-in. aperture in a circular guide of $\frac{15}{16}$ in. diameter are shown in Fig. 5·16-2. The data, which are rough, refer to the outside terminal planes T. A plot of Z_0/X_a as a function of d for $t = \frac{1}{32}$ in. is shown in Fig. 5·5-3.

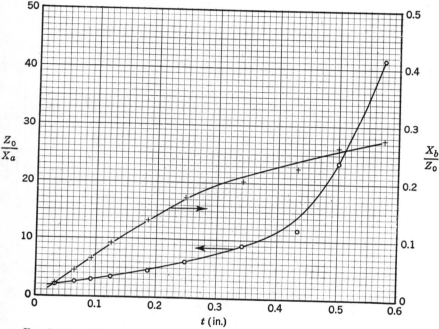

Fig. 5·16-2.—Measured circuit parameters of circular aperture in circular guide.

5·17. Resonant Ring in Circular Guide.—A thin wire of elliptical, circular, or rectangular cross section in the form of a circular ring concentric with the axis of a circular guide (H_{11}-mode in circular waveguide).

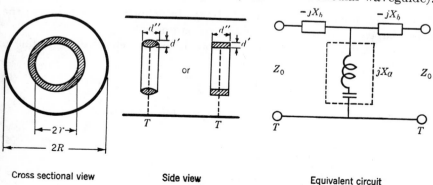

Fig. 5·17-1.

Equivalent-circuit Parameters.—At the reference plane T for the case of a ring with mean radius r formed with circular wire of radius r_0

$$\frac{X_a}{Z_0} = \frac{1}{2}\frac{R^2}{\lambda_g r}\frac{\chi_1'^2 - 1}{\chi_1'^2}\left[\frac{J_1(\chi_1')}{J_1'\left(\frac{\chi_1' r}{R}\right)}\right]^2 \left\{\left[\left(\frac{\lambda}{2\pi r}\right)^2 - 1\right]\ln\frac{8r}{r_0} - A\right\}, \quad (1)$$

$$\frac{X_b}{Z_0} \approx 0,$$

where

$$A = \left(\frac{\lambda}{2\pi r}\right)^2 \left\{\frac{2\pi r}{R}\frac{\chi_1'}{\chi_1'^2 - 1}\left[\frac{J_1'\left(\frac{\chi_1' r}{R}\right)}{J_1(\chi_1')}\right]^2 + \frac{8}{3}\right.$$

$$+ 2\frac{r}{R}\int_0^\infty \frac{K_1(\alpha)\left[I_1'\left(\alpha\frac{r}{R}\right)\right]^2}{I_1'(\alpha)}\left[\frac{2r}{R}\frac{I_1''\left(\alpha\frac{r}{R}\right)}{I_1'\left(\alpha\frac{r}{R}\right)} - \frac{I_1''(\alpha)}{I_1'(\alpha)}\right]d\alpha$$

$$- \frac{2\pi r}{R}\sum_{n=2}^\infty\left[\frac{1}{\sqrt{\chi_n'^2 - \left(\frac{2\pi R}{\lambda}\right)^2}} - \frac{1}{\chi_n'}\right]\frac{\chi_n'^2}{\chi_n'^2 - 1}\left[\frac{J_1'\left(\chi_n'\frac{r}{R}\right)}{J_1(\chi_n')}\right]^2\right\}$$

$$- 2 - \frac{2r}{R}\int_0^\infty \frac{K_1(\alpha)\left[I_1\left(\alpha\frac{r}{R}\right)\right]^2}{I_1(\alpha)}d\alpha$$

$$+ \frac{2\pi r}{R}\sum_{n=1}^\infty \frac{\chi_n - \sqrt{\chi_n^2 - \left(\frac{2\pi R}{\lambda}\right)^2}}{\chi_n^2}\left[\frac{J_1\left(\chi_n\frac{r}{R}\right)}{J_0(\chi_n)}\right]^2,$$

and

$$J_1(\chi_n) = 0, \quad J_1'(\chi_n') = 0, \quad \lambda_g = \frac{\lambda}{\sqrt{1 - \left(\frac{\lambda}{3.41R}\right)^2}}. \quad (2)$$

The Bessel functions $J_1(x)$, $K_1(x)$, $I_1(x)$ are defined in the glossary; the prime superscripts indicate a first or second derivative.

For a wire of elliptical cross section the equivalent circular radius r_0 is given by

$$2r_0 = \frac{d' + d''}{2}, \quad (3)$$

and for a wire of rectangular cross section (cf. Sec. 5·11c)

$$2r_0 = \frac{d'}{2}\frac{1}{E(\alpha) - \alpha'^2 F(\alpha)}, \quad (4)$$

$$2r_0 \approx \frac{d'}{2}\left[1 + \frac{d''}{\pi d'}\ln\left(4\pi e\frac{d'}{d''}\right)\cdots\right], \quad \frac{d''}{d'} \ll 1.$$

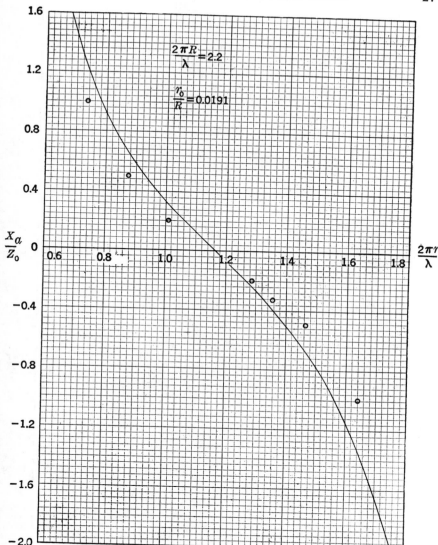

Fig. 5·17-2.—Shunt reactance of resonant ring in circular guide.

Restrictions.—The equivalent circuit is applicable in the wavelength range $1.84 < (2\pi R/\lambda) < 3.83$. The circuit parameters have been evaluated by an integral equation method and are restricted to wires of small cross section, $r_0/R < 0.2$, and to rings having radii within the range $0.2 < r/R < 0.8$. The equivalent circular radii r_0 have been computed by conformal mapping methods and are likewise subject to the aforementioned restrictions. No estimate of accuracy is available within the above range.

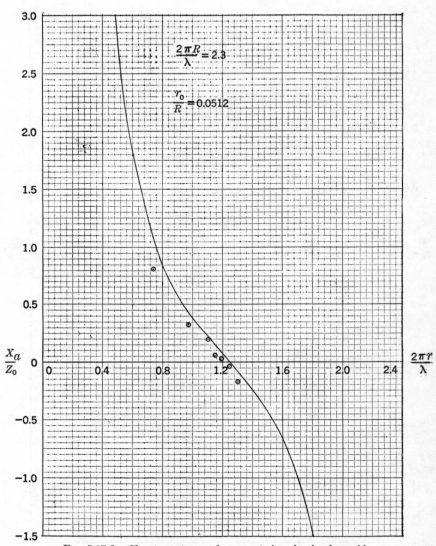

Fig. 5·17-3.—Shunt reactance of resonant rings in circular guide.

Numerical Results.—The relative shunt reactance X_a/Z_0 is plotted against $2\pi r/\lambda$ in Figs. 5·17-2 to 4 for several values of the parameters $2\pi R/\lambda$ and r_0/R. The equivalent circular diameters $d_0 = 2r_0$ for wires of rectangular cross section may be obtained as a function of d''/d' from the graph of Fig. 5·11-7.

Experimental Results.—Reactance measurements were performed at $\lambda = 3.20$ cm on circular rings formed with wire of cross-sectional dimen-

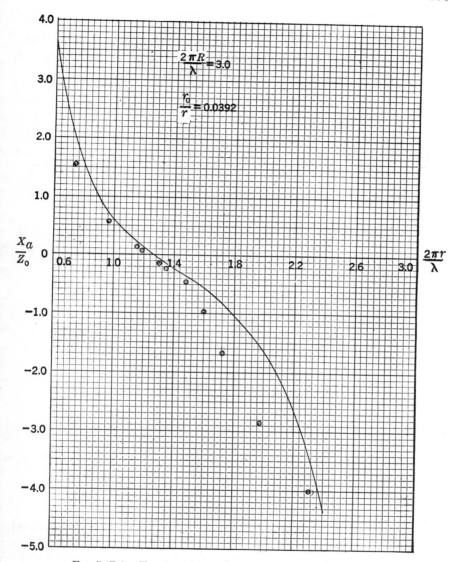

Fig. 5·17-4.—Shunt reactance of resonant ring in circular guide.

sions 0.040 by 0.040 in. The measured data, corresponding to circular guides for which $2\pi R/\lambda = 2.32$ and 2.95, are indicated by the circled points in Figs. 5·17-3 and 5·17-4, respectively. Additional measurements on wires of circular cross section are indicated by the circled points in Fig. 5·17-2; these data were taken in circular guide for which $2\pi R/\lambda = 2.19$ at a wavelength of approximately 10 cm.

GRATINGS AND ARRAYS IN FREE SPACE

5·18. Capacitive Strips.—An infinitely extended plane grating formed by metallic strips of zero thickness with edges parallel to the magnetic field (plane wave in free space incident at angle θ).

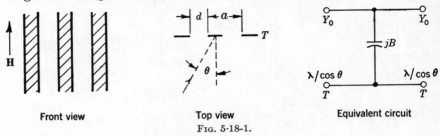

Fig. 5·18-1.

Equivalent-circuit Parameters.—At the reference plane T

$$\frac{B}{Y_0} = \frac{4a \cos \theta}{\lambda} \Bigg\{ \ln \csc \frac{\pi d}{2a}$$

$$+ \frac{1}{2} \frac{(1-\beta^2)^2 \left[\left(1 - \frac{\beta^2}{4}\right)(A_+ + A_-) + 4\beta^2 A_+ A_- \right]}{\left(1 - \frac{\beta^2}{4}\right) + \beta^2 \left(1 + \frac{\beta^2}{2} - \frac{\beta^4}{8}\right)(A_+ + A_-) + 2\beta^6 A_+ A_-} \Bigg\}, \quad (1a)$$

$$\frac{B}{Y_0} \approx \frac{4a \cos \theta}{\lambda} \left[\ln \frac{2a}{\pi d} + \frac{1}{2}(3 - 2\cos^2 \theta) \left(\frac{a}{\lambda}\right)^2 \right], \quad \frac{d}{a} \ll 1, \quad \frac{a}{\lambda} \ll 1, \quad (1b)$$

where

$$A_\pm = \frac{1}{\sqrt{1 \pm \frac{2a}{\lambda} \sin \theta - \left(\frac{a \cos \theta}{\lambda}\right)^2}} - 1,$$

$$\beta = \sin \frac{\pi d}{2a}.$$

Restrictions.—The equivalent circuit is valid for wavelengths and angles of incidence θ in the range $a(1 + \sin \theta)/\lambda < 1$. The quantity B/Y_0 has been computed by an integral equation method in which the first two diffraction modes are correctly treated to order β^2. Equation (1a) is estimated to be in error by less than 10 per cent for the range of values plotted in the accompanying figure. For the case $\theta = 0$, a more accurate expression for B/Y_0 is given by Eq. (2a) of Sec. 5·1a provided b therein is replaced by a and λ_g by $\lambda/\cos \theta$. This latter result, valid for all apertures d, indicates that at least for small θ Eq. (1a) may be justifiably employed for values of d/a larger than those plotted. Equation (1b) is

SEC. 5·18] CAPACITIVE STRIPS 281

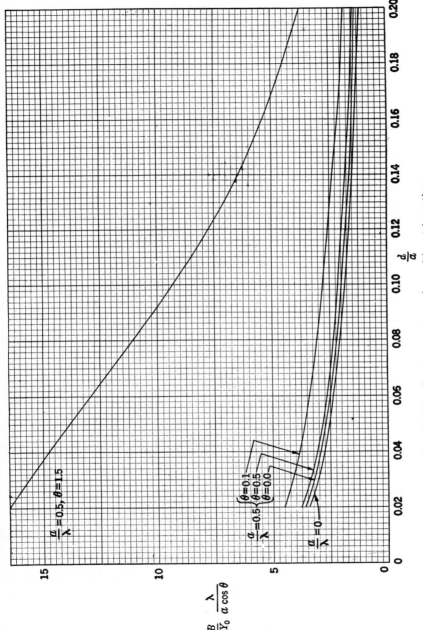

FIG. 5·18-2.—Shunt susceptance of capacitive strip grating.

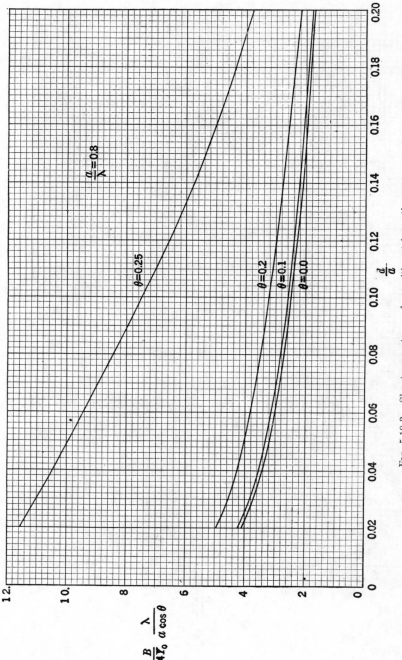

FIG. 5·18-3.—Shunt susceptance of capacitive strip grating.

SEC. 5·18] CAPACITIVE STRIPS

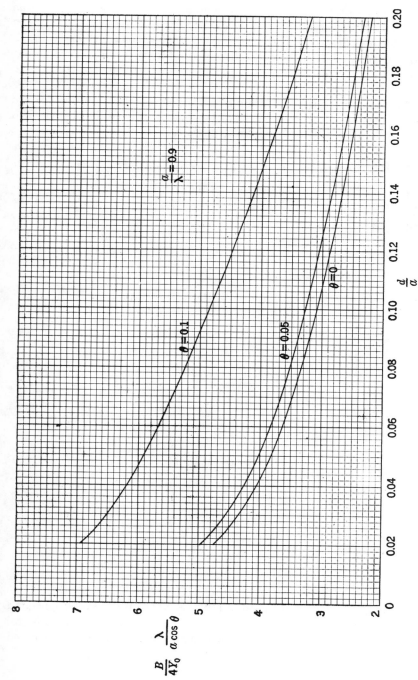

FIG. 5·18-4.—Shunt susceptance of capacitive strip grating.

an approximation for small apertures and agrees with Eq. (1a) to within 10 per cent in the range $d/a < 0.2$, $a/\lambda < 0.5$, and $\theta < 0.5$.

Numerical Results.—The quantity $B\lambda/4Y_0 a \cos\theta$ of Eq. (1a) is plotted in Figs. 5·18-2, 5·18-3, and 5·18-4 as a function of d/a in the range $d/a < 0.2$ for various values of θ (in radians) and for $a/\lambda = 0$, 0.5, 0.8, and 0.9.

5·19. Inductive Strips.—An infinitely extended plane grating formed by metallic strips of zero thickness with edges parallel to the electric field (plane wave in free space incident at angle θ).

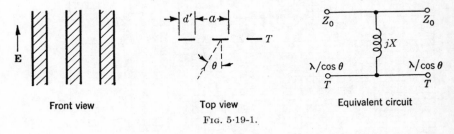

Fig. 5·19-1.

Equivalent-circuit Parameters.—At the terminal plane T

$$\frac{X}{Z_0} = \frac{a \cos\theta}{\lambda} \left\{ \ln \csc \frac{\pi d'}{2a} \right.$$

$$\left. + \frac{1}{2} \frac{(1 - \beta^2)^2 \left[\left(1 - \frac{\beta^2}{4}\right)(A_+ + A_-) + 4\beta^2 A_+ A_- \right]}{\left(1 - \frac{\beta^2}{4}\right) + \beta^2 \left(1 + \frac{\beta^2}{2} - \frac{\beta^4}{8}\right)(A_+ + A_-) + 2\beta^6 A_+ A_-} \right\}, \quad (1a)$$

$$\frac{X}{Z_0} \approx \frac{a \cos\theta}{\lambda} \left[\ln \frac{2a}{\pi d'} + \frac{1}{2}(3 - 2\cos^2\theta)\left(\frac{a}{\lambda}\right)^2 \right], \quad \frac{d'}{a} \ll 1, \quad \frac{a}{\lambda} \ll 1, \quad (1b)$$

where

$$A_\pm = \frac{1}{\sqrt{1 \pm \frac{2a \sin\theta}{\lambda} - \left(\frac{a \cos\theta}{\lambda}\right)^2}} - 1,$$

$$\beta = \sin\frac{\pi d'}{2a}.$$

Restrictions.—The equivalent circuit is valid for wavelengths and angles of incidence in the range $a(1 + \sin\theta)/\lambda < 1$. An integral equation method in which the first two diffraction modes are treated accurately to order β^2 has been employed to evaluate X/Z_0. Equation (1a) is estimated to be in error by less than 10 per cent for the range of values plotted below. For the case $\theta = 0$, a more accurate expression for X/Z_0 is $B/4Y_0$ obtained from Eq. (2a) of Sec. 5·1a with b therein replaced by a, d by d', and λ_g by $\lambda/\cos\theta$. This latter result, valid for the entire range of strip widths d', indicates that at least for small θ Eq. (1a) may be justifiably used for values of d'/a in excess of those plotted. Equation (1b) agrees with Eq. (1a) to within 10 per cent in the range $d'/a < 0.2$, $a/\lambda < 0.5$, and $\theta < 0.5$.

Numerical Results.—The term $X\lambda/Z_0 a \cos\theta$ of Eq. (1a) may be obtained as a function of d'/a from Figs. 5·18-2, 5·18-3, and 5·18-4, provided the replacements $X\lambda/Z_0 a \cos\theta$ for $B\lambda/4Y_0 a \cos\theta$ and d' for d are made. The angle θ is given in radians. The large values of reactance in the vicinity of $a \approx \lambda$ is indicative of almost perfect transmission through the grating in this range.

5·20. Capacitive Posts.—An infinitely extended grating formed by metallic obstacles of circular or rectangular cross section with axes parallel to the magnetic field (plane wave in free space incident normally).

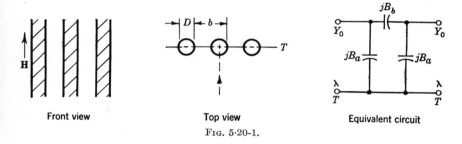

Front view   Top view   Equivalent circuit

Fig. 5·20-1.

Equivalent-circuit Parameters.—For obstacles of circular cross section the circuit parameters are the same as in Sec. 5·13 with λ_g therein replaced by λ, and $y/b = 0.5$. The results of Sec. 5·9c can be employed for the case of rectangular obstacles provided λ_g therein is likewise replaced by λ.

Restrictions.—Same as in Secs. 5·13 and 5·9c.

Numerical Results.—The variation of $B_a\lambda/Y_0 2b$ and $Y_0\lambda/B_b 2b$ with D/b and $2b/\lambda$ is indicated in Figs. 5·13-2 and 5·13-3 provided λ_g therein is replaced by λ. The results for rectangular obstacles may be obtained from Sec. 5·9c.

5·21. Inductive Posts.—An infinitely extended grating formed by small metallic obstacles of elliptical, circular, or rectangular cross section with their axes parallel to the electric field (plane wave in free space incident at an angle θ).

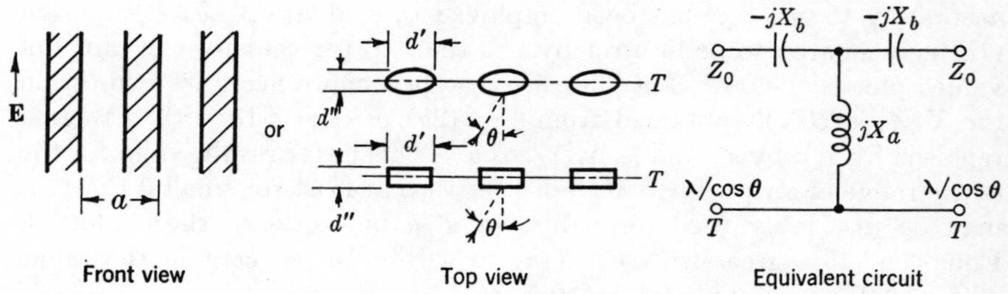

Front view | **Top view** | **Equivalent circuit**

Fig. 5·21-1.

Equivalent-circuit Parameters.—At the terminal plane T

$$\frac{X_a}{Z_0} = \frac{a \cos \theta}{\lambda} \left\{ \ln \frac{a}{2\pi r_0} \right.$$

$$\left. + \frac{1}{2} \sum_{\substack{m=-\infty \\ m \neq 0}}^{\infty} \left[\frac{1}{\sqrt{m^2 + \frac{2ma}{\lambda} \sin \theta - \left(\frac{a \cos \theta}{\lambda}\right)^2}} - \frac{1}{|m|} \right] \right\}, \quad (1a)$$

$$\frac{X_a}{Z_0} \approx \frac{a \cos \theta}{\lambda} \left[\ln \frac{a}{2\pi r_0} + 0.601(3 - 2\cos^2 \theta) \left(\frac{a}{\lambda}\right)^2 \right], \quad \frac{a}{\lambda} \ll 1, \quad (1b)$$

$$\frac{X_b}{Z_0} = \frac{a \cos \theta}{\lambda} \left(\frac{2\pi r_1}{a}\right)^2, \quad (2a)$$

where, if $d = (d' + d'')/2$,

$2r_0 = d,$ $2r_1 = \sqrt{dd''}$ for elliptical cross section,
$2r_0 = d,$ $2r_1 = d$ for circular cross section,
$2r_0 = \frac{d'}{2} f_0 \left(\frac{d''}{d'}\right),$ $2r_1 = \frac{d''}{\sqrt{2}} f_1 \left(\frac{d''}{d'}\right)$ for rectangular cross section.

The functions f_0 and f_1 are defined in Eqs. (8) and (9) of Sec. 5·11c (with $d_0 = 2r_0$, $d_1 = 2r_1$).

Restrictions.—The equivalent circuit is valid for wavelengths and incident angles in the range $a(1 + \sin \theta)/\lambda < 1$. Equations (1a) and (2a) were calculated by a variational method assuming for the obstacle current an angular distribution that is a combination of an even constant function and an odd sine function. The equivalent radii r_0 have been obtained by an equivalent static method. These results, valid only in the small-obstacle range, are estimated to be in error by less than 10 per cent for the range plotted in the accompanying figures.

Numerical Results.—The circuit parameter $X_a\lambda/Z_0 a \cos \theta$ of Eq. (1a) is plotted in two parts; in Fig. 5·21-2 the term $\ln (a/2\pi r_0)$ is presented as a

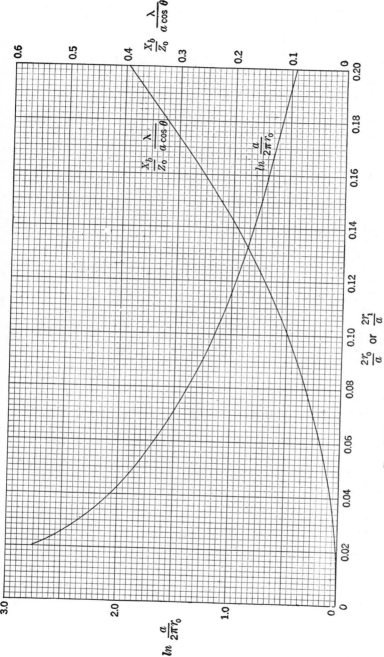

Fig. 5·21-2.—Circuit parameters for inductive post grating.

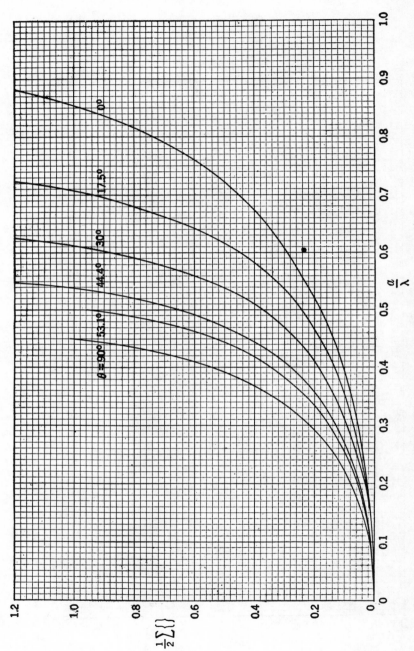

Fig. 5·21-3.—Sum terms for inductive post grating.

function of $2r_0/a$ and in Fig. 5·21-3 the quantity $\frac{1}{2}\Sigma\{\quad\}$ is given as a function of a/λ for various values of θ. The parameter $X_b\lambda/Z_0a \cos\theta$ of Eq. (2a) is plotted as a function of $2r_1/a$ in Fig. 5·21-2. The functions $f_0(d''/d')$ and $f_1(d''/d')$ appear in graphical form in the curves of Fig. 5·11-7 and Figs. 5·9-4 and 5·9-5, (see Sec. 5·11c).

5·22. Array of Semi-infinite Planes, E-plane.—An array consisting of an infinite number of semi-infinite metallic obstacles of zero thickness with edges parallel to the magnetic field (in space, plane wave incident at angle θ; in parallel-plate region, TEM-modes with a relative phase of $(2\pi b/\lambda)\sin\theta$ in adjacent guides).

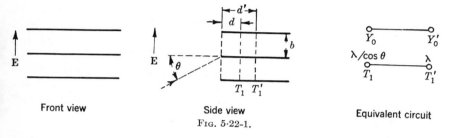

Fig. 5·22-1.

Equivalent-circuit Parameters.—At reference planes T_1 and T_1' the equivalent network is a simple junction of two uniform transmission lines whose characteristic admittance ratio is

$$\frac{Y_0'}{Y_0} = \cos\theta. \qquad (1)$$

The input and output planes are located at distances d and d' given by

$$\frac{2\pi d}{\lambda/\cos\theta} = 2x \ln 2 + \sin^{-1}\frac{2x}{\sqrt{1-4y^2}} - \sin^{-1}\frac{x}{\sqrt{1+2y}}$$
$$- \sin^{-1}\frac{x}{\sqrt{1-2y}} + S_2(2x;2y,0) - S_2(x;y,-y) - S_2(x;y,y), \qquad (2a)$$

$$\frac{2\pi d}{\lambda/\cos\theta} \approx 2x \ln 2 + \sin^{-1}\frac{2x}{\sqrt{1-4y^2}} - \sin^{-1}\frac{x}{\sqrt{1+2y}}$$
$$- \sin^{-1}\frac{x}{\sqrt{1-2y}} + \cdots, \qquad (2b)$$

$$\frac{2\pi d'}{\lambda} = 2x' \ln 2 + \sin^{-1} 2x' - \sin^{-1}\frac{x'}{1+y} - \sin^{-1}\frac{x'}{1-y}$$
$$+ S_2(2x';0,0) - S_2(x';0,-y) - S_2(x';0,y), \qquad (3a)$$

$$\frac{2\pi d'}{\lambda} \approx 2x' \ln 2 + \sin^{-1} 2x' - \sin^{-1} \frac{x'}{1+y} - \sin^{-1} \frac{x'}{1-y} + \cdots,$$
(3b)

where

$$x = \frac{b}{\lambda} \cos \theta, \qquad y = \frac{b}{\lambda} \sin \theta, \qquad x' = \frac{b}{\lambda},$$

$$S_2(x;\alpha,\beta) = \sum_{n=2}^{\infty} \left[\sin^{-1} \frac{x}{\sqrt{(n-\beta)^2 - \alpha^2}} - \frac{x}{n} \right].$$

Restrictions.—The equivalent circuit is valid in the wavelength range $\lambda > 2b$. The equations have been obtained by the transform method and are rigorous in the above range. The approximate Eqs. (2b) and (3b) agree with Eqs. (3a) and (3b), respectively, to within 4 per cent for $b/\lambda < 0.5$ and $\theta < 60°$. It is to be noted that the relative phase of the fields in adjacent guides of the parallel-plate region is $(2\pi b/\lambda) \sin \theta$; the average wavefront in the parallel-plate region is therefore the same as that in the outer space.

Numerical Results.—In Figs. 5·22-2 and 5·22-3 the reference plane distances $\pi d/b$ and $\pi d'/b$ are plotted as a function of b/λ for various values of the angle of incidence θ. The S_2 functions are tabulated in the Appendix.

5·23. Array of Semi-infinite Planes, H-plane.—An array consisting of an infinite number of semi-infinite metallic obstacles of zero thickness

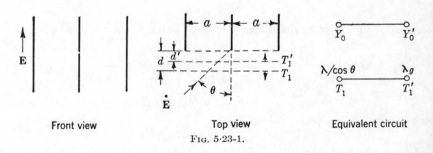

Fig. 5·23-1.

with edges parallel to the electric field (in space, plane wave incident at angle θ; in parallel-plate region, H_{10}-modes with a relative phase of $(2\pi a/\lambda) \sin \theta$ in adjacent guides).

SEC. 5·22] ARRAY OF SEMI-INFINITE PLANES, E-PLANE

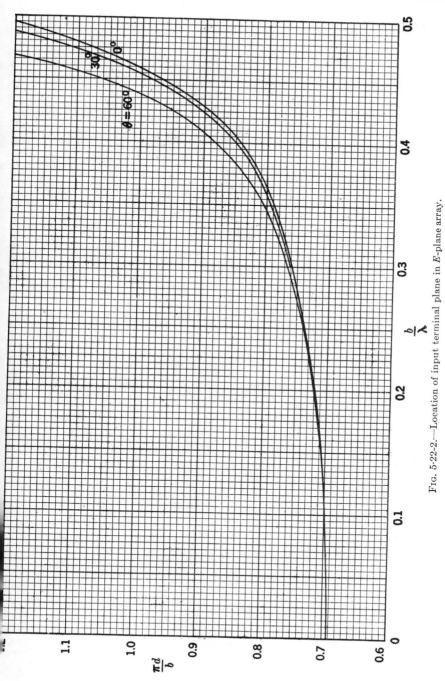

FIG. 5·22-2.—Location of input terminal plane in E-plane array.

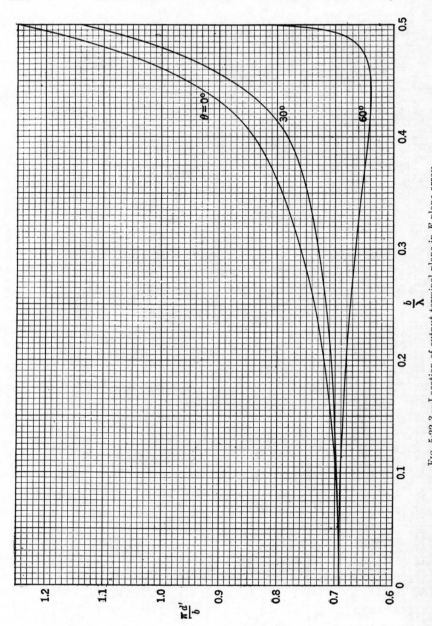

Fig. 5·22-3.—Location of output terminal plane in E-plane array.

Equivalent-circuit Parameters.—At reference planes T_1 and T_1' the equivalent network is a simple junction of two uniform transmission lines whose characteristic admittance ratio is

$$\frac{Y_0'}{Y_0} = \frac{\lambda}{\lambda_g \cos \theta} \tag{1}$$

The input and output planes are located at distances d and d' given by

$$\frac{2\pi d}{\lambda/\cos \theta} = \sin^{-1} \frac{x}{\sqrt{1-2y}} + \sin^{-1} \frac{x}{\sqrt{1+2y}} - 2x \ln 2 + S_2(x;y,y) + S_2(x;y,-y) - S_2(2x;2y,0), \tag{2}$$

$$\frac{2\pi d'}{\lambda_g} = \sin^{-1} \frac{x'}{\sqrt{(1-y)^2 - (0.5)^2}} + \sin^{-1} \frac{x'}{\sqrt{(1+y)^2 - (0.5)^2}} - 2x' \ln 2 + S_2(x';0.5,y) + S_2(x';0.5,-y) - S_2(2x';1,0), \tag{3}$$

where

$$x = \frac{a \cos \theta}{\lambda}, \quad x' = \frac{a}{\lambda_g}, \quad y = \frac{a \sin \theta}{\lambda},$$

$$S_2(x;\alpha,\beta) = \sum_{n=2}^{\infty} \left[\sin^{-1} \frac{x}{\sqrt{(n-\beta)^2 - \alpha^2}} - \frac{x}{n} \right], \quad \lambda_g = \frac{\lambda}{\sqrt{1 - \left(\frac{\lambda}{2a}\right)^2}}.$$

Restrictions.—The equivalent circuit is valid in the range

$$2a > \lambda > a(1 + \sin \theta).$$

The circuit parameters have been obtained by the transform method and are rigorous in the above range. It is to be noted that the relative phase of the modes in adjacent guides of the parallel-plate region is $(2\pi a/\lambda) \sin \theta$; the average wavefront in the parallel-plate region is therefore the same as that in the outer space.

Numerical Results.—The reference plane distances $\pi d/a$ and $\pi d'/a$ are indicated in Figs. 5·23-2 and 5·23-3 as a function of a/λ with the angle of incidence θ as a parameter. A table of the S_2 functions appears in the Appendix.

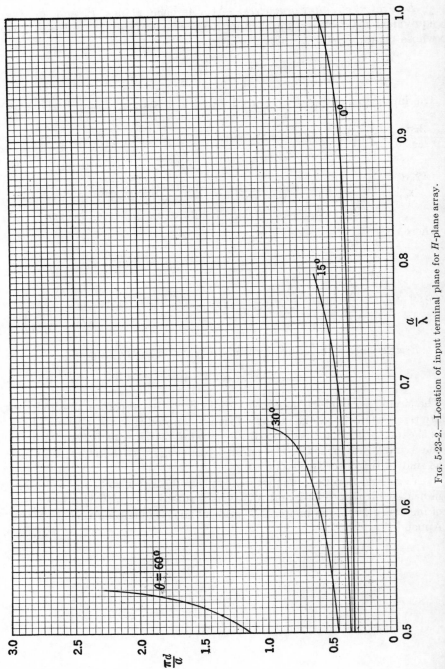

Fig. 5·23-2.—Location of input terminal plane for H-plane array.

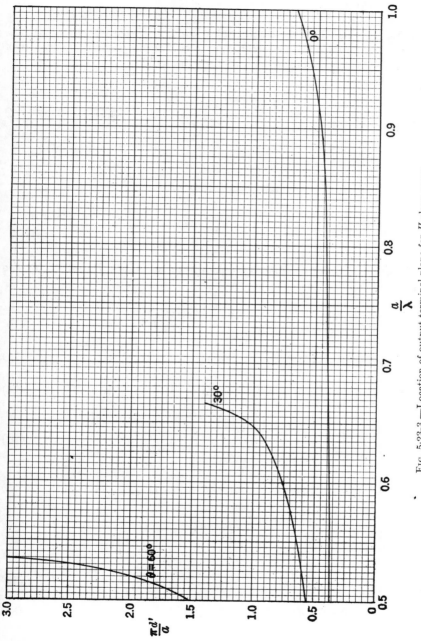

FIG. 5·23-3.—Location of output terminal plane for *H*-plane array.

ASYMMETRIC STRUCTURES; COUPLING OF TWO GUIDES

5·24. Junction of Two Rectangular Guides, H-plane. *a. Symmetrical Case.*—A junction of two rectangular guides of unequal widths but of equal heights (H_{10}-mode in each rectangular guide).

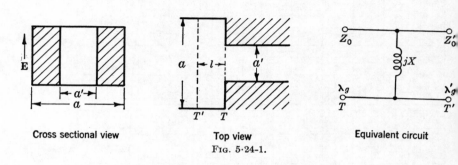

Cross sectional view   Top view   Equivalent circuit
FIG. 5·24-1.

Equivalent-circuit Parameters.—At reference planes T and T' for the larger and smaller guide, respectively,

$$\frac{Z_0'}{Z_0} = \frac{\lambda_g' a'}{\lambda_g a} \frac{X_{11}}{\alpha^2(X_{22} - X_0)} \left[1 + \left(\frac{2a' X_0}{\lambda_g'}\right)^2 \right], \tag{1a}$$

$$\frac{Z_0'}{Z_0} \approx \frac{\lambda_g' a'}{\lambda_g a} (1.750)(1 - 0.58\alpha^2), \qquad \alpha \ll 1, \tag{1b}$$

$$\frac{Z_0'}{Z_0} \approx \frac{\lambda_g' a'}{\lambda_g a} \left(1 + \beta + \frac{\beta^2}{2}\right), \qquad \beta = (1 - \alpha) \ll 1. \tag{1c}$$

$$\frac{X}{Z_0} = \frac{2a}{\lambda_g} X_{11} \frac{1 + \left(\frac{2a' X_0}{\lambda_g'}\right)^2}{1 + \left(\frac{2a'}{\lambda_g}\right)^2 X_0 X_{22}}, \tag{2a}$$

$$\frac{X}{Z_0} \approx \frac{2a}{\lambda_g} 2.330\alpha^2(1 + 1.56\alpha^2)(1 + 6.75\alpha^2 Q), \qquad \alpha \ll 1, \tag{2b}$$

$$\frac{Z_0}{X} \approx \frac{\lambda_g}{2a} \frac{\beta^2(1 + \beta) \ln \frac{2}{\beta}}{1 - \frac{\beta}{2}} \left(1 - \frac{27}{8} \frac{Q + Q'}{1 + 8 \ln \frac{2}{\beta}}\right), \qquad \beta \ll 1. \tag{2c}$$

$$\frac{l}{a} = \frac{\alpha X_0}{\pi}, \tag{3a}$$

$$\frac{l}{a} \approx 0.0084\alpha(1 - 1.5\alpha^2), \qquad \alpha \ll 1, \tag{3b}$$

$$\frac{l}{a} \approx 0, \qquad \beta \ll 1 \tag{3c}$$

SEC. 5·24] JUNCTION OF TWO RECTANGULAR GUIDES, H-PLANE 297

where

$$X_{11} = \frac{A}{1 - \frac{1}{4A}[(A+1)^2 N_{11} + 2(A+1)CN_{12} + C^2 N_{22}]},$$

$$X_{22} = \frac{A'}{1 - \frac{1}{4A'}[(A'+1)^2 N_{22} + 2(A'+1)CN_{12} + C^2 N_{11}]},$$

$$X_0 = X_{22} - \frac{X_{12}^2}{X_{11}} \approx A' - \frac{C^2}{A},$$

and

$$A = \frac{(1+R_1)(1-R_2) + T^2}{(1-R_1)(1-R_2) - T^2}, \qquad R_1 = -\left(\frac{1-\alpha}{1+\alpha}\right)^\alpha,$$

$$A' = \frac{(1-R_1)(1+R_2) + T^2}{(1-R_1)(1-R_2) - T^2}, \qquad R_2 = \left(\frac{1-\alpha}{1+\alpha}\right)^{1/\alpha},$$

$$C = \frac{2T}{(1-R_1)(1-R_2) - T^2}, \qquad T = \frac{4\alpha}{1-\alpha^2}\left(\frac{1-\alpha}{1+\alpha}\right)^{\frac{1}{2}(\alpha+\frac{1}{\alpha})},$$

$$N_{11} = 2\left(\frac{a}{\lambda}\right)^2 \left\{ 1 + \frac{16 R_1}{\pi(1-\alpha^2)}[E(\alpha) - \alpha'^2 F(\alpha)][E(\alpha') - \alpha^2 F(\alpha')] \right.$$
$$\left. - R_1^2 - \alpha^2 T^2 \right\} + \frac{12\alpha^4}{(1-\alpha^2)^2}\left(\frac{1-\alpha}{1+\alpha}\right)^{4\alpha}\left[Q - \frac{1}{2}\left(\frac{2a}{3\lambda}\right)^2\right]$$
$$+ \frac{48\alpha^2}{(1-\alpha^2)^2}\left(\frac{1-\alpha}{1+\alpha}\right)^{\alpha+\frac{3}{\alpha}}\left[Q' - \frac{1}{2}\left(\frac{2}{3}\frac{a'}{\lambda}\right)^2\right],$$

$$N_{22} = 2\left(\frac{a}{\lambda}\right)^2 \left\{ \alpha^2 + \frac{16\alpha^2 R_2 E(\alpha)}{\pi(1-\alpha^2)}[F(\alpha') - E(\alpha')] - \alpha^2 R_2^2 - T^2 \right\}$$
$$+ \frac{48\alpha^2}{(1-\alpha^2)^2}\left(\frac{1-\alpha}{1+\alpha}\right)^{3\alpha+\frac{1}{\alpha}}\left[Q - \frac{1}{2}\left(\frac{2}{3}\frac{a}{\lambda}\right)^2\right]$$
$$+ \frac{12}{(1-\alpha^2)^2}\left(\frac{1-\alpha}{1+\alpha}\right)^{4/\alpha}\left[Q' - \frac{1}{2}\left(\frac{2}{3}\frac{a'}{\lambda}\right)^2\right],$$

$$N_{12} = 2\left(\frac{a}{\lambda}\right)^2 \left\{ \alpha'^2 - \frac{4E(\alpha)}{\pi}[E(\alpha') - \alpha^2 F(\alpha')] - R_1 + \alpha^2 R_2 \right\} T$$
$$+ \frac{24\alpha^3}{(1-\alpha^2)^2}\left(\frac{1-\alpha}{1+\alpha}\right)^{\frac{1}{2}(7\alpha+\frac{1}{\alpha})}\left[Q - \frac{1}{2}\left(\frac{2a}{3\lambda}\right)^2\right]$$
$$+ \frac{24\alpha}{(1-\alpha^2)^2}\left(\frac{1-\alpha}{1+\alpha}\right)^{\frac{1}{2}(\alpha+\frac{7}{\alpha})}\left[Q' - \frac{1}{2}\left(\frac{2}{3}\frac{a'}{\lambda}\right)^2\right].$$

$$\alpha = \frac{a'}{a} = 1 - \beta, \qquad \alpha' = \sqrt{1-\alpha^2}.$$

$$Q = 1 - \sqrt{1 - \left(\frac{2a}{3\lambda}\right)^2}. \qquad Q' = 1 - \sqrt{1 - \left(\frac{2}{3}\frac{a'}{\lambda}\right)^2}.$$

The functions $F(\alpha)$ and $E(\alpha)$ are complete elliptical integrals of the first and second kinds, respectively.

Restrictions.—The equivalent circuit is valid in the range

$$0.5 < \frac{a}{\lambda} < 1.5$$

and $2a'/\lambda > 1$. Equations (1a), (2a), and (3a) have been obtained by the equivalent static method using one static mode incident in each guide and are estimated to be in error by less than 1 per cent over most of the wavelength range; the error may rise to as much as 5 per cent at the limit $a/\lambda = 1.5$. Equations (1b), (2b), and (3b), valid in the small-aperture range, are approximations that differ from the more accurate Eqs. (1a), (2a), and (3a) by less than 5 per cent for $0.5 < (a/\lambda) < 1$ and $\alpha < 0.4$; for $\alpha < 0.5$ the difference is less than 10 per cent. Equations (1c), (2c), and (3c) are approximations in the wide-aperture range and are in error by less than 6 per cent for $0.5 < (a/\lambda) < 1$ and $\beta < 0.3$. The constants X_{11}, X_{12}, and X_{22} are the impedance matrix elements of the T network referred to a common reference plane T and correspond to a characteristic impedance choice, $Z_0'/Z_0 = \lambda_g' a/\lambda_g a'$.

Numerical Results.—Figure 5·24-2 contain plots of $Z_0' \lambda_g a / Z_0 \lambda_g' a'$ and l/a as functions of α for various values of the parameter a/λ. The quantities $X\lambda_g/Z_0 2a$ and $Z_0 2a/X\lambda_g$ are plotted in Fig. 5·24-3 as a function of α for various values of the parameter a/λ. These results have been plotted from data obtained by use of the equivalent static method with two modes incident in each guide and are somewhat more accurate than the analytic results given in Eqs. (1a), (2a), and (3a).

b. Asymmetrical Case.—A junction of two rectangular guides of unequal widths but of equal heights (H_{10}-mode in each rectangular guide).

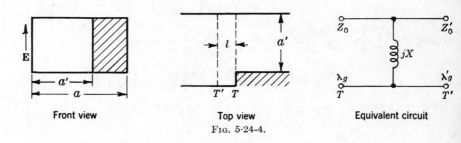

Front view   Top view   Equivalent circuit

FIG. 5·24-4.

Equivalent-circuit Parameters.—At reference planes T and T' for the larger and smaller guide, respectively.

SEC. 5·24] JUNCTION OF TWO RECTANGULAR GUIDES, H-PLANE

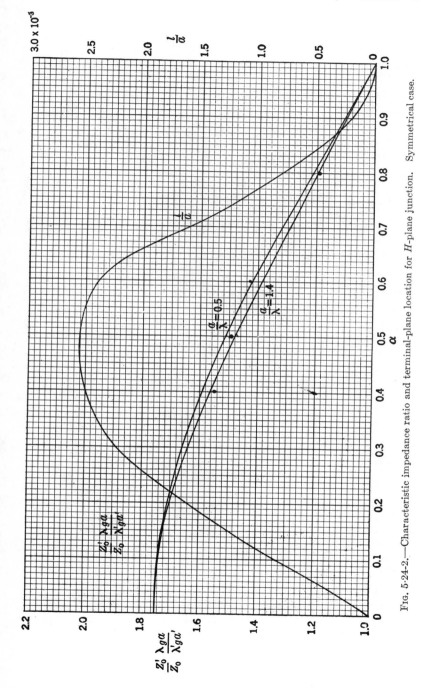

FIG. 5·24-2.—Characteristic impedance ratio and terminal-plane location for H-plane junction. Symmetrical case.

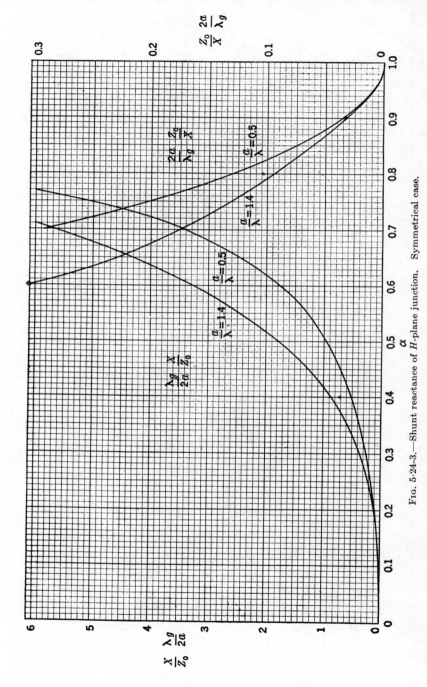

Fig. 5·24-3.—Shunt reactance of H-plane junction. Symmetrical case.

SEC. 5·24]   JUNCTION OF TWO RECTANGULAR GUIDES, H-PLANE   301

$$\frac{Z_0'}{Z_0} = \text{same as in Eq. (1a), Sec. 5·24a,} \tag{4a}$$

$$\frac{Z_0'}{Z_0} \approx \frac{\lambda_g' a'^3}{\lambda_g a^3} 5.046(1 - 1.44\alpha^2), \qquad \alpha \ll 1, \tag{4b}$$

$$\frac{Z_0'}{Z_0} \approx \frac{\frac{\lambda_g'}{a'}}{\frac{\lambda_g}{a}}\left(1 - \beta - \frac{7\beta^2}{4}\right), \qquad \beta = (1 - \alpha) \ll 1. \tag{4c}$$

$$\frac{X}{Z_0} = \text{same as Eq. (2a), Sec. 5·24a,} \tag{5a}$$

$$\frac{X}{Z_0} \approx \frac{2a}{\lambda_g} 5.764\alpha^4(1 - 1.33\alpha^2)(1 + 41.3\alpha^4 Q), \qquad \alpha \ll 1 \tag{5b}$$

$$\frac{Z_0}{X} \approx \frac{\lambda_g}{2a} \frac{\beta^2}{1 - \frac{\beta}{2} - \beta^2\left(\ln\frac{2}{\beta} + \frac{3}{2}\right)} \left\{\frac{7}{4}\ln\frac{2}{\beta}\left[1 + \beta - \frac{5}{4}\beta^2 \ln\frac{2}{\beta}\right]\right.$$

$$\left. \left[1 + \frac{4}{7}(1-\beta)^2\left(\frac{a}{\lambda}\right)^2\right] - \frac{8}{9}\left[1 + 3\beta \ln\frac{2}{\beta}\right][Q + Q']\right\}. \tag{5c}$$

$$\frac{l}{a} = \text{same as in Eq. (3a), Sec. 5·24a,} \tag{6a}$$

$$\frac{l}{a} \approx 0.0259\alpha(1 - 0.6\alpha^2), \qquad \alpha \ll 1, \tag{6b}$$

$$\frac{l}{a} \approx 0, \qquad \alpha \approx 1. \tag{6c}$$

The parameters X_{11}, X_{12}, X_{22}, X_0, A, A', C, α, α' and β are defined as in Sec. 5·24a but with

$$R_1 = -\left(\frac{1 + 3\alpha^2}{1 - \alpha^2}\right)\left(\frac{1-\alpha}{1+\alpha}\right)^{2\alpha},$$

$$R_2 = \left(\frac{3 + \alpha^2}{1 - \alpha^2}\right)\left(\frac{1-\alpha}{1+\alpha}\right)^{2/\alpha},$$

$$T = \frac{16\alpha^2}{(1-\alpha^2)^2}\left(\frac{1-\alpha}{1+\alpha}\right)^{\alpha + (1/\alpha)},$$

$$N_{11} = 2\left(\frac{a}{\lambda}\right)^2\left[1 + \frac{8\alpha(1-\alpha^2)}{1 + 3\alpha^2} R_1 \ln\frac{1-\alpha}{1+\alpha} + \frac{16\alpha^2}{1 + 3\alpha^2} R_1 - R_1^2 - \alpha^2 T^2\right]$$

$$+ 2\left[\frac{32}{3}\frac{\alpha^4}{(1-\alpha^2)^2}\left(\frac{1-\alpha}{1+\alpha}\right)^{3\alpha}\right]^2 \left[Q - \frac{1}{2}\left(\frac{a}{\lambda}\right)^2\right]$$

$$+ 2\left[\frac{32\alpha^2}{(1-\alpha^2)^2}\left(\frac{1-\alpha}{1+\alpha}\right)^{\alpha+(2/\alpha)}\right]^2 \left[Q' - \frac{1}{2}\left(\frac{a'}{\lambda}\right)^2\right],$$

$$N_{22} = 2\left(\frac{a}{\lambda}\right)^2\left[\alpha^2 - \frac{8\alpha(1-\alpha^2)}{3+\alpha^2}R_2\ln\frac{1-\alpha}{1+\alpha} + \frac{16\alpha^2}{1+3\alpha^2}R_2 - \alpha^2 R_2^2 - T^2\right]$$

$$+ 2\left[\frac{32\alpha^2}{(1-\alpha^2)^2}\left(\frac{1-\alpha}{1+\alpha}\right)^{2\alpha+(1/\alpha)}\right]^2\left[Q - \frac{1}{2}\left(\frac{a}{\lambda}\right)^2\right]$$

$$+ 2\left[\frac{32}{3}\frac{1}{(1-\alpha^2)^2}\left(\frac{1-\alpha}{1+\alpha}\right)^{3/\alpha}\right]^2\left[Q' - \frac{1}{2}\left(\frac{a'}{\lambda}\right)^2\right],$$

$$N_{12} = 2\left(\frac{a}{\lambda}\right)^2\left[\frac{(1-\alpha^2)^2}{2\alpha}\ln\frac{1-\alpha}{1+\alpha} - R_1 - \alpha^2 R_2\right]T$$

$$+ \frac{2}{3}\left(\frac{1-\alpha}{1+\alpha}\right)^{5\alpha+(1/\alpha)}\left[\frac{32\alpha^3}{(1-\alpha^2)^2}\right]^2\left[Q - \frac{1}{2}\left(\frac{a}{\lambda}\right)^2\right]$$

$$+ \frac{2}{3}\left(\frac{1-\alpha}{1+\alpha}\right)^{\alpha+(5/\alpha)}\left[\frac{32\alpha}{(1-\alpha^2)^2}\right]^2\left[Q' - \frac{1}{2}\left(\frac{a'}{\lambda}\right)^2\right],$$

$$Q = 1 - \sqrt{1 - \left(\frac{a}{\lambda}\right)^2}, \qquad Q' = 1 - \sqrt{1 - \left(\frac{a'}{\lambda}\right)^2}.$$

Restrictions.—The equivalent circuit is valid in the range

$$0.5 < \frac{a}{\lambda} < 1.0.$$

Equations (4a), (5a), and (6a) have been obtained by the equivalent static method employing one static mode incident in each guide. These equations are estimated to be in error by less than 1 per cent over most of the wavelength range and by as high as 5 per cent at the limit $a/\lambda = 1.0$. Equations (4b), (5b), and (6b), valid in the small-aperture range, are approximations that differ from Eqs. (4a), (5a), and (6a) by less than 5 per cent for $\alpha < 0.4$ and $0.5 < a/\lambda < 1$. Equations (4c), (5c), and ((6c) are approximations in the large-aperture range and are in error by less than 10 per cent for $\beta < 0.2$ and $0.5 < a/\lambda < 0.8$. The parameters X_{11}, X_{12}, and X_{22} are the impedance matrix elements of the T network referred to the common reference plane T and correspond to a characteristic impedance choice $Z_0'/Z_0 = \lambda_g'a/\lambda_g a'$.

Numerical Results.—In Fig. 5·24-5 there are indicated curves for $Z_0'\lambda_g a^3/Z_0\lambda_g' a'^3$ and l/a as functions of α for various values of the parameter λ/a. Figure 5·24-6 contains a plot of $X\lambda_g/Z_0 2a$ as a function of α for various values of λ/a. These results have been plotted from data that are obtained by use of the equivalent static method with two modes incident in each guide and thus are somewhat more accurate than the analytic results indicated in Eqs. (4a), (5a), and (6a).

5·25. Bifurcation of a Rectangular Guide, H-plane.—A bifurcation of a rectangular guide by a metallic wall of zero thickness aligned parallel to the electric field (H_{10}-modes in rectangular guides 1 and 2, no propagation in guide 3).

SEC. 5·24] JUNCTION OF TWO RECTANGULAR GUIDES, H-PLANE

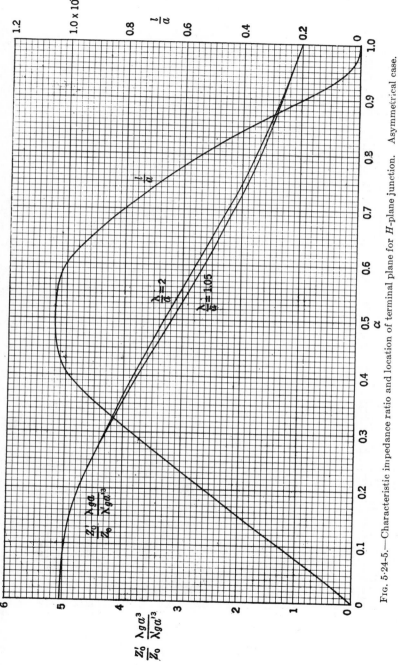

FIG. 5·24-5.—Characteristic impedance ratio and location of terminal plane for H-plane junction. Asymmetrical case.

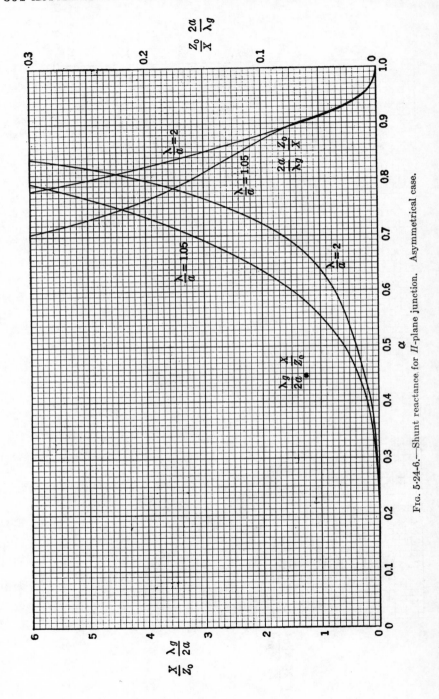

Fig. 5·24-6.—Shunt reactance for H-plane junction. Asymmetrical case.

5·25. BIFURCATION OF A RECTANGULAR GUIDE, H-PLANE

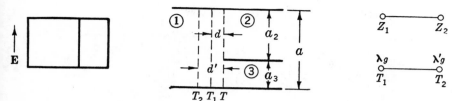

Cross sectional view   Top view   Equivalent circuit
FIG. 5·25-1.

Equivalent-circuit Parameters.—At reference planes T_1 and T_2 in guides 1 and 2, respectively, the equivalent network is a simple junction of two uniform transmission lines whose characteristic impedance ratio is

$$\frac{Z_2}{Z_1} = \frac{\lambda'_g}{\lambda_g}. \tag{1}$$

The terminals T_1 and T_2 are located at distances d and d' given by

$$\theta = \frac{2\pi d}{\lambda_g} = x \left(\frac{a_2}{a} \ln \frac{a}{a_2} + \frac{a_3}{a} \ln \frac{a}{a_3} \right) - \sin^{-1} \frac{x_3}{\sqrt{1 - \left(\frac{a_3}{a}\right)^2}}$$

$$+ S_2(x;1,0) - S_2\left(x_2; \frac{a_2}{a}, 0\right) - S_2\left(x_3; \frac{a_3}{a}, 0\right), \tag{2}$$

$$\theta' = \frac{2\pi d'}{\lambda'_g} = x' \left(\frac{a_2}{a} \ln \frac{a}{a_2} + \frac{a_3}{a} \ln \frac{a}{a_3} \right) - \sin^{-1} \frac{x'_3}{\sqrt{1 - \left(\frac{a_3}{a_2}\right)^2}}$$

$$+ S_2\left(x'; \frac{a}{a_2}, 0\right) - S_2(x'_2;1,0) - S_2\left(x'_3; \frac{a_3}{a_2}, 0\right), \tag{3}$$

where

$$x = \frac{2a}{\lambda_g}, \quad x_2 = \frac{2a_2}{\lambda_g}, \quad x_3 = \frac{2a_3}{\lambda_g}, \quad x = x_2 + x_3,$$

$$x' = \frac{2a}{\lambda'_g}, \quad x'_2 = \frac{2a_2}{\lambda'_g}, \quad x'_3 = \frac{2a_3}{\lambda'_g}, \quad x' = x'_2 + x'_3,$$

$$S_2(x;\alpha,0) = \sum_{n=2}^{\infty} \left[\sin^{-1} \frac{x}{\sqrt{n^2 - \alpha^2}} - \frac{x}{n} \right]$$

and the wavelengths in guides 1 and 2, respectively, are

$$\lambda_g = \frac{\lambda}{\sqrt{1 - \left(\frac{\lambda}{2a}\right)^2}}, \quad \lambda'_g = \frac{\lambda}{\sqrt{1 - \left(\frac{\lambda}{2a_2}\right)^2}}.$$

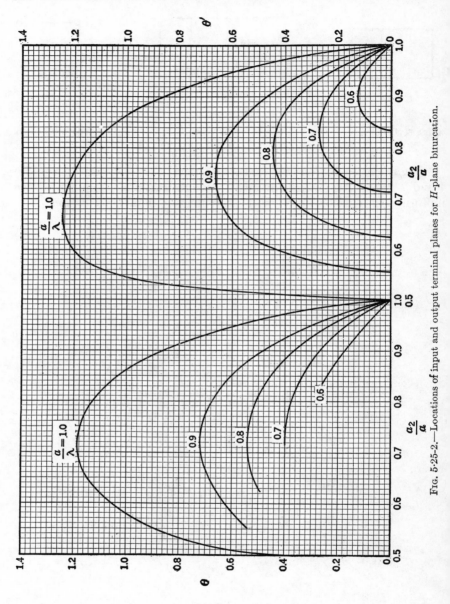

Fig. 5·25-2.—Locations of input and output terminal planes for H-plane bifurcation.

SEC. 5·26] CHANGE IN HEIGHT OF RECTANGULAR GUIDE 307

An alternative representation at the reference plane T is provided by a T-type network having the reactance parameters

$$\frac{X_{11}}{Z_1} = \frac{\frac{\lambda_g'}{\lambda_g} + \tan\theta \tan\theta'}{\frac{\lambda_g'}{\lambda_g} \tan\theta - \tan\theta'},$$

$$\frac{X_{12}}{Z_1} = \frac{\frac{\lambda_g'}{\lambda_g} \sec\theta \sec\theta'}{\frac{\lambda_g'}{\lambda_g} \tan\theta - \tan\theta'},$$

$$\frac{X_{22}}{Z_1} = \frac{\lambda_g'}{\lambda_g} \frac{1 + \frac{\lambda_g'}{\lambda_g} \tan\theta \tan\theta'}{\frac{\lambda_g'}{\lambda_g} \tan\theta - \tan\theta'}.$$

Restrictions.—The equivalent circuits are valid in the range $a < \lambda < 2a$ and $\lambda < 2a_2$. The circuit parameters have been obtained by the transform method and are rigorous in the above range.

Numerical Results.—In Fig. 5·25-2, θ and θ' are plotted as functions of a_2/a for various values of a/λ in the range of validity of the formulas. The S_2 functions are tabulated in the Appendix.

5·26. Change in Height of Rectangular Guide. *a. Symmetrical Case.* An axially symmetrical junction of two rectangular guides of equal widths and unequal heights (H_{10}-mode in rectangular guides).

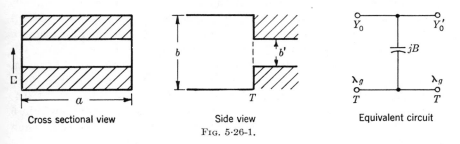

Cross sectional view   Side view   Equivalent circuit
FIG. 5·26-1.

Equivalent-circuit Parameters.—At the terminal plane T

$$\frac{Y_0}{Y_0'} = \frac{b'}{b} = \alpha = 1 - \delta, \tag{1}$$

$$\frac{B}{Y_0} = \frac{2b}{\lambda_g}\left[\ln\left(\frac{1-\alpha^2}{4\alpha}\right)\left(\frac{1+\alpha}{1-\alpha}\right)^{\frac{1}{2}(\alpha+\frac{1}{\alpha})} + 2\frac{A+A'+2C}{AA'-C^2} \right.$$
$$\left. + \left(\frac{b}{4\lambda_g}\right)^2\left(\frac{1-\alpha}{1+\alpha}\right)^{4\alpha}\left(\frac{5\alpha^2-1}{1-\alpha^2} + \frac{4}{3}\frac{\alpha^2 C}{A}\right)^2\right], \tag{2a}$$

$$\frac{B}{Y_0} \approx \frac{2b}{\lambda_g}\left[\ln\left(\frac{1-\alpha^2}{4\alpha}\right)\left(\frac{1+\alpha}{1-\alpha}\right)^{\frac{1}{2}(\alpha+\frac{1}{\alpha})} + \frac{2}{A}\right], \qquad (2b)$$

$$\frac{B}{Y_0} \approx \frac{2b}{\lambda_g}\left[\ln\frac{e}{4\alpha} + \frac{\alpha^2}{3} + \frac{1}{2}\left(\frac{b}{\lambda_g}\right)^2(1-\alpha^2)^4\right], \qquad \alpha \ll 1, \qquad (2c)$$

$$\frac{B}{Y_0} \approx \frac{2b}{\lambda_g}\left(\frac{\delta}{2}\right)^2\left[\frac{2\ln\frac{2}{\delta}}{1-\delta} + 1 + \frac{17}{16}\left(\frac{b}{\lambda_g}\right)^2\right], \qquad \delta \ll 1, \qquad (2d)$$

where

$$A = \left(\frac{1+\alpha}{1-\alpha}\right)^{2\alpha}\frac{1+\sqrt{1-\left(\dfrac{b}{\lambda_g}\right)^2}}{1-\sqrt{1-\left(\dfrac{b}{\lambda_g}\right)^2}} - \frac{1+3\alpha^2}{1-\alpha^2},$$

$$A' = \left(\frac{1+\alpha}{1-\alpha}\right)^{2/\alpha}\frac{1+\sqrt{1-\left(\dfrac{b'}{\lambda_g}\right)^2}}{1-\sqrt{1-\left(\dfrac{b'}{\lambda_g}\right)^2}} + \frac{3+\alpha^2}{1-\alpha^2},$$

$$C = \left(\frac{4\alpha}{1-\alpha^2}\right)^2, \qquad e = 2.718.$$

Restrictions.—The equivalent circuit is valid in the range $b/\lambda_g < 1$. Equation (2a) has been obtained by the equivalent static method employing a static aperture field due to the incidence of the two lowest modes and is correct to within 1 per cent in the range $b/\lambda_g < 1$. The approximate Eq. (2b) is correct to within 3 per cent for $(b/\lambda_g) < 0.7$; for $(b/\lambda_g) \geqq 0.7$ it is still correct to within 3 per cent if $\alpha < 0.7$. Equation (2c) is an asymptotic expansion of Eq. (2a) correct to within 5 per cent if $\alpha < 0.6$ and $b/\lambda_g < 0.5$ and to within 2 per cent if $\alpha < 0.4$ and $b/\lambda_g < 0.4$. Equation (2d) is an asymptotic expansion of Eq. (2a) correct to within 5 per cent when $\delta < 0.5$ and $b/\lambda_g < 0.5$ and to within 3 per cent when $\delta < 0.4$ and $b/\lambda_g < 0.4$.

Numerical Results.—In Fig. 5·26-3 $B\lambda_g/Y_0 b$, as obtained from Eq. (2a), is plotted as a function of b'/b with b/λ_g as a parameter.

Experimental Results.—The above results have been verified experimentally at least for the cases $b/\lambda_g = 0.23$ and $\alpha > 0.15$ to within a few per cent.

b. Asymmetrical Case.—An axially asymmetrical junction of two rectangular guides of equal widths but unequal heights (H_{10}-modes in rectangular guides).

SEC. 5·26]   CHANGE IN HEIGHT OF RECTANGULAR GUIDE   309

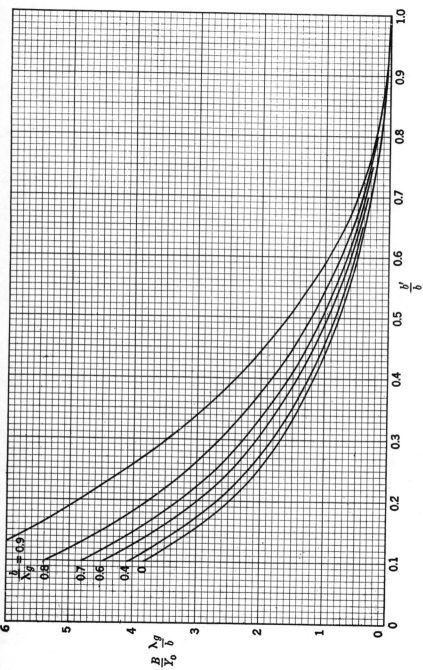

FIG. 5·26-3.—Shunt susceptance for change in height of rectangular guide.

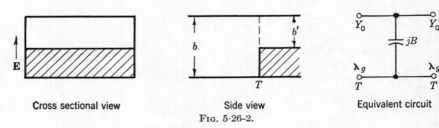

Fig. 5·26-2.

Cross sectional view — Side view — Equivalent circuit

Equivalent-circuit Parameters.—Same as Sec. 5·26a except λ_g is replaced by $\lambda_g/2$.

Restrictions.—Same as Sec. 5·26a except λ_g is replaced by $\lambda_g/2$.

Numerical Results.—If the λ_g in Fig. 5·26-3 is replaced by $\lambda_g/2$, one obtains a plot of $B\lambda_g/2Y_0 b$ as a function of b'/b with $2b/\lambda_g$ as a parameter.

5·27. Change in Radius of Coaxial Guide. *a. Equal Outer Radii.*— A junction of two coaxial guides of unequal inner but equal outer radii (principal mode in coaxial guides).

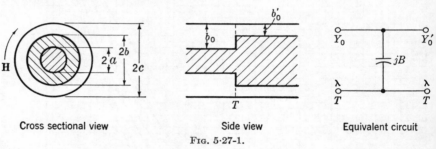

Fig. 5·27-1.

Cross sectional view — Side view — Equivalent circuit

Equivalent-circuit Parameters.—At the reference plane T

$$\frac{Y_0'}{Y_0} = \frac{\ln\dfrac{c}{a}}{\ln\dfrac{c}{b}}, \tag{1}$$

$$\frac{B}{Y_0} = \frac{2b_0 A_1}{\lambda}\left[2\ln\left(\frac{1-\alpha^2}{4\alpha}\right)\left(\frac{1+\alpha}{1-\alpha}\right)^{\frac{1}{2}\left(\alpha+\frac{1}{\alpha}\right)} + 4\frac{A+A'+2C}{AA'-C^2} \right.$$
$$\left. +\frac{1}{2}\left(\frac{b_0}{\lambda}\right)^2\left(\frac{1-\alpha}{1+\alpha}\right)^{4\alpha}\left(\frac{5\alpha^2-1}{1-\alpha^2} + \frac{4}{3}\frac{\alpha^2 C}{A}\right)^2 + \frac{A_2}{2}\right], \tag{2a}$$

$$\frac{B}{Y_0} \approx \frac{2b_0 A_1}{\lambda}\left[2\ln\frac{e}{4\alpha} + \frac{2\alpha^2}{3} + 4\left(\frac{b_0}{\lambda}\right)^2(1-\alpha^2)^4 + \frac{A_2}{2}\right], \quad \alpha \ll 1, \tag{2b}$$

$$\frac{B}{Y_0} \approx \frac{2b_0 A_1}{\lambda}\left(\frac{\delta}{2}\right)^2\left[\frac{4\ln\dfrac{2}{\delta}}{1-\delta} + 2 + \frac{17}{2}\left(\frac{b_0}{\lambda}\right)^2 + \frac{A_2}{2}\right], \quad \delta \ll 1, \tag{2c}$$

where

$$\alpha = 1 - \delta = \frac{c-b}{c-a}, \qquad b_0 = c - a, \qquad b_0' = c - b, \qquad e = 2.718.$$

The terms A, A', and C are defined in Sec. 5·26a, provided the λ_g, b, and b' therein are replaced by $\lambda/2$, b_0, and b_0' of this section. The terms A_1 and A_2 are defined in Sec. 5·3a.

Restrictions.—The equivalent circuit is valid in the wavelength range $\lambda > 2(c-a)/\gamma_1$, provided the field is rotationally symmetrical. The circuit parameter has been evaluated by a variational method treating the first higher E-mode correctly and all higher modes by plane-parallel approximations. Equation (2a) is estimated to be correct to within a few per cent for $c/a < 5$ and for wavelengths not too close to $2(c-a)/\gamma_1$. Equation (2b) is a small-aperture approximation and, for $c/a = 1$, agrees with Eq. (2a) to within 5 per cent for $\alpha < 0.6$ and $2b_0/\lambda < 0.5$ and to within 2 per cent for $\alpha < 0.4$ and $2b_0/\lambda < 0.4$. Similarly, Eq. (2c) is an asymptotic expansion of Eq. (2a) and, for $c/a = 1$, is correct to within 5 per cent if $\delta < 0.5$ and $2b_0/\lambda < 0.5$ and to within 3 per cent if $\delta < 0.4$ and $2b_0/\lambda < 0.4$. For $c/a > 1$ the agreement of Eqs. (2b) and (2c) with (2a) is presumably of the same order of magnitude in the indicated ranges.

Numerical Results.—For $c/a = 1$, the graph of $B\lambda/Y_0 2b_0$ as a function of α for various values of $2b_0/\lambda$ may be obtained from Fig. 5·26-3 if the λ_g therein is replaced everywhere by $\lambda/2$ and b by b_0. For $c/a > 1$, $B\lambda/Y_0 2b_0$ may be obtained from its value for $c/a = 1$ by addition of the term $A_2/2$ and multiplication of the resulting sum by A_1. The quantities A_1 and A_2 are plotted in Figs. 5·3-2 to 5·3-4. The term $\pi\gamma_1 = \left(\dfrac{c}{a} - 1\right)\chi_0$ is tabulated in Table 2·3 as a function of the ratio c/a (note that the c of Table 2·3 is the c/a of this section).

b. Equal Inner Radii.—A junction of two coaxial guides of equal inner but unequal outer radii (principal mode in coaxial guides).

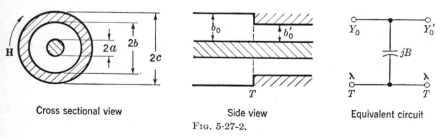

Fig. 5·27-2.

Equivalent-circuit Parameters.—At the reference plane T

$$\left. \begin{array}{l} \dfrac{Y_0'}{Y_0} = \dfrac{\ln \dfrac{c}{a}}{\ln \dfrac{b}{a}}, \\[1em] \dfrac{B}{Y_0} = \text{same as Eqs. (2a) to (2c) of Sec. 5·27}a, \end{array} \right\} \quad (1)$$

except that now

$$\alpha = 1 - \delta = \frac{b-a}{c-a}, \qquad b_0 = c - a, \qquad b_0' = b - a.$$

The terms A, A', and C are defined in Sec. 5·26a provided the λ_g, b, and b' therein are replaced by $\lambda/2$, b_0, and b_0', respectively, of this section. The terms A_1 and A_2 are the same as in Sec. 5·3b.

Restrictions.—Same as in Sec. 5·27a.

Numerical Results.—For $c/a = 1$, the graph of $B\lambda/Y_0 2b_0$ as a function of α for various values of $2b_0/\lambda$ may be obtained from Fig. 5·26-3 if the λ_g therein is replaced everywhere by $\lambda/2$. For $c/a > 1$, $B\lambda/Y_0 2b_0$ may be obtained from its value for $c/a = 1$ by addition of the term $A_2/2$ and multiplication of the resulting sum by A_1. The quantities A_1 and A_2 are plotted in Figs. 5·3-6 to 5·3-8.

5·28. E-plane Corners. *a. Right-angle Bends.*—A right-angle E-plane junction of two rectangular guides of equal dimensions (H_{10}-mode in rectangular guides).

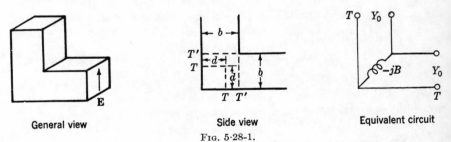

General view   Side view   Equivalent circuit

Fig. 5·28-1.

Equivalent-circuit Parameters.—At reference planes T the equivalent circuit is pure shunt. The relative shunt susceptance and location of the terminal planes are given by

$$\frac{B}{Y_0} = \frac{\left(\dfrac{B_a}{Y_0}\right)^2 + 1}{\dfrac{B_b}{Y_0}} - 2\frac{B_a}{Y_0}, \quad (1)$$

Sec. 5·28] E-PLANE CORNERS 313

$$\frac{2\pi(b-d)}{\lambda_g} = \cot^{-1}\left(2\frac{B_b}{Y_0} - \frac{B_a}{Y_0}\right), \quad (2)$$

where

$$\frac{B_a}{Y_0} = \frac{2b}{\lambda_g}\left\{-\frac{\cot \pi x}{x} + \frac{1}{\pi x^2} + \frac{\pi}{6} - \ln 2\right.$$

$$\left. - \left[\frac{A_0 e^{-\frac{\pi}{2}} + (A_1 - A_2)e^{-\pi} + (1 + 5e^{-\pi})\frac{A_0^2}{16}}{1 - (1 + 5e^{-\pi})\left(\frac{A_1 - A_2}{4}\right)}\right]\right\}, \quad (3a)$$

$$\frac{B_a}{Y_0} \approx \frac{2b}{\lambda_g}\left[0.878 + 0.498\left(\frac{2b}{\lambda_g}\right)^2\right], \quad \frac{2b}{\lambda_g} \ll 1, \quad (3b)$$

$$\frac{2B_b}{Y_0} - \frac{B_a}{Y_0} = \frac{\lambda_g}{2\pi b}\left\{1 + \pi x \cot \pi x - \pi x^2\left[5 \ln 2 - \frac{7\pi}{6} - 8\sum_{n=1}^{\infty}\frac{1}{n(e^{2\pi n} - 1)}\right]\right.$$

$$\left. + \pi x^2 \left[\frac{A_0' e^{-\frac{\pi}{2}} - (A_1 + A_2)e^{-\pi} + (1 - 3e^{-\pi})\frac{A_0'^2}{16}}{1 + (1 - 3e^{-\pi})\frac{A_1 + A_2}{4}}\right]\right\}, \quad (4a)$$

$$\frac{B_b}{Y_0} \approx \frac{\lambda_g}{2\pi b}\left[1 - 0.114\left(\frac{2b}{\lambda_g}\right)^2\right], \quad \frac{2b}{\lambda_g} \ll 1, \quad (4b)$$

and

$$x = \frac{2b}{\lambda_g},$$

$$A_0 = \frac{4}{\pi}\frac{x^2}{1 - x^2}, \qquad A_0' = A_0 - \frac{8}{\sinh \pi},$$

$$A_1 = \frac{1}{\pi}\frac{x^2}{1 - 0.5x^2}, \qquad A_2 = 4\left[\frac{1}{\sqrt{1-x^2}(1 - e^{-2\pi\sqrt{1-x^2}})} - \frac{1}{1 - e^{-2\pi}}\right].$$

An alternative circuit at the reference planes T' is shown in Fig. 5·28-2.

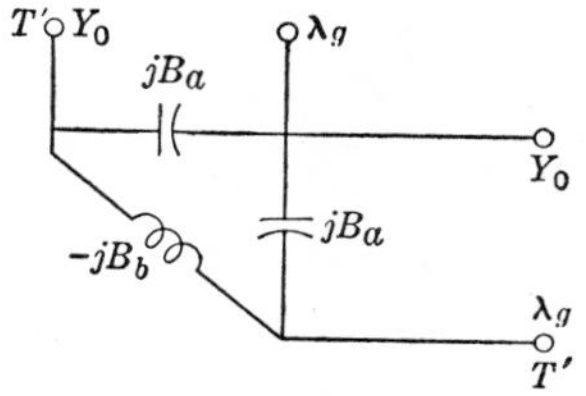

Restrictions.—The equivalent circuits are valid in the range $2b/\lambda_g < 1$. The circuit parameters obtained by the equivalent static method, employing two lowest modes incident in each guide, are accurate to within 1 per cent in the above range. Equations (3b)

Fig. 5·28-2.

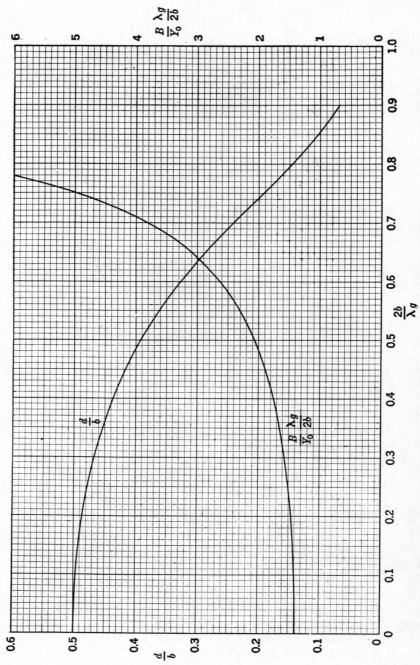

Fig. 5·28-3.—Shunt susceptance and location of terminal planes T for right-angle bend. E-plane.

SEC. 5·28] E-PLANE CORNERS

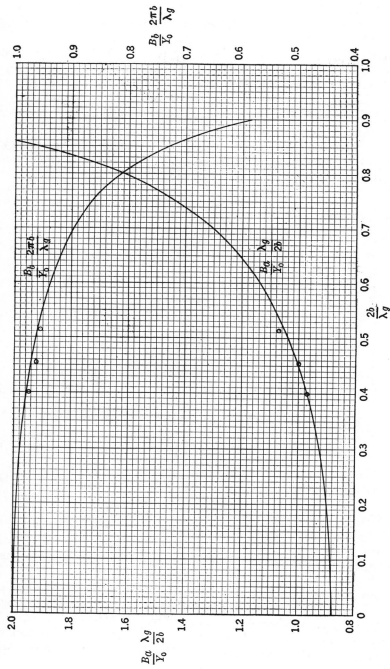

FIG. 5·28-4.—Circuit parameters of right-angle bend at terminals T'. E-plane.

and (4b) are asymptotic approximations to B_a and B_b that agree with the values from Eqs. (3a) and (4a) to within 8 per cent for $2b/\lambda_g < 0.6$.

Numerical Results.—In Fig. 5·28-3, d/b and $B\lambda_g/Y_0 2b$ are plotted as functions of $2b/\lambda_g$. The alternative circuit parameters $B_a\lambda_g/Y_0 2b$ and $B_b 2\pi b/Y_0 \lambda_g$ are shown in Fig. 5·28-4, as functions of $2b/\lambda_g$.

Experimental Results.—A few measured values for B_a and B_b are indicated by the circled points in Fig. 5·28-4. These data were taken at wavelengths in the vicinity of 3 cm.

b. Arbitrary Angle Bends.—A symmetrical, arbitrary angle, E-plane junction of two rectangular guides of equal dimensions (H_{10}-mode in rectangular guides).

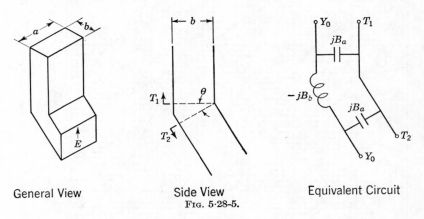

General View   Side View   Equivalent Circuit
Fig. 5·28-5.

Equivalent-circuit Parameters.—At the reference planes T

$$\frac{B_a}{Y_0} = \frac{2b}{\lambda_g}\left\{\Psi\left[-\frac{1}{2}\left(1 - \frac{\theta}{\pi}\right)\right] - \Psi\left[-\frac{1}{2}\right]\right\}, \tag{1}$$

$$\frac{B_b}{Y_0} = \frac{\lambda_g}{2\pi b}\cot\frac{\theta}{2}, \tag{2}$$

where $\Psi(x)$ is the derivative of the logarithm of $x!$

Restrictions.—The equivalent circuit is applicable in the wavelength range $2b/\lambda_g < 1$. The circuit parameters have been obtained by a simple equivalent static method and are rigorous only in the static limit $b/\lambda_g \approx 0$.

Numerical Results.—The circuit parameters $B_a\lambda_g/Y_0 2b$ and $B_b b/Y_0 \lambda_g$ are plotted as a function of θ in Fig. 5·28-6. The solid curves represent the static values of the parameters as given in Eqs. (1) and (2).

Experimental Results.—Measured values of the circuit parameters are indicated by the circled points in Fig. 5·28-6. These data apply

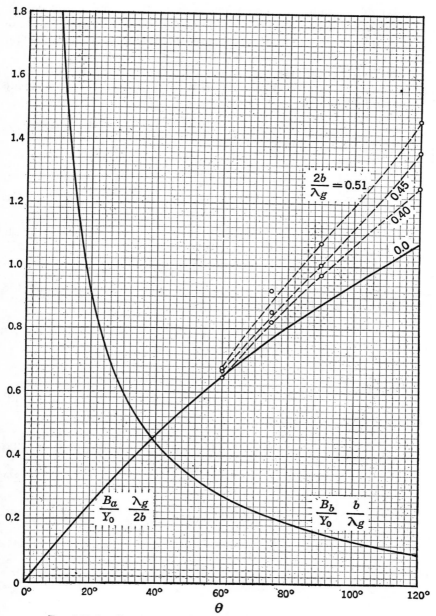

Fig. 5·28-6.—Circuit parameters for E-plane bends of arbitrary angle.

to rectangular guides of dimensions $a = 0.900$ in. and $b = 0.400$ in. and were measured for several values of b/λ_g. The measured values of $B_b b/Y_0 \lambda_g$ are within a few per cent of the static values given in Eq. (2) for the bend angles $\theta = 60$, 75, and 90° if $2b/\lambda_g < 0.51$.

5·29. H-plane Corners. *a. Right-angle Bends.*—An H-plane right-angle junction of two rectangular guides with equal dimensions (H_{10}-mode in rectangular guides).

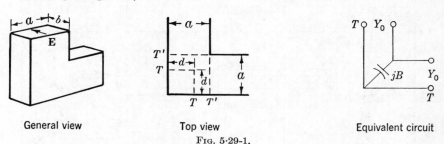

General view   Top view   Equivalent circuit
Fig. 5·29-1.

Equivalent-circuit Parameters.—At the terminal planes T the equivalent circuit is pure shunt. The relative shunt susceptance and the location of the terminal planes are given by

$$\frac{B}{Y_0} = \frac{1 + \left(\dfrac{B_a}{Y_0}\right)^2}{\dfrac{B_b}{Y_0}} + 2\,\frac{B_a}{Y_0}, \tag{1}$$

$$\frac{2\pi(a - d)}{\lambda_g} = -\cot^{-1}\frac{B_a + 2B_b}{Y_0}, \tag{2}$$

where

$$\frac{B_a}{Y_0} = \frac{\lambda_g}{2a}\left[A_1 - x \cot \pi x + \frac{1}{\pi}\left(\frac{1 + x^2}{1 - x^2}\right) + A_2\right](1 + A_3), \tag{3a}$$

$$\frac{B_a}{Y_0} \approx 0.42 + \frac{0.64}{1 - x}, \qquad x \approx 1, \tag{3b}$$

$$\frac{B_a + 2B_b}{Y_0} = \frac{\lambda_g}{2a}\left[-A_1' - x \cot \pi x - \frac{1}{\pi}\left(\frac{1 + x^2}{1 - x^2}\right) + A_2'\right](1 + A_3'), \tag{4a}$$

$$\frac{B_a + 2B_b}{Y_0} \approx -2.04 + 1.63x, \qquad x \approx 1, \tag{4b}$$

and

$$x = \frac{2a}{\lambda_g}, \qquad \lambda_g = \frac{\lambda}{\sqrt{1 - \left(\dfrac{\lambda}{2a}\right)^2}}$$

$$A_1 = \frac{2}{1 - e^{-2\pi}} - \frac{4(1 + 63e^{-2\pi})}{[2(1 + 5e^{-\pi})(1 + 63e^{-2\pi}) - (\frac{64}{3})^2 e^{-3\pi}]} = 0.3366,$$

$$A_1' = -\frac{2}{1-e^{-2\pi}} + \frac{4(1-33e^{-2\pi})}{[2(1-3e^{-\pi})(1-33e^{-2\pi}) - (\tfrac{32}{3})^2 e^{-3\pi}]} = 0.3071,$$

$$A_2 = \frac{\left[0.1801\left(\frac{1+x^2}{4-x^2}\right) + 0.2021\right]^2}{\frac{\sqrt{3-x^2}}{1-e^{-2\pi\sqrt{3-x^2}}} - 0.1858 - 0.1592\left(\frac{1+x^2}{7-x^2}\right)},$$

$$A_2' = \frac{\left[0.1801\left(\frac{1+x^2}{4-x^2}\right) + 0.1669\right]^2}{\frac{\sqrt{3-x^2}}{1-e^{-2\pi\sqrt{3-x^2}}} + 0.1432 + 0.1592\left(\frac{1+x^2}{7-x^2}\right)},$$

$$A_3 = 0.01 - 0.0095x \quad \text{for } 0 \leq x \leq 1,$$

$$A_3 = -\frac{0.00416}{1.5-x} + 0.0040 + 0.0019x + 0.0020x^2 \quad \text{for } 1 \leq x \leq 1.436,$$

$$A_3' = -\frac{0.00146}{1.26-x} - 0.0016 + 0.0034x - 0.0026x^2 \quad \text{for } 0 \leq x \leq 1.2,$$

$$A_3' = \frac{0.00136}{x-1.26} + 0.0076 - 0.0066x \quad \text{for } 1.3 \leq x \leq 1.5.$$

An alternative equivalent circuit at reference planes T' is in Fig. 5·29-2.

Restrictions.—The equivalent circuit is valid in the wavelength range $a < \lambda < 2a$. The circuit parameters have been derived by the equivalent static method with two static modes incident in each guide. The error in the susceptance values is estimated to be of the order of magnitude of 1 per cent in the range $0 \leq 2a/\lambda_g \leq 1.44$, excluding the range $1.2 \leq 2a/\lambda_g \leq 1.3$ in which the variational correction A_3' diverges.

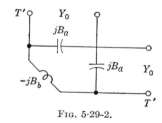

Fig. 5·29-2.

Equations (3b) and (4b) are approximations to Eqs. (3a) and (4a), respectively, and agree with these to within 10 per cent in the resonant range $0.7 \leq 2a/\lambda_g \leq 1.2$. The quantity A_2 is less than 10 per cent of B_a/Y_0 for $2a/\lambda_g < 1.3$; A_2' is less than 10 per cent of $(B_a + 2B_b)/Y_0$ for $2a/\lambda_g < 1.0$.

Numerical Results.—The quantities $B2a/Y_0\lambda_g$ and d/a are plotted as a function of $2a/\lambda_g$ in Fig. 5·29-3. A similar plot for the parameters $Y_0\lambda_g/B_b 2a$ and $(B_a + 2B_b)2a/Y_0\lambda_g$ appears in Fig. 5·29-4.

Experimental Results.—For comparison with the theoretical values a number of experimental data are indicated by circled points in Fig. 5·29-4. These data were measured in rectangular guides of dimensions $a = 0.900$ in. and $b = 0.400$ in.

b. *Arbitrary Angle H-plane Corners.*—A symmetrical, arbitrary angle, H-plane junction of two rectangular guides of identical cross-sectional dimensions (H_{10}-mode in rectangular guides).

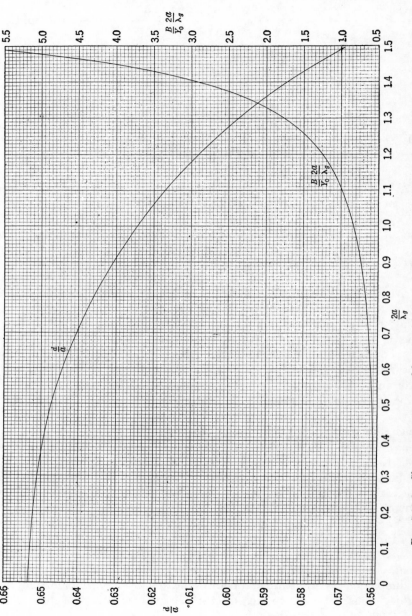

FIG. 5·29-3.—Shunt susceptance and location of terminal planes T for right-angle bend. H-plane.

SEC. 5·29] H-PLANE CORNERS

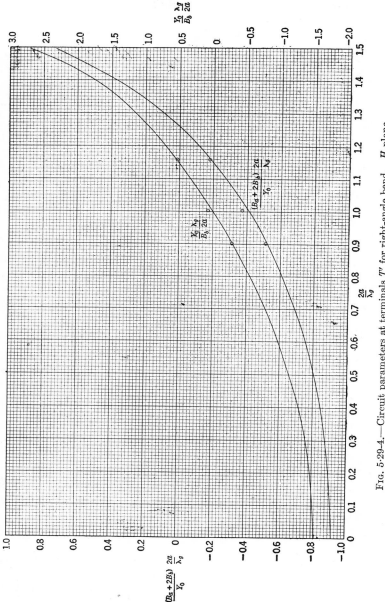

FIG. 5·29-4.—Circuit parameters at terminals T' for right-angle bend. H-plane.

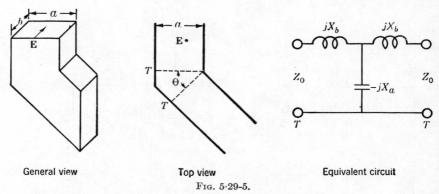

General view   Top view   Equivalent circuit
Fig. 5·29-5.

Experimental Results.—The equivalent circuit of Fig. 5·29-5 is valid in the wavelength range $a < \lambda < 2a$. The circuit parameters Z_0/X_a and X_b/Z_0, relative to the reference planes T, were measured in rectangular waveguide of dimensions $a = 0.900$ in. and $b = 0.400$ in. as a function of the bend angle θ at various wavelengths. The resulting data are:

θ	$\lambda = 3.0$ cm ($\lambda_g = 1.560''$)		$\lambda = 3.2$ cm ($\lambda_g = 1.762''$)		$\lambda = 3.4$ cm ($\lambda_g = 1.999''$)	
	Z_0/X_a	X_b/Z_0	Z_0/X_a	X_b/Z_0	Z_0/X_a	X_b/Z_0
30°	0.849	0.493	0.782	0.420	0.718	0.365
45°					0.995	0.567
50°	1.102	0.943			1.068	0.650
60°	1.088	1.27	1.158	0.998	1.176	0.817
75°	0.864	2.14				
90°	0.324	6.46	0.739	2.75	1.055	1.76
120°			−2.17	−4.58	0.044	47.9

5·30. Junction of a Rectangular and a Radial Guide, E-plane.—An E-plane junction of a rectangular guide and a radial guide of angular aperture θ (H_{10}-mode in rectangular guide; principal E-type mode in radial guide).

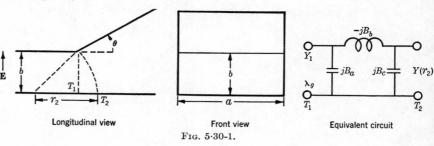

Longitudinal view   Front view   Equivalent circuit
Fig. 5·30-1.

SEC. 5·31] COUPLING OF A COAXIAL AND A CIRCULAR GUIDE

Equivalent-circuit Parameters.—At the reference planes T_1 for the rectangular guide and T_2 for the radial guide

$$\frac{Y(r_2)}{Y_1} = \frac{b}{r_2\theta} = \frac{\sin\theta}{\theta}, \quad (1)$$

$$\frac{B_a}{Y_1} = \frac{2\pi b}{\lambda_g \theta} \ln\frac{\theta}{\sin\theta}, \quad (2)$$

$$\frac{B_c}{Y_1} = \frac{2b}{\lambda_g}\left[0.577 + \Psi\left(\frac{\theta}{\pi}\right)\right], \quad (3)$$

$$\frac{B_b}{Y_1} = \frac{\lambda_g}{\pi b}\frac{\sin\theta}{\theta}\frac{\sin\theta}{1 - \frac{\sin 2\theta}{2\theta}}, \quad (4)$$

where $\Psi(x)$ is the logarithmic derivative of $x!$ and θ is measured in radians.

It is to be noted that the characteristic admittance of the radial line is variable and the propagation constant of the radial line is $\kappa = 2\pi/\lambda_g$ with

$$\lambda_g = \frac{\lambda}{\sqrt{1 - \left(\frac{\lambda}{2a}\right)^2}}.$$

The above formulas also apply to the case of a radial guide with downward taper, i.e., θ negative.

Restrictions.—The equivalent circuit is applicable in the wavelength range $2b/\lambda_g < 1$. The circuit parameters have been obtained by a simple equivalent static method and are strictly valid only in the range $2b/\lambda_g \ll 1$. The error is estimated to lie within a few per cent for $2b/\lambda_g < 0.1$.

Numerical Results.—Numerical values of the Ψ function may be obtained from Jahnke-Emde, Tables of Functions, 3d ed. (revised), page 16.

5·31. Coupling of a Coaxial and a Circular Guide.—A coaxial guide having a hollow center conductor of zero wall thickness coupled to a circular guide (principal mode in coaxial guide, E_{01}-mode in circular guide).

Cross sectional view

Side view
FIG. 5·31-1.

Equivalent circuit

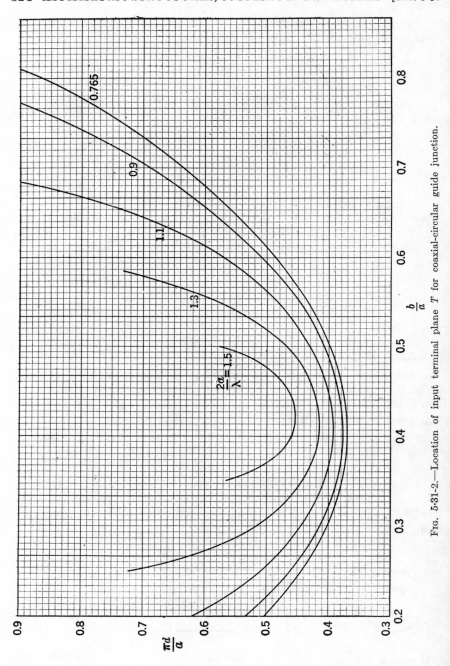

Fig. 5·31-2.—Location of input terminal plane T for coaxial-circular guide junction.

SEC. 5·31] COUPLING OF RECTANGULAR AND CIRCULAR GUIDE   325

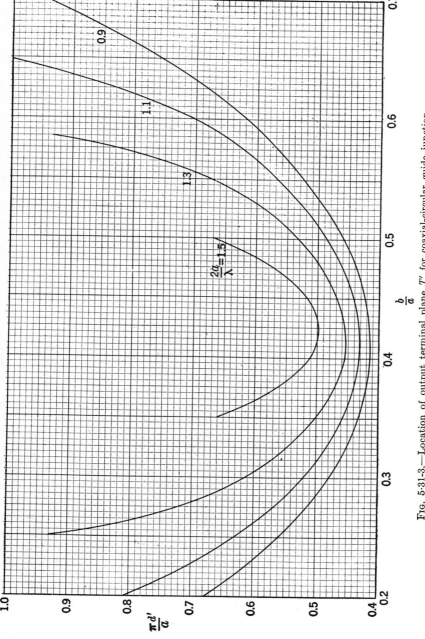

FIG. 5·31-3.—Location of output terminal plane T' for coaxial-circular guide junction.

Equivalent-circuit Parameters.—At the reference planes T and T' the equivalent network is a simple junction of two transmission lines whose characteristic impedance ratio is

$$\frac{Z_0'}{Z_0} = \frac{\lambda}{\lambda_g}. \tag{1}$$

The locations of the input and output planes are given by

$$\frac{2\pi d}{\lambda} = \sin^{-1}\frac{\alpha x}{\beta_1} + \sin^{-1}\frac{(1-\alpha)x}{\gamma_1} - \frac{2b}{\lambda}[-(1-\alpha)\ln(1-\alpha) - \alpha\ln\alpha]$$
$$- S_2^{J_0}(x;0) + S_2^{J_0}(\alpha x;0) + S_2^{Z_0}[(1-\alpha)x;0,c], \tag{2}$$

$$\frac{2\pi d'}{\lambda_g} = \sin^{-1}\frac{\alpha x'}{\beta_1\sqrt{1-\alpha^2}} + \sin^{-1}\frac{(1-\alpha)x'}{\sqrt{\gamma_1^2 - \beta_1^2(1-\alpha)^2}}$$
$$- \frac{2b}{\lambda_g}[-(1-\alpha)\ln(1-\alpha) - \alpha\ln\alpha] - S_2^{J_0}(x';\beta_1) + S_2^{J_0}(\alpha x';\alpha\beta_1)$$
$$+ S_2^{Z_0}[(1-\alpha)x';(1-\alpha)\beta_1,c], \tag{3}$$

where

$$x = \frac{2a}{\lambda}, \quad x' = \frac{2a}{\lambda_g}, \quad \alpha = \frac{b}{a} = \frac{1}{c}, \quad \lambda_g = \frac{\lambda}{\sqrt{1-\left(\frac{\lambda}{2.61a}\right)^2}},$$

$$S_2^{J_0}(x;y) = \sum_{n=2}^{\infty}\left[\sin^{-1}\frac{x}{\sqrt{\beta_n^2 - y^2}} - \frac{x}{n}\right], \quad J_0(\pi\beta_n) = 0,$$

$$S_2^{Z_0}(x;y,c) = \sum_{n=2}^{\infty}\left(\sin^{-1}\frac{x}{\sqrt{\gamma_n^2 - y^2}} - \frac{x}{n}\right),$$

where $\chi_{0n} = \pi\gamma_n/(c-1)$ is the nth nonvanishing root of

$$J_0(\chi)N_0(\chi c) - N_0(\chi)J_0(\chi c) = 0.$$

Restrictions.—The equivalent circuit is valid in the wavelength range $1.14a < \lambda < 2.61a$, provided $\lambda > 2.61b$ and the fields are rotationally symmetrical. The locations of the reference planes have been determined by the transform method and are rigorous in the above range.

Numerical Results.—The quantities $\pi d/a$ and $\pi d'/a$ are plotted in Figs. 5·31-2 and 5·31-3 as a function of b/a for various values of the parameter $2a/\lambda$. The roots $\pi\beta_n$ and $\chi_{0n}(c)$ may be obtained from the tables in Secs. 2·3a and 2·4a. The summation functions $S_2^{J_0}$ and $S_2^{Z_0}$ are tabulated in the Appendix.

5·32. Coupling of Rectangular and Circular Guide.—A junction of a rectangular and circular guide with a common symmetry axis (H_{10}-mode in rectangular guide, nonpropagating H_{11}-mode in circular guide).

5·32] COUPLING OF RECTANGULAR AND CIRCULAR GUIDE

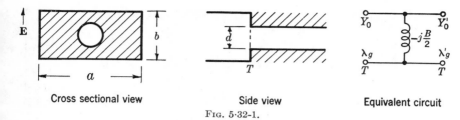

Cross sectional view   Side view   Equivalent circuit
Fig. 5·32-1.

Equivalent-circuit Parameters.—At the reference plane T

$$\frac{Y_0'}{Y_0} = -j\,\frac{0.446ab\lambda_g}{d^3}\sqrt{1 - \left(\frac{1.706d}{\lambda}\right)^2}\left[\frac{1 - \left(0.853\dfrac{d}{a}\right)^2}{2J_1'\left(\dfrac{\pi d}{2a}\right)}\right]^2, \quad (1a)$$

$$\frac{Y_0'}{Y_0} \approx -j\,\frac{0.446ab\lambda_g}{d^3}\sqrt{1 - \left(\frac{1.706d}{\lambda}\right)^2}, \qquad \frac{d}{a} \ll 1. \quad (1b)$$

$$\frac{B}{2Y_0} = \text{one-half the quantity } \frac{B}{Y_0} \text{ given in Eq. (1a) of Sec. 5·4a,} \quad (2a)$$

$$\frac{B}{2Y_0} \approx \frac{3}{4\pi}\frac{ab\lambda_g}{d^3}. \quad (2b)$$

where

$$\lambda_g = \frac{\lambda}{\sqrt{1 - \left(\dfrac{\lambda}{2a}\right)^2}}, \qquad \lambda_g' = j\,\frac{1.706d}{\sqrt{1 - \left(\dfrac{1.706d}{\lambda}\right)^2}}$$

The corresponding parameters for a two-terminal representation are given in Sec. 4·4.

Restrictions.—The equivalent circuit is an approximation in the range $2a > \lambda > 1.706d$ provided neither the H_{11}- nor the H_{30}-mode propagates in the rectangular guide. The circuit parameters have been determined by variational methods, Y_0'/Y_0 with the assumption that the aperture field is that of the H_{11}-mode in the circular guide and B/Y_0 as indicated in Sec. 5·4a. For $d < b$ and λ not too close to the cutoff wavelength of the next propagating mode, the error is estimated to be less than 10 per cent. The asymptotic Eqs. (1b) and (2b) agree with Eqs. (1a) and (2a) to within about 5 per cent if $d/a < 0.3$. For the equivalent circuit to be strictly valid, the output terminal plane must be located more accurately than indicated above.

Numerical Results.—The quantity $jY_0'a/Y_0b$ is plotted in Fig. 5·32-2 as a function of d/a for several values of λ/a. A similar plot for $B/2Y_0$ may be obtained by taking one-half of the quantity B/Y_0 plotted in Fig. 5·4-2.

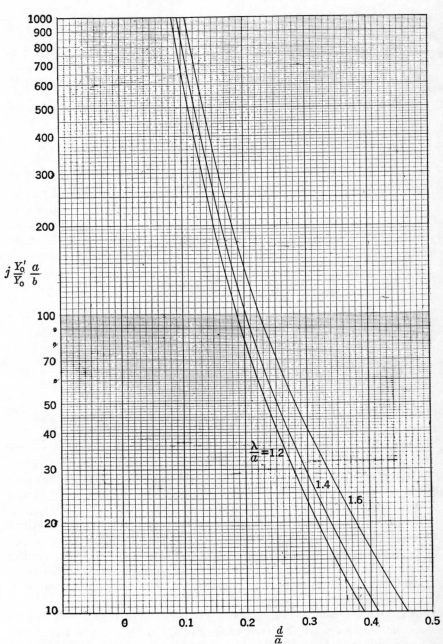

Fig. 5·32-2.—Characteristic admittance ratio for rectangular-circular guide junction.

Experimental Results.—Several experimentally determined data for the case of an infinitely long circular guide are indicated in Fig. 4·4-2.

5·33. Aperture Coupling of Two Guides. *a. Junction of Two Rectangular Guides.*—A junction of two rectangular guides of different cross sections coupled by a small elliptical or circular aperture in a transverse plate of zero thickness (H_{10}-mode in rectangular guides).

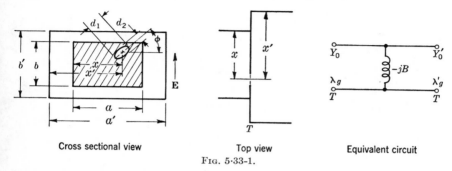

Cross sectional view   Top view   Equivalent circuit
Fig. 5·33-1.

Equivalent-circuit Parameters.—For the case of an elliptical aperture oriented as shown in the above figure

$$\frac{B}{Y_0} = \frac{\lambda_g ab}{4\pi \sin^2 \frac{\pi x}{a}} \left[\frac{1}{M} - 2\pi \left(\frac{\sin^2 \frac{\pi x}{a}}{a^2 b} + \frac{\sin^2 \frac{\pi x'}{a'}}{a'^2 b'} \right) \right], \quad (1a)$$

$$\frac{B}{Y_0} \approx \frac{\lambda_g ab}{4\pi M \sin^2 \frac{\pi x}{a}}, \quad \left. \begin{array}{c} d_1 \\ d_2 \end{array} \right\} \ll a, b, \quad (1b)$$

$$\frac{Y_0'}{Y_0} = \frac{\lambda_g ab}{\lambda_g' a' b'} \frac{\sin^2 \frac{\pi x'}{a'}}{\sin^2 \frac{\pi x}{a}}, \quad (2)$$

where
$$M = M_1 \cos^2 \phi + M_2 \sin^2 \phi.$$

The coefficients M_1 and M_2 are given by Eqs. (4) and (5) of Sec. 5·4b. For the case of a circular aperture ($d = d_1 = d_2$)

$$M = \frac{d^3}{6}.$$

Restrictions.—The equivalent circuit is applicable in the wavelength range $a' < \lambda < 2a$ ($a' > a$). Comments are the same as in Sec. 5·4b.

Numerical Results.—The coefficients M_1 and M_2 are shown in Fig. 5·4-4 as a function of d_2/d_1.

b. Junction of Two Circular Guides.—Two circular guides having a common axis of symmetry but of different radii coupled by a small elliptical or circular aperture in a transverse plate of zero thickness (H_{11}-mode in circular guides).

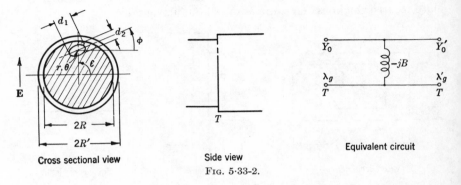

Cross sectional view   Side view   Equivalent circuit
Fig. 5·33-2.

Equivalent-circuit Parameters.—For the case of an elliptical aperture with orientation $\phi = 0$ or $\pi/2$.

$$\frac{B}{Y_0} = 0.952 \frac{\lambda_g}{4\pi} \frac{\pi R^2}{\left[\frac{2J_1(\alpha_1 r)}{\alpha_1 r}\right]^2} \left\{ \frac{1}{M} - 7.74 \left[\frac{1}{2\pi R^3} \left(\frac{2J_1(\alpha_1 r)}{\alpha_1 r}\right)^2 \right. \right.$$
$$\left. \left. + \frac{1}{2\pi R'^3} \left(\frac{2J_1(\alpha_2 r)}{\alpha_2 r}\right)^2 \right] \right\}, \quad (3a)$$

$$\frac{B}{Y_0} \approx 0.952 \frac{\lambda_g}{4\pi M} \frac{\pi R^2}{\left[\frac{2J_1(\alpha_1 r)}{\alpha_1 r}\right]^2}, \quad \left. \begin{array}{l} d_1 \\ d_2 \end{array} \right\} \ll a, r, \quad (3b)$$

$$\frac{Y_0'}{Y_0} = \frac{\lambda_g J_1^2(\alpha_2 r)}{\lambda_g' J_1^2(\alpha_1 r)} = \frac{\lambda_g R^2}{\lambda_g' R'^2} \left[\frac{\frac{J_1(\alpha_2 r)}{\alpha_2 r}}{\frac{J_1(\alpha_1 r)}{\alpha_1 r}} \right]^2, \quad (4)$$

where
$$M = M_1 \cos^2 \phi + M_2 \sin^2 \phi,$$
$$\alpha_1 = \frac{1.841}{R}, \quad \alpha_2 = \frac{1.841}{R'},$$
$$\lambda_g = \frac{\lambda}{\sqrt{1 - \left(\frac{\lambda}{3.41R}\right)^2}}, \quad \lambda_g' = \frac{\lambda}{\sqrt{1 - \left(\frac{\lambda}{3.41R'}\right)^2}}.$$

The arrangement of the various terms has been chosen to emphasize the similarity between Eqs. (3) and (4) and Eqs. (1) and (2). If the aperture center is located on a diameter perpendicular to that indicated in the above figure, Eqs. (3) and (4) are still applicable provided $J_1(\alpha r)/\alpha r$ is replaced by $J_1'(\alpha r)$.

The coefficients M_1 and M_2 are defined in Sec. 5·4b. For the case of a circular aperture ($d = d_1 = d_2$)

$$M = \frac{d^3}{6}.$$

Restrictions.—The equivalent circuit is applicable in the wavelength range $2.61R' < \lambda < 3.41R$. The restriction of the aperture orientation to $\phi = 0$ or $\pi/2$ is necessary to ensure that only the H_{11}-mode of the indicated polarization can be propagated in the circular guide. Otherwise, comments are the same as in Sec. 5·4b except that B/Y_0 is dependent on the aperture position along the direction of the electric field.

Numerical Results.—The coefficients M_1 and M_2 are shown in Fig. 5·4-4 as a function of d_2/d_1.

c. Junction of Two Coaxial Guides.—Two concentric coaxial guides of unequal inner and outer radii coupled by a small elliptical or circular aperture in a transverse plate of zero thickness (principal mode in coaxial guides).

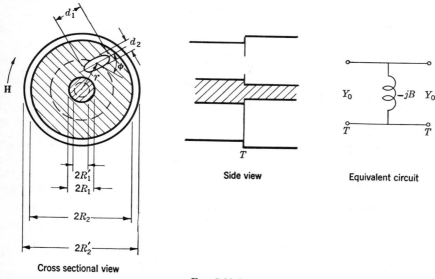

FIG. 5·33-3.

Equivalent-circuit Parameters.—For an elliptical aperture oriented as shown in the above figure

$$\frac{B}{Y_0} = \frac{\lambda r^2 \ln \frac{R_2}{R_1}}{M}, \quad (5)$$

$$\frac{Y'_0}{Y_0} = \frac{\ln \frac{R_2}{R_1}}{\ln \frac{R'_2}{R'_1}}, \tag{6}$$

where

$$M = M_1 \cos^2 \phi + M_2 \sin^2 \phi.$$

The coefficients M_1 and M_2 are defined in Sec. 5·4b. For the case of a circular aperture ($d = d_1 = d_2$)

$$M = M_1 = M_2 = \frac{d^3}{6}.$$

Restrictions.—The equivalent circuit is applicable for all wavelengths greater than the cutoff wavelength of the H_{10}-mode. Comments same as in Sec. 5·4b.

Numerical Results.—The coefficients M_1 and M_2 are plotted in Fig. 5·4-4 as a function of d_2/d_1.

d. Junction of a Rectangular and Circular Guide.—A rectangular and a circular guide coupled by a small elliptical or circular aperture in a transverse plate of zero thickness (H_{10}-mode in rectangular guide, H_{11}-mode in circular guide).

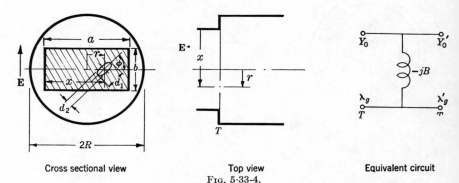

Cross sectional view   Top view   Equivalent circuit
FIG. 5·33-4.

Equivalent-circuit Parameters.—For an elliptical aperture oriented as shown in the above figure and with $\phi = 0$ or $\pi/2$

$$\frac{B}{Y_0} = \frac{\lambda_g a b}{4\pi \sin^2 \frac{\pi x}{a}} \left\{ \frac{1}{M} - \left[\frac{2\pi}{a^2 b} \sin^2 \frac{\pi x}{a} + \frac{7.74}{2\pi R^3} 4 J_1'^2(\alpha_2 r) \right] \right\}, \tag{7a}$$

$$\frac{B}{Y_0} \approx \frac{\lambda_g a b}{4\pi M \sin^2 \frac{\pi x}{a}}, \quad \left. \begin{array}{c} d_1 \\ d_2 \end{array} \right\} \ll a, b, r, \tag{7b}$$

$$\frac{Y_0'}{Y_0} = \frac{1}{0.952} \frac{\lambda_g ab}{\lambda_g' \pi R^2} \left[\frac{2 J_1'(\alpha_2 r)}{\sin \frac{\pi x}{a}} \right]^2, \tag{8}$$

where

$$M = M_1 \cos^2 \phi + M_2 \sin^2 \phi,$$

$$\alpha_2 = \frac{1.841}{R},$$

$$\lambda_g' = \frac{\lambda}{\sqrt{1 - \left(\frac{\lambda}{3.41R}\right)^2}}, \qquad \lambda_g = \frac{\lambda}{\sqrt{1 - \left(\frac{\lambda}{2a}\right)^2}}.$$

The coefficients M_1 and M_2 are defined in Sec. 5·4b. For a circular aperture ($d = d_1 = d_2$)

$$M = \frac{d^3}{6}.$$

If the aperture center is located on a diameter perpendicular to that shown in the above figure, the function $J_1'(\alpha_2 r)$ should be replaced by $J_1(\alpha_2 r)/\alpha_2 r$.

Restrictions.—The equivalent circuit is applicable in the wavelength range in which only the H_{10}-mode in rectangular and the H_{11}-mode in circular guide can be propagated. Other restrictions are the same as in Sec. 5·4b. The location of the aperture center is restricted to points where the direction of the dominant-mode magnetic field is the same in both guides; this is not necessary and is employed for simplification of the above formulas. The restriction of the angle ϕ to 0 or $\pi/2$ is necessary to avoid excitation of an H_{11}-mode with polarization perpendicular to that indicated in Fig. 5·33-4. The description of this latter mode necessitates the use of a six-terminal network.

Numerical Results.—The coefficients M_1 and M_2 are plotted in Fig. 5·4-4 as a function of d_2/d_1.

5·34. Circular Bends. E-plane.—An E-plane junction of a rectangular guide and a uniform circular bend of rectangular cross section (H_{10}-mode in rectangular guide and in circular bend).

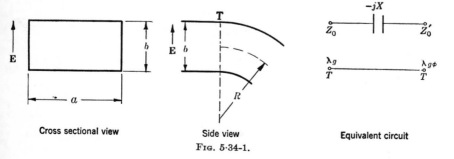

Cross sectional view   Side view   Equivalent circuit
                            Fig. 5·34-1.

Equivalent-circuit Parameters.—At the reference plane T

$$\frac{Z'_0}{Z_0} = 1 + \frac{1}{12}\left(\frac{b}{R}\right)^2\left[\frac{1}{2} - \frac{1}{5}\left(\frac{2\pi b}{\lambda_g}\right)^2\right], \tag{1}$$

$$\frac{X}{Z_0} = \frac{32}{\pi^7}\left(\frac{2\pi b}{\lambda_g}\right)^3\left(\frac{b}{R}\right)^2 \sum_{n=1,3,\cdots}^{\infty}\frac{1}{n^7}\sqrt{1 - \left(\frac{2b}{n\lambda_g}\right)^2}. \tag{2}$$

where

$$\lambda_g = \frac{\lambda}{\sqrt{1 - \left(\frac{\lambda}{2a}\right)^2}},$$

$$\lambda_{g\phi} = \lambda_g\left\{1 - \frac{1}{12}\left(\frac{b}{R}\right)^2\left[-\frac{1}{2} + \frac{1}{5}\left(\frac{2\pi b}{\lambda_g}\right)^2 \cdots\right]\right\}.$$

The circular bend is a uniform angular guide with propagation wavelength $\lambda_{g\phi}$, the latter being measured along the arc of mean radius R. The electrical length corresponding to an angular distance ϕ or to a (mean) arc distance s is

$$\frac{2\pi s}{\lambda_{g\phi}} = m\phi, \qquad \text{where } m = \frac{2\pi R}{\lambda_{g\phi}}.$$

Restrictions.—The equivalent circuit is applicable in the wavelength range $2b/\lambda_g < 1$. Equations (1) and (2) have been derived by a variational method employing the form of the lowest mode in the angular guide for the electric-field distribution in the junction plane. The circuit parameters have been evaluated to order $(b/R)^2$, but no estimate of the range of accuracy is available.

5·35. Circular Bends. H-Plane.—An H-plane junction of a rectangular guide and a uniform circular bend, i.e., angular guide, of rectangular cross section (H_{10}-mode in rectangular and angular guides).

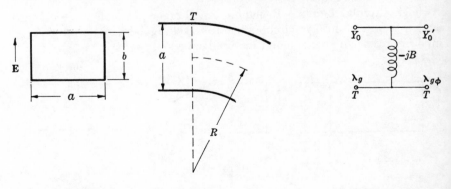

Cross sectional view   Top view   Equivalent circuit

Fig. 5·35-1.

Equivalent-circuit Parameters.—At the reference plane T

$$\frac{Y_0'}{Y_0} = 1 + \frac{\pi^2}{120}\left(\frac{a}{R}\right)^2 \left[1 + \frac{\frac{15}{\pi^4}}{\left(\frac{2a}{\lambda}\right)^2 - 1} - \left(2 - \frac{5}{\pi^2} - \frac{60}{\pi^4}\right)\left(\frac{2a}{\lambda}\right)^2 \right.$$
$$\left. + \left(1 + \frac{15}{6\pi^2} - \frac{105}{\pi^4}\right)\left(\frac{2a}{\lambda}\right)^4 \cdots \right], \quad (1)$$

$$\frac{B}{Y_0} = \frac{\lambda_g}{a}\frac{8}{\pi^4}\left(\frac{a}{R}\right)^2 \sum_{n=2,4,}^{\infty} \left\{\frac{n^3}{(n^2-1)^4}\left[1 - \frac{\left(\frac{4a}{\lambda}\right)^2}{n^2-1}\right]^2 \sqrt{1 - \left(\frac{2a}{n\lambda}\right)^2}\right\}, \quad (2)$$

where

$$\lambda_g = \frac{\lambda}{\sqrt{1 - \left(\frac{\lambda}{2a}\right)^2}}$$

$$\frac{1}{\lambda_{g\phi}^2} = \frac{1}{\lambda_g^2} + \frac{1}{24R^2}\left[1 - \left(\frac{12+\pi^2}{2\pi^2}\right)\left(\frac{2a}{\lambda}\right)^2 + \left(\frac{15-\pi^2}{2\pi^2}\right)\left(\frac{2a}{\lambda}\right)^4 \cdots\right].$$

The propagation wavelength $\lambda_{g\phi}$ in the uniform angular guide (circular bend) is measured along the arc of mean radius R. The electrical length corresponding to the angular distance ϕ or to a mean arc distance S is

$$\frac{2\pi s}{\lambda_{g\phi}} = m\phi, \quad \text{where } m = \frac{2\pi R}{\lambda_{g\phi}}.$$

With this interpretation the angular guide may be treated as a uniform line with characteristic admittance Y_0' and wavelength $\lambda_{g\phi}$.

Restrictions.—The equivalent circuit is applicable in the wavelength range $a < \lambda < 2a$. The circuit parameters have been derived by a variational method employing a sine function for the distribution of electric field in the junction plane. The results have been evaluated to order $(a/R)^2$ but no estimate of the range of accuracy is available.

CHAPTER 6

SIX-TERMINAL STRUCTURES

An arbitrary junction of three accessible waveguides, each propagating only a single mode, is designated as a six-terminal or three-terminal-pair waveguide structure. The over-all description of the propagating modes in such a structure is accomplished by representation of the waveguide regions as transmission lines and by representation of the junction region as a six-terminal lumped-constant equivalent circuit. The three transmission lines together with the lumped-constant circuit comprise a six-terminal network with the aid of which reflection, transmission, standing-wave, etc., properties of the over-all structure can be determined by conventional network calculations. The quantitative description of the transmission lines is effected by indication of their characteristic impedances and propagation wavelengths. The description of the six-terminal lumped-constant circuit requires, in general, the specification of six circuit parameters and the locations of the three corresponding terminal planes. If the structure possesses geometrical symmetries, a reduction in the number of required circuit parameters is possible.

Since only relative impedances are physically significant in microwave calculations, all impedances in the following networks will be referred to the characteristic impedance of one of the transmission lines. The choice of characteristic impedances of the transmission lines as well as the locations of the relevant terminal planes will be indicated explicitly in the equivalent-network diagrams of the various structures. As discussed in Sec. 3·3 the equivalent circuit of a microwave structure can be simplified considerably by an appropriate choice of terminal planes. Simplifications of this sort become increasingly important for multiterminal structures. However, the equivalent networks for most of the structures described below will be represented at terminal planes which are simple in so far as the derivation of the circuit parameters are concerned. For the important cases of the open E- and H-plane T structures of Secs. 6·1 and 6·5 the terminal-plane transformations to simple networks will be presented in some detail.

6·1. Open T-junction, E-plane.

A right-angle T-type junction of two rectangular guides of unequal heights but equal widths (H_{10}-mode in rectangular guides).

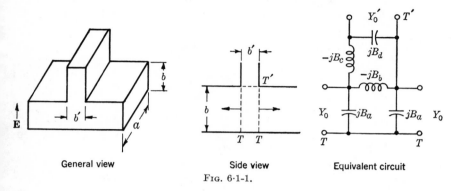

General view   Side view   Equivalent circuit
Fig. 6·1-1.

Equivalent-circuit Parameters.—At the reference planes T and T' for the main and stub guides, respectively:

$$\frac{Y_0'}{Y_0} = \frac{b}{b'} = \frac{1}{2\alpha}, \tag{1}$$

$$\frac{B_a}{Y_0} = \frac{2b'}{\lambda_g}\left(\tan^{-1}\frac{1}{\alpha} + \frac{\ln\sqrt{1+\alpha^2}}{\alpha}\right), \tag{2a}$$

$$\frac{B_a}{Y_0} \approx \frac{\pi b'}{\lambda_g}\left[1 - \frac{b'}{2\pi b} + \frac{8}{\pi^2}\left(\frac{2b'}{\lambda_g}\right)^2\left(1 - 0.368\frac{b'}{b}\right)\right], \tag{2b}$$

$$\frac{B_c}{Y_0} = \frac{\lambda_g}{2\pi b'}, \tag{3}$$

$$\frac{B_a - 2B_b}{Y_0} = \frac{2b}{\lambda_g}\left(\frac{\pi\alpha}{3} + A_1\right), \tag{4a}$$

$$\frac{B_b}{Y_0} \approx 1.100\frac{b'}{\lambda_g}\left[1 - 0.227\frac{b'}{b} + 0.008\left(\frac{b'}{b}\right)^2\right], \tag{4b}$$

$$\frac{B_d}{Y_0} = \frac{b}{\lambda_g}\left(\frac{\pi}{3\alpha} + A_2\right), \tag{5a}$$

$$\frac{B_d}{Y_0} \approx \frac{b}{\lambda_g}\left[2\ln\frac{eb}{2b'} + 1.100\frac{b'}{b} - 0.167\left(\frac{b'}{b}\right)^2 + 0.008\left(\frac{b'}{b}\right)^3\right] \tag{5b}$$

where

$$A_1 = -\frac{2\alpha}{\pi}e^{-\frac{2\tan^{-1}\alpha}{\alpha}}\left\{1 + \frac{5+\alpha^2}{4(1+\alpha^2)}e^{-\frac{2\tan^{-1}\alpha}{\alpha}}\right.$$
$$\left. + \left[\frac{4}{1+\alpha^2} + \left(\frac{5+\alpha^2}{1+\alpha^2}\right)^2\right]\frac{e^{-\frac{4\tan^{-1}\alpha}{\alpha}}}{9}\right\}$$

$$\frac{A_1 + A_2}{2} = \alpha \tan^{-1}\frac{1}{\alpha} + \frac{\tan^{-1}\alpha}{\alpha} + \ln\frac{1+\alpha^2}{4\alpha} - \frac{\pi(1+\alpha^2)}{6\alpha}.$$

An alternative equivalent circuit at the reference planes T_1 and T_1' is shown in Fig. 6·1-2. The corresponding parameters are given by

$$\frac{2\pi}{\lambda_g}\left(\frac{b'}{2} - d\right) = \tan^{-1} A_0, \tag{6}$$

$$\frac{2\pi d'}{\lambda_g} = \tan^{-1}\frac{A_0 A_c + A_b}{A_0 - A_a}, \tag{7}$$

$$n^2 = \frac{2b}{b'}\frac{(1 + A_0^2)(A_b + A_a A_c)}{(A_a - A_0)^2 + (A_b + A_0 A_c)^2}, \tag{8}$$

$$\frac{X}{2Z_0} = \frac{A_0(1 + A_c^2 - A_a^2 - A_b^2) - (1 - A_0^2)(A_a - A_b A_c)}{(A_a - A_0)^2 + (A_b + A_0 A_c)^2}, \tag{9}$$

where

$$A_0 = \frac{B_a}{Y_0}, \qquad A_a = \frac{B_a - 2B_b - 2B_c}{Y_0},$$

$$A_c = \frac{b'}{b}\frac{(B_d - B_c)}{Y_0}, \qquad A_b = \frac{2b'}{b}\left(\frac{B_c}{Y_0}\right)^2 - A_a A_c.$$

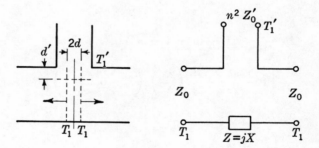

Side view Equivalent circuit

Fig. 6·1-2.

Another equivalent circuit at reference planes T_2 and T_2' is indicated in Fig. 6·1-3. In this case the circuit parameters are

$$\frac{2\pi l}{\lambda_g} = \frac{2\pi d}{\lambda_g}, \tag{10}$$

$$\frac{2\pi l'}{\lambda_g} = \tan^{-1}\frac{A_0 A_a + 1}{A_0 A_b - A_c}, \tag{11}$$

$$m^2 = \frac{2b}{b'}\frac{(1 + A_0 A_a)^2 + (A_c - A_0 A_b)^2}{(1 + A_0)^2(A_b + A_a A_c)}, \tag{12}$$

$$\frac{2B}{Y_0} = \frac{(1 - A_0^2)(A_a - A_b A_c) + A_0(A_a^2 + A_b^2 - A_c^2 - 1)}{(1 + A_0 A_a)^2 + (A_c - A_0 A_b)^2} \tag{13}$$

Restrictions.—The equivalent circuit is valid in the range $2b < \lambda_g$ or $2b' < \lambda_g$ according as $b > b'$ or $b < b'$, respectively. The circuit parameter B_a has been obtained by the equivalent static method, employing two incident static modes in the main guide, and is accurate to within 1 per cent over most of the wavelength range. For the case $b/\lambda_g = 0$ an analytical expression for B_a is given in Eq. (2a); the asymptotic Eq. (2b) agrees with the plotted results to within 10 per cent in the range $b'/b < 1$ and $b/\lambda_g < 0.5$. The circuit parameters B_b, B_c, and B_d indicated in the accompanying graphs have been obtained by a variational method and are estimated to be in error by only a few per cent over most of the wavelength range. For the case $b/\lambda_g = 0$ analytical expressions for these

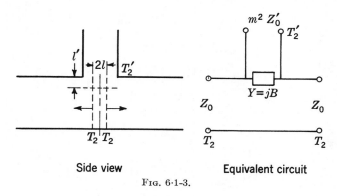

Fig. 6·1-3.

parameters obtained by the equivalent static method are given by Eqs. (3), (4a), and (5a); these equations agree with the plotted results to within a few per cent. The asymptotic relations Eqs. (4b) and (5b) agree with the results of Eqs. (4a) and (5a) to within 10 per cent for $b/\lambda_g = 0$. The alternative circuit parameters given by Eqs. (6) to (13) have been obtained by reference plane transformations (*cf.* Sec. 3·3).

Numerical Results.—The circuit parameters $B_a\lambda_g/Y_0b$, $B_b\lambda_g/Y_0b$, $B_cb/Y_0\lambda_g$, and $B_d\lambda_g/Y_0b$ are indicated in Figs. 6·1-4 to 6·1-7 as functions of b'/b with b/λ_g as a parameter. The alternative circuit parameters d/b', d'/b, n^2, and $X\lambda_g/Z_0b$, computed from the preceding results, are shown in Figs. 6·1-8 to 6·1-11. The circuit parameters l'/b, m^2, and $B\lambda_g/Y_0b$ are given in Figs. 6·1-12 to 6·1-14 as functions of b'/b. The multiplicity of representations is presented primarily to illustrate the differences in frequency dependence of circuit elements in various equivalent circuit representations.

Experimental Results.—The theoretical results indicated above have been verified within an experimental error of a few per cent for a number of cases in the range $b'/b < 1$ and $b/\lambda_g < 0.8$.

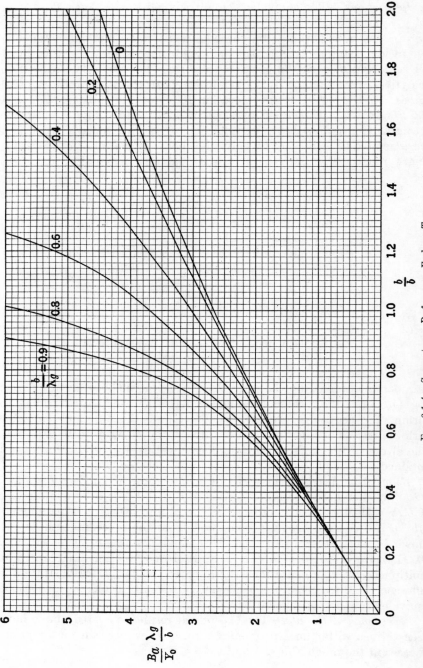

Fig. 6.1-4.—Susceptance B_a for open E-plane T.

SEC. 6·1] OPEN T-JUNCTION, E-PLANE 341

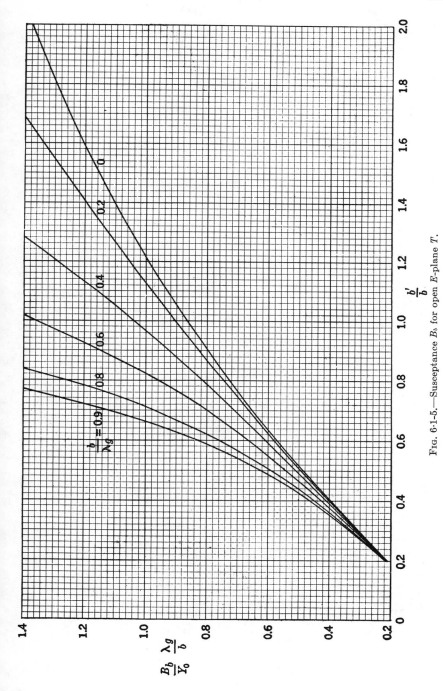

FIG. 6·1-5.—Susceptance B_b for open E-plane T.

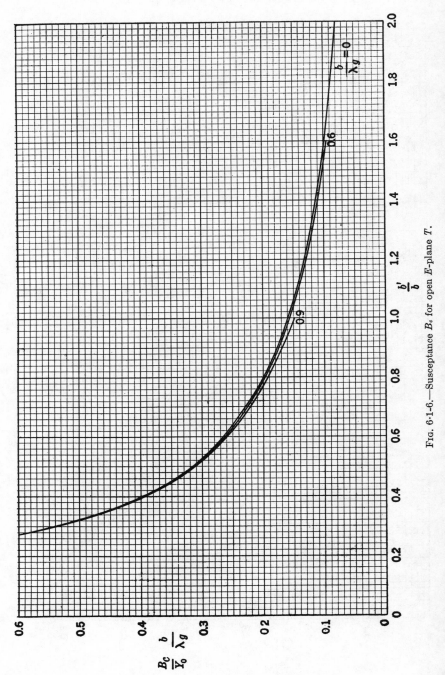

Fig. 6·1-6.—Susceptance B_c for open E-plane T.

SEC. 6·1] OPEN T-JUNCTION, E-PLANE

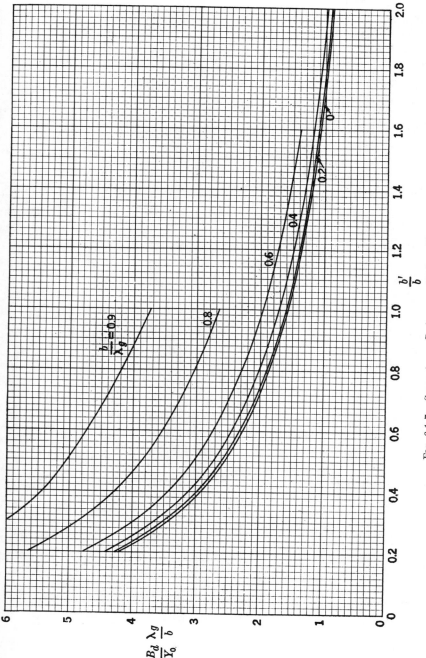

FIG. 6·1-7.—Susceptance B_d for open E-plane T.

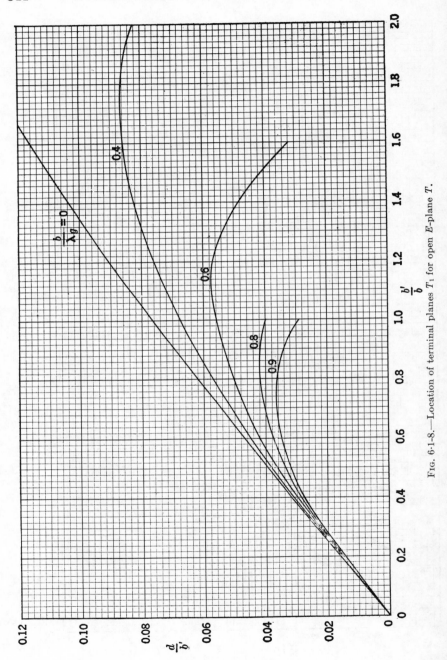

Fig. 6·1-8.—Location of terminal planes T_1 for open E-plane T.

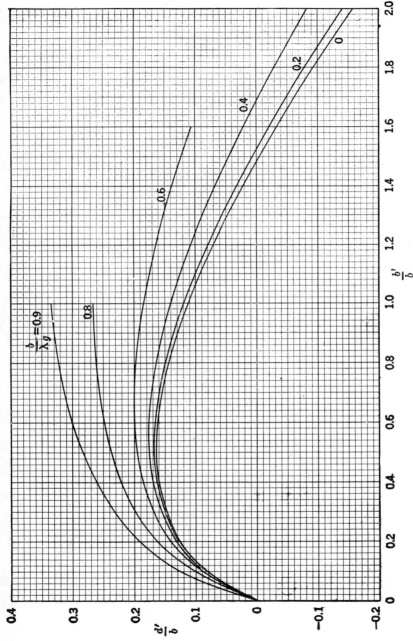

Fig. 6.1-9.—Location of terminal plane T'_1 for open E-plane T.

346   E-PLANE JUNCTIONS, RECTANGULAR GUIDES   [SEC. 6.1

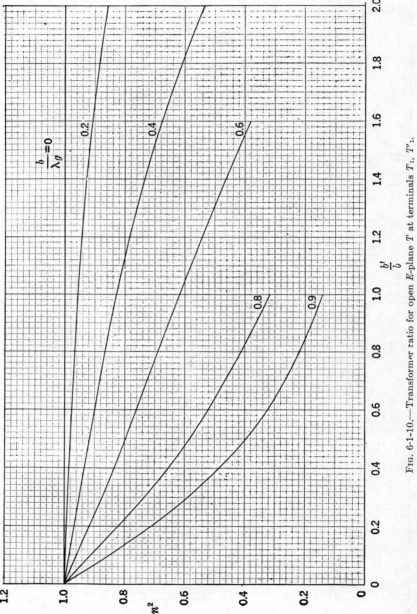

FIG. 6.1-10.—Transformer ratio for open E-plane T at terminals T_1, T'_1.

SEC. 6·1] OPEN T-JUNCTION, E-PLANE

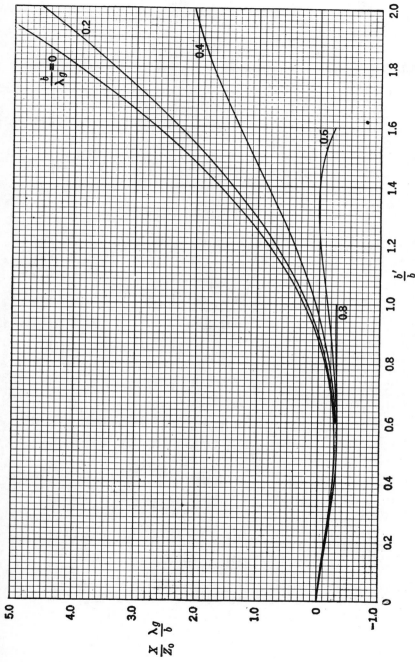

FIG. 6·1-11.—Series reactance of open E-plane T at terminals T_1, T'_1.

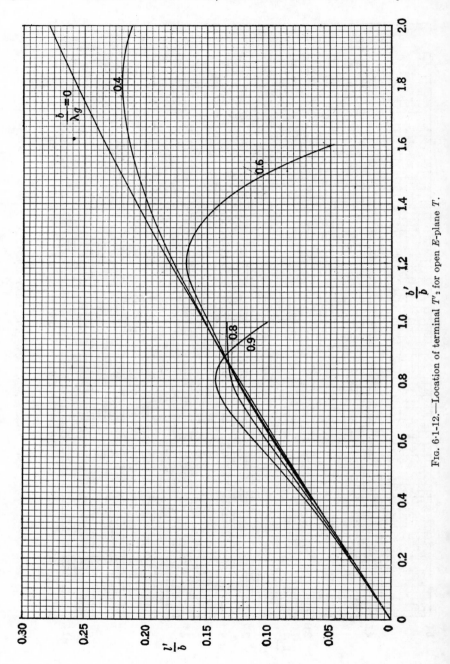

Fig. 6·1-12.—Location of terminal T'_2 for open E-plane T.

SEC. 6·1] OPEN T-JUNCTION, E-PLANE 349

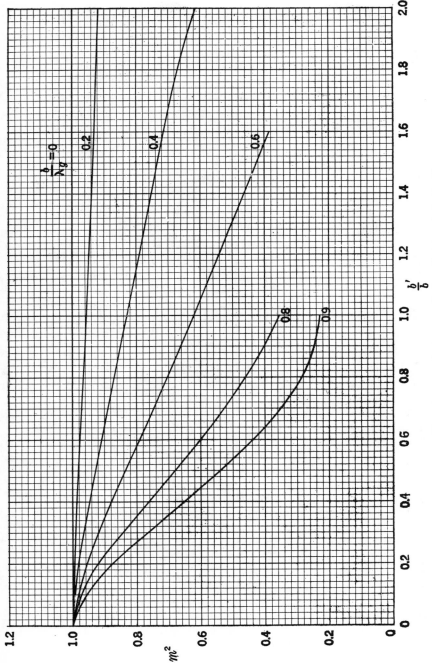

FIG. 6·1·13.—Transformer ratio for open E-plane T at terminals T_2, T'_2.

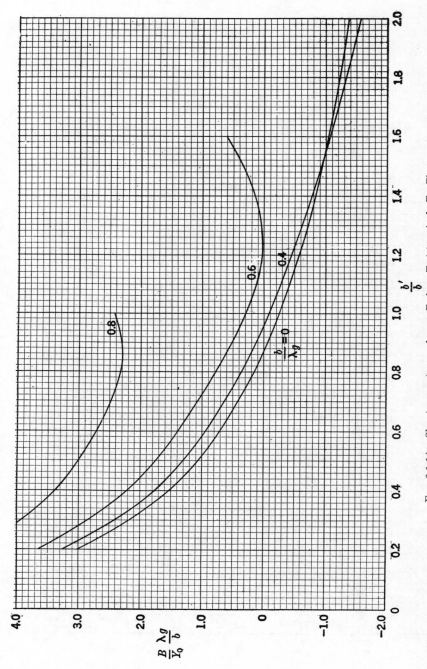

Fig. 6·1-14.—Shunt susceptance of open E-plane T at terminals T_2, T'_2.

6·2. Slit-coupled T-junctions in Rectangular Guide, E-Plane.

—A symmetrical right-angle T-type junction of two rectangular guides of unequal heights, but equal widths, coupled by a small slit in a wall of zero thickness. Sides of slit perpendicular to electric field (H_{10}-mode in rectangular guides).

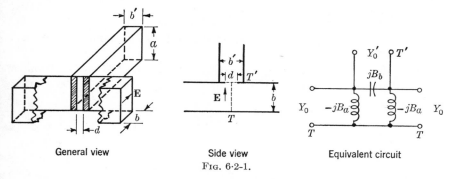

General view   Side view   Equivalent circuit
Fig. 6·2-1.

Equivalent-circuit Parameters.—At the reference planes T and T'

$$\frac{Y_0'}{Y_0} = \frac{b}{b'}, \tag{1}$$

$$\frac{B_a}{Y_0} = \frac{\frac{b}{\lambda_g}\left(\frac{\pi d}{4b}\right)^2}{1 + \frac{1}{6}\left(\frac{\pi d}{4b}\right)^2 \left\{\left(\frac{b}{b'}\right)^2 + \frac{1}{2} + 6\left[1 - \sqrt{1 - \left(\frac{2b}{\lambda_g}\right)^2}\right]\right.}$$
$$\left. + 12\left(\frac{b}{b'}\right)^2 \left[\frac{1}{\sqrt{1 - \left(\frac{2b'}{\lambda_g}\right)^2}} - 1\right]\right\}, \tag{2a}$$

$$\frac{B_a}{Y_0} \approx \frac{b}{\lambda_g}\left(\frac{\pi d}{4b}\right)^2, \qquad d \ll b \tag{2b}$$

$$\frac{B_b}{Y_0} - \frac{B_a}{2Y_0} = \frac{4b}{\lambda_g}\left\{\ln\frac{2\sqrt{2bb'}}{\pi d} + \frac{1}{2}\sum_{n=1}^{\infty}\left[\frac{1}{\sqrt{n^2 - \left(\frac{2b}{\lambda_g}\right)^2}} - \frac{1}{n}\right]\right\}, \tag{3a}$$

$$\frac{B_b}{Y_0} \approx \frac{4b}{\lambda_g}\left[\ln\frac{2\sqrt{2bb'}}{\pi d} + \left(\frac{b}{\lambda_g}\right)^2\right], \qquad b \ll \lambda_g. \tag{3b}$$

Restrictions.—The equivalent circuit is applicable in the wavelength range $2b/\lambda_g < 1$ provided the slit dimension d is small compared with the wavelength. The equivalent-circuit parameters have been derived by an integral equation method. The derivation of the parameter

B_a/Y_0 neglects terms of the third order in the aperture coordinates. To be rigorous the circuit element B_b should be replaced by a π network or its equivalent. However, in the small-aperture range, $\pi d/b' < 1$ ($b' < b$), the susceptance of the series arm of this π network is approximately infinite. The sum of the susceptances of the shunt arms is represented to the second order in the aperture coordinates by Eq. (3a). It is estimated that Eqs. (2a) and (3a) are correct to within a few per cent for $\pi d/b' < 1$. The effect of wall thickness is important and will be considered in Sec. 8·11.

6·3. 120° Y-Junction, E-plane.—A symmetrical 120° Y-junction of three identical rectangular guides in the E-plane (H_{10}-mode in rectangular guides).

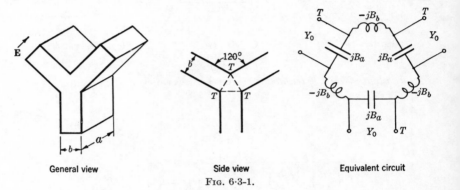

General view   Side view   Equivalent circuit

FIG. 6·3-1.

Equivalent-circuit Parameters.—At the reference planes T

$$\frac{B_a}{Y_0} = \frac{2b}{\lambda_g} 0.6455, \tag{1}$$

$$\frac{B_b}{Y_0} = \frac{\lambda_g}{b} \frac{2\sqrt{3}}{\pi}. \tag{2}$$

Restrictions.—The equivalent circuit is applicable in the wavelength range $2b/\lambda_g < 1$. Equations (1) and (2) are static approximations and have been obtained by conformal mapping methods, etc. They are estimated to be accurate to within a few per cent in the range $b/\lambda_g < 0.1$.

Experimental Results.—Measurement taken at a wavelength of $\lambda = 3.20$ cm with rectangular guides of dimensions $a = 0.901$ in. and $b = 0.404$ in. yield the values

$$\frac{B_a}{Y_0} = 0.29, \qquad \frac{B_b}{Y_0} = 4.68.$$

These are to be compared with the values obtained by Eqs. (1) and (2)

$$\frac{B_a}{Y_0} = 0.296, \qquad \frac{B_b}{Y_0} = 4.81.$$

6·4. E-plane Bifurcation.—A bifurcation of a rectangular guide by a partition of zero thickness perpendicular to the direction of the electric field (H_{10}-mode in each of the rectangular guides).

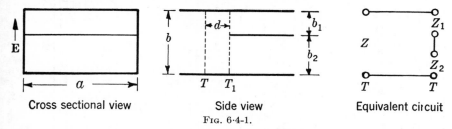

Cross sectional view Side view Equivalent circuit
FIG. 6·4-1.

Equivalent-circuit Parameters.—At the reference plane T the equivalent circuit is simply that of three guides in series, the characteristic impedances of the various guides being

$$\frac{Z_1}{Z_0} = \frac{b_1}{b}, \qquad \frac{Z_2}{Z_0} = \frac{b_2}{b}. \tag{1}$$

The location of T is given by

$$\frac{2\pi d}{\lambda_g} = \frac{2b}{\lambda_g}\left(\frac{b_1}{b}\ln\frac{b}{b_1} + \frac{b_2}{b}\ln\frac{b}{b_2}\right) + S_1\left(\frac{2b}{\lambda_g};0,0\right) - S_1\left(\frac{2b_1}{\lambda_g};0,0\right) - S_1\left(\frac{2b_2}{\lambda_g};0,0\right), \tag{2a}$$

$$\frac{2\pi d}{\lambda_g} \approx \frac{2b}{\lambda_g}\left(\frac{b_1}{b}\ln\frac{b}{b_1} + \frac{b_2}{b}\ln\frac{b}{b_2}\right), \tag{2b}$$

where

$$S_1(x;0,0) = \sum_{n=1}^{\infty}\left(\sin^{-1}\frac{x}{n} - \frac{x}{n}\right).$$

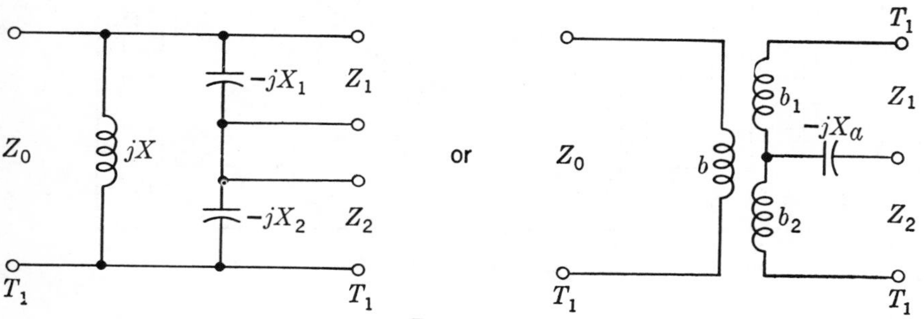

FIG. 6·4-2.

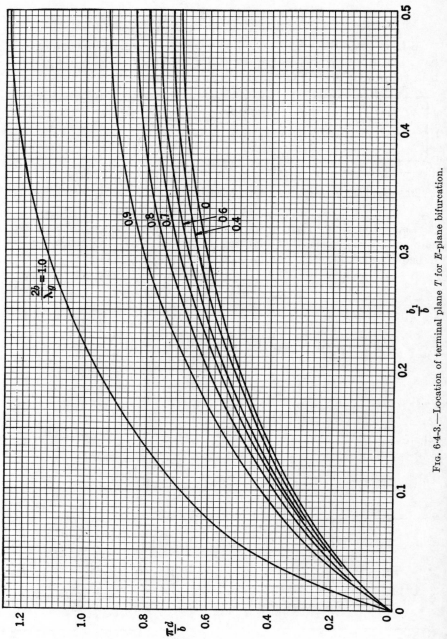

FIG. 6·4·3.—Location of terminal plane T for E-plane bifurcation.

At the reference plane T_1 alternative equivalent circuits (*cf.* Sec. 3·3, N-Terminal Pair Representations) are shown in Fig. 6·4-2. The corresponding circuit parameters are

$$\left. \begin{array}{ll} \dfrac{X}{Z_0} = \cot \dfrac{2\pi d}{\lambda_g}, & \dfrac{X_2}{Z_0} = \dfrac{b_2}{b} \cot \dfrac{2\pi d}{\lambda_g}, \\ \dfrac{X_1}{Z_0} = \dfrac{b_1}{b} \cot \dfrac{2\pi d}{\lambda_g}, & \dfrac{X_a}{Z_0} = \dfrac{b_1 b_2}{b^2} \cot \dfrac{2\pi d}{\lambda_g}. \end{array} \right\} \quad (3)$$

As indicated in Fig. 6·4-2 the turns ratios of the ideal three-winding transformer are $b:b_1:b_2$ (*cf.* Fig. 3·18*b*).

Restrictions.—The equivalent circuit is applicable in the wavelength range $2b < \lambda_g < \infty$. The circuit parameters have been obtained by the transform method and are rigorous. The approximation (2*b*) is the asymptotic form of Eq. (2*a*) in the range $2b/\lambda_g \ll 1$.

Numerical Results.—Curves of $\pi d/b$ as a function of b_1/b with $2b/\lambda_g$ as a parameter are shown in Fig. 6·4-3. The summation function $S_1(x;0,0)$ may be obtained from the summation functions tabulated in the Appendix.

6·5. Open T-junction, H-plane.—An *H*-plane, T-type junction of three rectangular guides of equal dimensions (H_{10}-mode in rectangular guides).

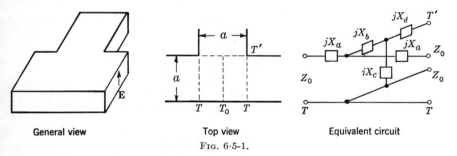

General view   Top view   Equivalent circuit
Fig. 6·5-1.

Equivalent-circuit Parameters.—At the reference planes T and T'

$$\frac{Z_0}{X_a} \equiv \frac{1}{X_1} \approx \frac{\lambda_g}{2a} \left[x \cot \frac{\pi x}{2} - 0.0103 - \frac{\left(\dfrac{4}{5\pi} \dfrac{1+x^2}{4-x^2} + 0.2614 \right)^2}{\left(\dfrac{\sqrt{3-x^2}}{1 - e^{-2\pi\sqrt{3-x^2}}} - 0.0694 \right)} \right], \quad (1)$$

$$\frac{X_a + 2X_b + 2X_c}{Z_0} \equiv X_2 \approx \frac{2a}{\lambda_g} \left[-\frac{A}{AC + B^2} - 0.057 + \frac{0.085}{1.62 - x^2} \right], \quad (2)$$

$$-\frac{2X_c}{Z_0} \equiv X_3 \approx \frac{2a}{\lambda_g} \left(\frac{B}{AC + B^2} \right), \quad (3)$$

$$\frac{2X_d}{Z_0} \equiv X_3 + X_4 \approx \frac{2a}{\lambda_g} \left(\frac{B + C}{AC + B^2} \right), \quad (4)$$

where

$$A = \frac{x}{2} \cot(\pi x) - 0.0322, \qquad x = \frac{2a}{\lambda_g},$$

$$B = \frac{1}{\pi} \frac{1+x^2}{1-x^2} + 0.03246,$$

$$C = x \tan\left(\frac{\pi x}{2}\right) + 0.0195.$$

At the reference planes T_0 and T' the equivalent circuit is that indicated above with T replaced by T_0 and X_a, X_b, ... by X'_a, X'_b, ..., where

$$\frac{X'_a}{Z_0} = \frac{X_1 - \tan\left(\frac{\pi x}{2}\right)}{1 + X_1 \tan\left(\frac{\pi x}{2}\right)}, \tag{5a}$$

$$\frac{X'_a}{Z_0} \approx \frac{2a}{\lambda_g}[0.16 - 0.06(x-1) + 0.04(x-1)^2], \tag{5b}$$

$$\frac{X'_a + 2X'_b + 2X'_c}{Z_0} = \frac{X_2 - \tan\left(\frac{\pi x}{2}\right)}{1 + X_2 \tan\left(\frac{\pi x}{2}\right)}, \tag{6a}$$

$$\frac{X'_a + 2X'_b + 2X'_c}{Z_0} \approx \frac{2a}{\lambda_g}[-1.59 + 2.42(x-1) + 5.06(x-1)^2], \tag{6b}$$

$$-\frac{2X'_c}{Z_0} = \frac{X_3 \sec\left(\frac{\pi x}{2}\right)}{1 + X_2 \tan\left(\frac{\pi x}{2}\right)}, \tag{7a}$$

$$-\frac{2X'_c}{Z_0} \approx +\frac{2a}{\lambda_g}[2.00 + 1.06(x-1) + 0.84(x-1)^2], \tag{7b}$$

$$\frac{2X'_d}{Z_0} = \frac{X_4 + (X_2 X_4 - X_3^2)\tan\left(\frac{\pi x}{2}\right) + X_3 \sec\left(\frac{\pi x}{2}\right)}{1 + X_2 \tan\left(\frac{\pi x}{2}\right)}, \tag{8a}$$

$$\frac{2X'_d}{Z_0} \approx \frac{2a}{\lambda_g}[2.00 + 1.04(x-1) + 1.36(x-1)^2]. \tag{8b}$$

An alternative equivalent circuit at the reference planes T_1 and T'_1 is shown in Fig. 6·5-2. The reference-plane locations and associated circuit parameters are given by

$$d = \frac{\lambda_g}{2\pi} \tan^{-1} X_1 - \frac{a}{2}, \tag{9}$$

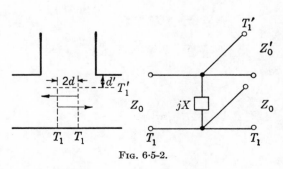

Fig. 6·5-2.

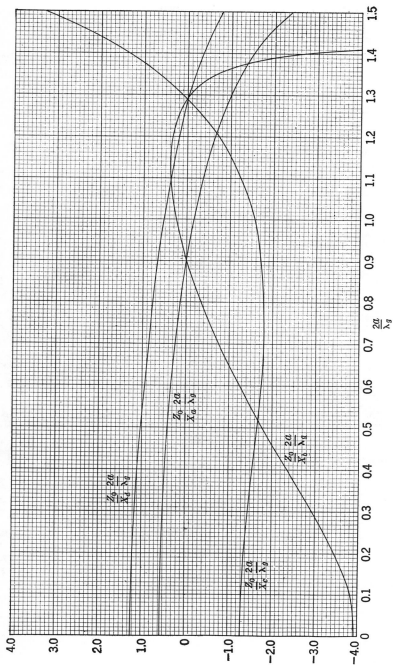

FIG. 6·5-3.—Circuit parameters for open H-plane T at terminals T, T'.

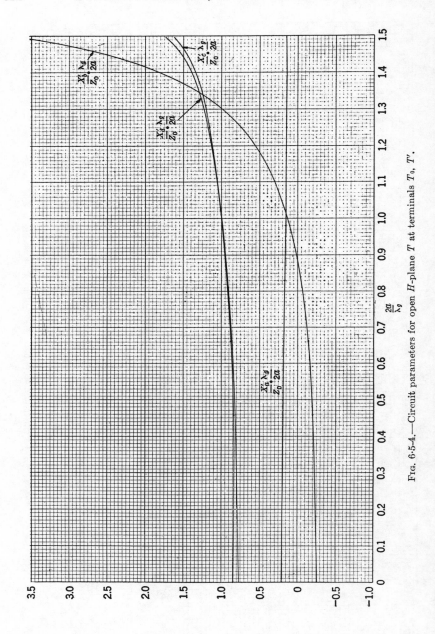

Fig. 6·5-4.—Circuit parameters for open H-plane T at terminals T_0, T'.

SEC. 6·5]   OPEN T-JUNCTION, H-PLANE

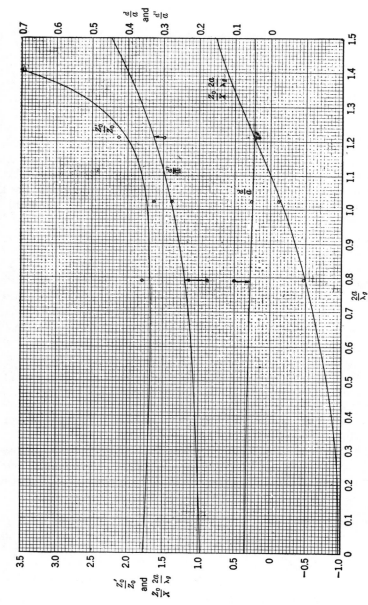

FIG. 6·5-5.—Locations of terminals T_1, T'_1 and associated circuit parameters for open H-plane T.

$$d' = \frac{\lambda_g}{2\pi} \tan^{-1} \frac{X_1 X_4 + X_0}{2(X_1 - X_2)}, \tag{10}$$

$$\frac{X}{Z_0} = \frac{1}{2} \frac{4(X_2 - X_1)^2 + (X_0 + X_1 X_4)^2}{(1 - X_1^2)(4X_2 - X_0 X_4) - X_1(4 + X_4^2 - 4X_2^2 - X_0^2)}, \tag{11}$$

$$\frac{Z_0'}{Z_0} = \frac{4(X_2 - X_1)^2 + (X_0 + X_1 X_4)^2}{4(1 + X_1^2) X_3^2}, \tag{12}$$

where

$$X_0 = X_3^2 - X_2 X_4.$$

Restrictions.—The circuit is valid in the wavelength range $a < \lambda < 2a$. The quantities plotted in the accompanying graphs have been obtained by the equivalent static method employing two static modes incident in each guide and are estimated to be in error by about 1 per cent. Equations (1) to (4) are approximate analytical results [except for the last two terms in Eq. (2)] for one static mode incident in each guide and are in error by less than 15 per cent for $0 < 2a/\lambda_g < 1$. The corresponding Eqs. (5a) to (8a) at the shifted reference system are correct to within 10 per cent for $0 < 2a/\lambda_g < 1.25$. The approximations (5b) to (8b) are in error by less than 4 per cent in the range $0.7 < \lambda_g < 1.25$.

Numerical Results.—For illustrative purposes the equivalent-circuit parameters have been plotted in a number of reference systems. In Fig. 6·5-3 the exact values of the elements $Z_0 2a/X_a \lambda_g, \ldots$ at the reference planes T and T' have been plotted as a function of $2a/\lambda_g$. In Fig. 6·5-4 the exact values of $X_a' \lambda_g/Z_0 2a, \ldots$ at the reference planes T_0 and T' have been plotted against $2a/\lambda_g$. The parameters of the shunt representation at the reference planes T_1 and T_1' are indicated in Fig. 6·5-5 where d/a, d'/a, $Z_0 2a/X\lambda_g$, and Z_0'/Z_0 are plotted against $2a/\lambda_g$. It is to be noted that the reference terminals in the main guide overlap one another.

Experimental Results.—A few experimental data taken at wavelengths of 3 and 10 cm are indicated by the circled points in Fig. 6·5-5.

6·6. Slit-coupled T-junction in Rectangular Guide, H-plane.—A T-junction of two rectangular guides of unequal widths but equal heights coupled by a small slit in a side wall of zero thickness. Electric field parallel to sides of slit (H_{10}-modes in rectangular guides).

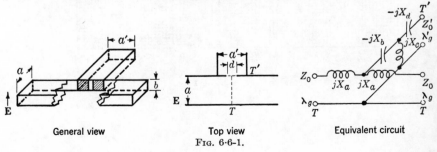

General view   Top view   Equivalent circuit

FIG. 6·6-1.

SEC. 6·6] SLIT-COUPLED T-JUNCTION IN RECTANGULAR GUIDE

Equivalent-circuit Parameters.—At the reference plane T and T' for the main and stub guides, respectively,

$$\frac{Z_0'}{Z_0} = \frac{\lambda_g' a'}{\lambda_g a}, \tag{1}$$

$$\frac{X_a}{Z_0} = \frac{\dfrac{a}{\lambda_g}\left(\dfrac{\pi d}{4a}\right)^4}{1 - \dfrac{1}{2}\left(\dfrac{\pi d}{4a}\right)^4}, \tag{2}$$

$$\frac{X_b}{Z_0} = \frac{a}{\lambda_g}\left[\left(\frac{4a}{\pi d}\right)^2 + 3 - 2\left(\frac{a}{a'}\right)^2\right], \tag{3a}$$

$$\frac{X_b}{Z_0} \approx \frac{a}{\lambda_g}\left(\frac{4a}{\pi d}\right)^2, \qquad d \ll a, \tag{3b}$$

$$\frac{X_c}{Z_0} = \frac{2a}{\lambda_g}, \tag{4}$$

$$\frac{X_d}{Z_0} = \frac{2a}{\lambda_g}, \tag{5}$$

where

$$\lambda_g = \frac{\lambda}{\sqrt{1 - \left(\dfrac{\lambda}{2a}\right)^2}}, \qquad \lambda_g' = \frac{\lambda}{\sqrt{1 - \left(\dfrac{\lambda}{2a'}\right)^2}}.$$

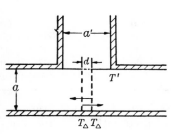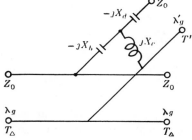

FIG. 6·6-2.

On shift of the reference planes T to T_Δ, a distance Δ in guide wavelengths, where

$$\Delta = \frac{1}{32\pi}\frac{a}{\lambda_g}\left(\frac{\pi d}{2a}\right)^4, \tag{6}$$

the alternative equivalent circuit of Fig. 6·6-2 is obtained wherein the series reactance X_a vanishes. As indicated in this figure the new reference planes in the main guide overlap. Hence, when the output impedance in the stub is so adjusted that the junction is reflectionless, the electrical length of the main guide is larger than the geometrical length by the amount 2Δ. Since the shift Δ is relatively small in the range of validity of the above formulas, the "shifted" circuit parameters X'_b, X'_c, and X'_d are approximately the same as those in the original reference system. More accurate values can be obtained by use of the transformation relations given in Eqs. (3·21).

Restrictions.—The equivalent circuit is applicable in the wavelength range $a < \lambda < 2a'$ (if $a' < a$). The formulas have been derived by an integral equation method employing the small-aperture approximations that $d < \lambda/\pi$ and the wavelength is not too close to cutoff of the second mode. It is to be emphasized that the above formulas apply to slits in walls of zero thickness. The power transmitted through a small slit of thickness t and width d is approximately 27.3 t/d db less than that computed from the above equations.

Experimental Results.—With the reference system as shown in Fig. 6·6-2 the equivalent-circuit parameters of the above junction were measured at a wavelength of $\lambda = 3.20$ cm for waveguides of dimensions $a = a' = 0.900$ in. and for various slit widths d. These measurements together with the corresponding theoretical values from Eqs. (2) to (5) are indicated in the following table:

d, in.	$\dfrac{X'_b}{Z_0}$		$\dfrac{X'_c}{Z_0}$		$\dfrac{X'_d}{Z_0}$		Δ (in λ_g)	
	Exp.	Theor.	Exp.	Theor.	Exp.	Theor.	Exp.	Theor.
0.189	22	19.3	1.14	1.02	1.00	1.02	.....	0.0000
0.299	8.12	8.05	1.09	1.02	1.07	1.02	0.001	0.0004
0.389	4.67	4.96	1.00	1.02	1.00	1.02	0.001	0.0010
0.486	3.29	3.35	1.06	1.02	1.06	1.02	0.003	0.0026
0.688	2.98	2.20	1.28	1.09	1.78	1.23	0.010	0.0106

The data apply to slits in a wall of 0.005 in. thickness. Because of mechanical difficulties (buckling, etc.) associated with a thickness of this magnitude, the experimental accuracy is not too good. The theoretical results are not applicable to the case of largest width and have been included merely to indicate the order of the error in this case.

In order to emphasize the importance of the wall thickness t the following experimental results for the case $t = 0.050$ in. are included:

d, in.	$\dfrac{X'_b}{Z_0}$	$\dfrac{X'_c}{Z_0}$	$\dfrac{X'_d}{Z_0}$	Δ (in λ_g)
0.300	3.23	4.14	14.3	0.051

6·7. 120° Y-junction, H-Plane.—A symmetrical 120° Y-junction of three identical rectangular guides in the H-plane. (H_{10}-mode in rectangular guides.)

SEC. 6·8] APERTURE-COUPLED T-JUNCTIONS, E-PLANE

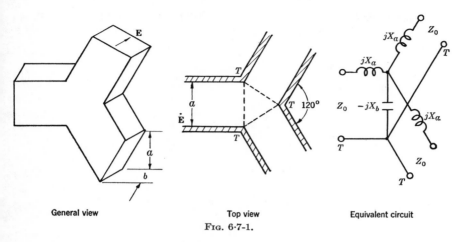

General view   Top view   Equivalent circuit
FIG. 6·7-1.

Experimental Results.—At the symmetrically chosen terminal planes T, measurements taken in rectangular guides of width $a \approx 0.90$ in. at a number of wavelengths yield the following values for the circuit parameters:

λ cm	λ_g in.	$\dfrac{X_a}{Z_0}$	$\dfrac{X_b}{Z_0}$
3.40	1.9996	1.499	0.698
3.20	1.7620	1.141	0.598
3.00	1.5606	0.900	0.537

6·8. Aperture-coupled T-junctions, E-Plane. *a. Rectangular Stub Guide.*—A right-angle T-type junction composed of two dissimilar rectangular guides coupled by a centered small elliptical or circular aperture in a metallic wall of zero thickness (H_{10}-mode in rectangular guides).

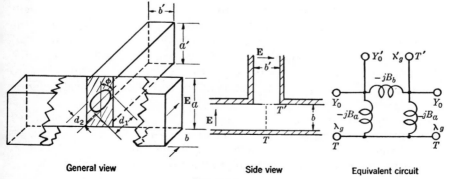

General view   Side view   Equivalent circuit
FIG. 6·8-1.

Equivalent-circuit Parameters.—At the reference planes T and T' for the case of coupling by a centered elliptical aperture oriented as shown in Fig. 6·8-1,

$$\frac{Y_0'}{Y_0} = \frac{\lambda_g ab}{\lambda_g' a'b'}, \tag{1}$$

$$\frac{B_a}{Y_0} = \frac{\dfrac{2\pi P}{\lambda_g ab}\left(\dfrac{\lambda_g}{\lambda}\right)^2}{1 - \dfrac{4\pi P}{\lambda^2 b}}, \tag{2a}$$

$$\frac{B_a}{Y_0} \approx \frac{2\pi P}{\lambda_g ab}\left(\frac{\lambda_g}{\lambda}\right)^2, \qquad \frac{d_1}{\lambda} \ll 1, \tag{2b}$$

$$\frac{B_b}{Y_0} = \frac{\lambda_g ab}{4\pi}\left[\frac{1}{M} - \left(\frac{\pi}{a^2 b} + \frac{2\pi}{a'^2 b'}\right)\right], \tag{3a}$$

$$\frac{B_b}{Y_0} \approx \frac{\lambda_g ab}{4\pi M}, \qquad \frac{d_1}{\lambda} \ll 1, \tag{3b}$$

where

$$M = M_1 \cos^2 \phi + M_2 \sin^2 \phi,$$

$$P = \frac{d_1 d_2^2}{12} \cdot \frac{\pi}{2} \frac{1}{E(e)},$$

$$P \approx \frac{\pi}{24} d_1 d_2^2, \qquad \frac{d_2}{d_1} \ll 1,$$

$$\lambda_g = \frac{\lambda}{\sqrt{1 - \left(\dfrac{\lambda}{2a}\right)^2}}, \qquad \lambda_g' = \frac{\lambda}{\sqrt{1 - \left(\dfrac{\lambda}{2a'}\right)^2}}.$$

The coefficients M_1 and M_2 are defined in Eqs. (4) and (5) of Sec. 5·4b. The function $E(e)$ is the complete elliptic integral of the second kind, $e = \sqrt{1 - (d_2/d_1)^2}$ being the eccentricity of the ellipse. For the case of a centered circular aperture, $d_1 = d_2 = d$

$$M = \frac{d^3}{6}, \qquad \text{and} \qquad P = \frac{d^3}{12}.$$

Restrictions.—In the small-aperture range $d_1/\lambda \ll 1$ the equivalent circuit of Fig. 6·8-1 is applicable at wavelengths for which only the dominant H_{10}-mode can be propagated in both guides. Equations (2a) and (3a) have been obtained by an integral equation method employing the small-aperture approximations (*cf.* Restrictions, Sec. 5·4b). It is to be emphasized that the above results apply to apertures in walls of zero thickness (*cf.* Chap. 8).

Numerical Results.—The coefficients M_1, M_2, and P may be obtained as a function of d_2/d_1 from the curves in Fig. 5·4-4.

b. *Circular Stub Guide.*—A symmetrical right-angle T-type junction of a rectangular and a circular guide coupled by a centered small elliptical or circular aperture in a metallic wall of zero thickness (H_{10}-mode in rectangular guide, H_{11}-mode of indicated polarization in circular guide).

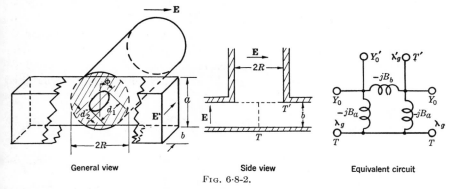

General view Side view Equivalent circuit
FIG. 6·8-2.

Equivalent-circuit Parameters.—At the reference planes T and T', for coupling by a centered elliptical aperture of orientation $\phi = 0$ or $\pi/2$,

$$\frac{Y_0'}{Y_0} = 1.051 \frac{\lambda_g ab}{\lambda_g' \pi R^2}, \tag{4}$$

$$\frac{B_a}{Y_0} = \frac{\dfrac{2\pi P}{\lambda_g ab}\left(\dfrac{\lambda_g}{\lambda}\right)^2}{1 - \dfrac{4\pi P}{\lambda^2 b}}, \tag{5a}$$

$$\frac{B_a}{Y_0} \approx \frac{2\pi P}{\lambda_g ab}\left(\frac{\lambda_g}{\lambda}\right)^2, \qquad \frac{d_1}{\lambda} \ll 1, \tag{5b}$$

$$\frac{B_b}{Y_0} = \frac{\lambda_g ab}{4\pi}\left[\frac{1}{M} - \left(\frac{\pi}{a^2 b} + \frac{7.74}{2\pi R^3}\right)\right], \tag{6a}$$

$$\frac{B_b}{Y_0} \approx \frac{\lambda_g ab}{4\pi M}, \qquad \frac{d_1}{\lambda} \ll 1, \tag{6b}$$

and

$$\lambda_g = \frac{\lambda}{\sqrt{1 - \left(\dfrac{\lambda}{2a}\right)^2}}, \qquad \lambda_g' = \frac{\lambda}{\sqrt{1 - \left(\dfrac{\lambda}{3.41R}\right)^2}},$$

where the remaining quantities are defined as in Sec. 6·8a.

For the special case of a centered circular aperture with $d_1 = d_2 = d$,

$$M = \frac{d^3}{6}, \qquad P = \frac{d^3}{12}.$$

Restrictions.—In the small-aperture range the equivalent circuit is applicable at wavelengths for which only a single dominant mode can be propagated in each waveguide. The angle ϕ must be restricted to 0 or $\pi/2$ to ensure that only a single H_{11}-mode of the above-indicated polarization can be propagated in the circular guide. For further comments note Restrictions in Secs. 6·8a and 5·4b.

Numerical Results.—The coefficients M_1, M_2, and P appear as functions of d_2/d_1 in the plots of Fig. 5·4-4.

6·9. Aperture-coupled T-junction in Rectangular Guide, H-plane.
a. Rectangular Stub Guide.—A right-angle T-type junction of two unequal rectangular guides coupled by a centered small elliptical or circular aperture in a metallic wall of zero thickness (H_{10}-mode in rectangular guides).

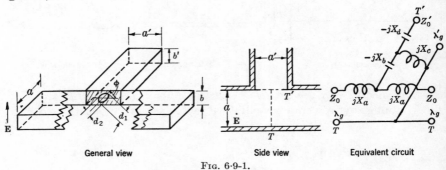

Fig. 6·9-1.

Equivalent-circuit Parameters.—For the case of an elliptical aperture, oriented as shown in the above figure, the circuit parameters at the reference planes T and T' are

$$\frac{Z_0'}{Z_0} = \frac{\lambda_g' a' b'}{\lambda_g ab}, \tag{1}$$

$$\frac{X_a}{Z_0} \approx 0 \tag{2}$$

$$\frac{X_b}{Z_0} = \frac{a}{\lambda_g}\left(\frac{a^2 b}{\pi M} + 1 + 2\sqrt{\frac{ab}{a'b'}} - 2\frac{a^2 b}{a'^2 b'}\right) \tag{3a}$$

$$\frac{X_b}{Z_0} \approx \frac{a}{\lambda_g}\left(\frac{a^2 b}{\pi M}\right), \qquad \frac{d_1}{a} \ll 1 \tag{3b}$$

$$\frac{X_c}{Z_0} = \frac{2a}{\lambda_g}, \tag{4}$$

$$\frac{X_d}{Z_0} = \frac{2a}{\lambda_g}, \tag{5}$$

where

$$\lambda_g = \frac{\lambda}{\sqrt{1 - \left(\dfrac{\lambda}{2a}\right)^2}}, \qquad \lambda_g' = \frac{\lambda}{\sqrt{1 - \left(\dfrac{\lambda}{2a'}\right)^2}},$$

$$M = M_1 \cos^2 \phi + M_2 \sin^2 \phi.$$

The coefficients M_1 and M_2 are defined in Sec. 5·4b.

For the case of a centered circular aperture ($d_1 = d_2 = d$)

$$M = M_1 = M_2 = \frac{d^3}{6},$$

and Eq. (2) can be written more accurately as

$$\frac{X_a}{Z_0} = \frac{4\pi^3}{45} \frac{a^2}{\lambda_g b} \left(\frac{d}{2a}\right)^5. \tag{6}$$

If the reference planes T are then shifted a distance Δ in guide wavelengths, where

$$\Delta = \frac{2\pi^2}{45} \frac{a^2}{\lambda_g b} \left(\frac{d}{2a}\right)^5, \tag{7}$$

the equivalent circuit at the new reference planes T_Δ has a vanishing series reactance X_a. Since the shift Δ is small, the circuit parameters at the reference planes T_Δ are approximately the same as those at T.

Restrictions.—The equivalent circuit is applicable in the wavelength range in which only the H_{10}-mode can be propagated. The circuit parameters have been derived by the integral equation method employing the small-aperture approximations that $\pi d_1 < \lambda$ with λ not too close to cutoff of the next higher mode. The restriction to zero wall thickness is to be emphasized. The power transmitted through a small circular hole of diameter d in a wall of thickness t is approximately $32.0 \, t/d$ db less than that computed by the above circuit.

Numerical Results.—The coefficients M_1 and M_2 are plotted as a function of d_2/d_1 in the graph of Fig. 5·4-4.

b. *Circular Stub Guide.*—A symmetrical H-plane, T-type junction of a rectangular and a circular guide coupled by a centered small elliptical or circular aperture in a metallic wall of zero thickness (H_{10}-mode in rectangular guide, H_{11}-mode of indicated polarization in circular guide).

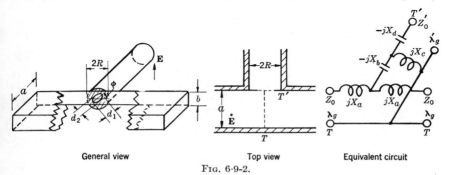

General view   Top view   Equivalent circuit

FIG. 6·9-2.

Equivalent-circuit Parameters.—At the reference plane T and T', for the case of an elliptical aperture with orientation $\phi = 0$ or $\pi/2$,

$$\frac{Z_0'}{Z_0} = 0.952 \frac{\lambda_g' \pi R^2}{\lambda_g ab}, \tag{8}$$

$$\frac{X_a}{Z_0} \approx 0, \tag{9}$$

$$\frac{X_b}{Z_0} = \frac{a}{\lambda_g}\left(\frac{a^2 b}{\pi M} + 1 + 2.05 \sqrt{\frac{ab}{\pi R^2}} - 2.46 \frac{a^2 b}{2\pi R^3}\right), \tag{10a}$$

$$\frac{X_b}{Z_0} \approx \frac{a}{\lambda_g} \frac{a^2 b}{\pi M}, \qquad \frac{d_1}{a} \ll 1, \tag{10b}$$

$$\frac{X_c}{Z_0} = \frac{2a}{\lambda_g}, \tag{11}$$

$$\frac{X_d}{Z_0} = \frac{2a}{\lambda_g}, \tag{12}$$

where

$$\lambda_g = \frac{\lambda}{\sqrt{1 - \left(\frac{\lambda}{2a}\right)^2}}, \quad \lambda_g' = \frac{\lambda}{\sqrt{1 - \left(\frac{\lambda}{3.41R}\right)^2}},$$

$$M = M_1 \cos^2 \phi + M_2 \sin^2 \phi.$$

For the special case of a circular aperture see comments in Sec. 6·9a.

Restrictions.—The equivalent circuit is valid in the wavelength range for which only a single dominant mode can be propagated in each guide. To ensure that only a single H_{11}-mode, with polarization as indicated in Fig. 6.9-2, can be propagated in the circular guide, the angle ϕ must be restricted to 0 or $\pi/2$. Also note Restrictions in Sec. 6·9a.

Numerical Results.—The coefficients M_1 and M_2 are plotted as a function of d_2/d_1 in the graph of Fig. 5·4-4.

6·10. Aperture-coupled T-junction in Coaxial Guide.—A right-angle T-type junction of a coaxial and a rectangular guide with coupling by a small centered elliptical or circular aperture in a metallic wall of zero thickness (principal mode in coaxial guide, H_{10}-mode in rectangular guide).

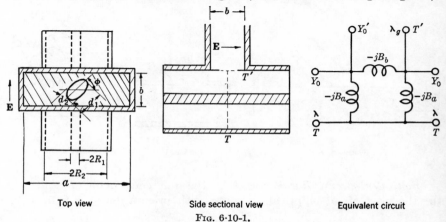

Top view   Side sectional view   Equivalent circuit
Fig. 6·10-1.

SEC. 6·11]   BIFURCATION OF A COAXIAL LINE   369

Equivalent-circuit Parameters.—At the reference planes T and T' for coupling by a centered elliptical aperture of orientation ϕ,

$$\frac{Y_0'}{Y_0} = \frac{\lambda 4\pi R_2^2 \ln\left(\frac{R_2}{R_1}\right)}{\lambda_g ab}, \tag{1}$$

$$\frac{B_a}{Y_0} = \frac{P}{2\lambda R_2^2 \ln\left(\frac{R_2}{R_1}\right)}, \tag{2}$$

$$\frac{B_b}{Y_0} = \frac{\lambda R_2^2 \ln\left(\frac{R_2}{R_1}\right)}{M}, \tag{3}$$

where

$$\lambda_g = \frac{\lambda}{\sqrt{1 - \left(\frac{\lambda}{2a}\right)^2}}.$$

The coefficients P and M are defined in Secs. 5·4b and 6·8a. For a centered circular aperture

$$M = \frac{d^3}{6}, \qquad P = \frac{d^3}{12}.$$

Restrictions.—Subject to the provision that $d_1 \ll \lambda$, the equivalent circuit is applicable in the wavelength range for which only the principal mode can be propagated in the coaxial guide and only the H_{10}-mode in the rectangular guide. The circuit parameters have been obtained by an integral equation method employing the small-aperture approximations mentioned in Sec. 5·4b. In addition it is assumed that the curvature of the surface containing the aperture has negligible effect. The restriction to zero thickness is to be emphasized (*cf.* Chap. 8).

Numerical Results.—The coefficients M_1, M_2, and P are plotted as a function of d_2/d_1 in Fig. 5·4-4.

6·11. Bifurcation of a Coaxial Line.—A bifurcation of a coaxial guide by a partition of zero thickness perpendicular to the electric field (principal mode in all guides).

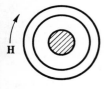
Cross sectional view

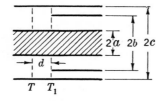
Side view
FIG. 6·11-1.

Equivalent circuit

Equivalent-circuit Parameters.—At the common reference plane T the equivalent circuit is a series junction of three transmission lines of characteristic impedances

$$\frac{Z_1}{Z_0} = \frac{\ln \frac{c}{b}}{\ln \frac{c}{a}}, \qquad \frac{Z_2}{Z_0} = \frac{\ln \frac{b}{a}}{\ln \frac{c}{a}}, \qquad Z_0 = Z_1 + Z_2. \tag{1}$$

The location of the reference plane T is given by

$$\frac{2\pi d}{\lambda} = \frac{2b_0}{\lambda}\left(\frac{b_1}{b_0}\ln\frac{b_0}{b_1} + \frac{b_2}{b_0}\ln\frac{b_0}{b_2}\right) + S_1^{Z_0}(x;0,\alpha) - S_1^{Z_0}(x_1;0,\alpha_1) - S_1^{Z_0}(x_2;0,\alpha_2), \tag{2a}$$

$$\frac{\pi d}{b_0} \approx \frac{b_1}{b_0}\ln\frac{b_0}{b_1} + \frac{b_2}{b_0}\ln\frac{b_0}{b_2}, \tag{2b}$$

where

$$x = \frac{2b_0}{\lambda} = \frac{2(c-a)}{\lambda}, \qquad x_1 = \frac{2b_1}{\lambda} = \frac{2(c-b)}{\lambda}, \qquad x_2 = \frac{2b_2}{\lambda} = \frac{2(b-a)}{\lambda},$$

$$\alpha = \frac{c}{a}, \qquad \alpha_1 = \frac{c}{b}, \qquad \alpha_2 = \frac{b}{a},$$

and

$$S_1^{Z_0}(x;0,\alpha) = \sum_{n=1}^{\infty}\left(\sin^{-1}\frac{x}{\gamma_n} - \frac{x}{n}\right),$$

$$\frac{\pi\gamma_n(\alpha)}{\alpha - 1} = \chi_{0n}(\alpha) = \chi_n, \qquad J_0(\chi_n)N_0(\chi_n\alpha) - N_0(\chi_n)J_0(\chi_n\alpha) = 0,$$

with similar definitions for α_1 and α_2.

Alternative equivalent circuits at the reference plane T_1 are identical with the alternative equivalent circuits shown in Sec. 6·4. The values of the corresponding circuit parameters are given in Eqs. (3) of Sec. 6·4 except that the $2\pi d/\lambda_g$ therein is replaced by the $2\pi d/\lambda$ of this section.

Restrictions.—The equivalent circuits are valid in the wavelength range $\lambda > 2(c-a)/\gamma_1(c/a)$ provided the fields are rotationally symmetrical. The circuit parameters have been determined by the transform method and are rigorous in the above range. The approximation (2b) for $\pi d/b_0$ is in error by less than 2 per cent provided $2(c-a)/\lambda < 0.3$. For $c/a < 3$ the results of this section differ by only a few per cent from those of Sec. 6·4 provided λ_g and b in the latter are replaced by the λ and b_0 of this section.

Numerical Results.—The quantity $\pi d/b_0$ is plotted in Fig. 6·11-2 as a function of $(c-b)/(c-a) = b_1/b_0$ with $2(c-a)/\lambda$ as a parameter and

SEC. 6·11] BIFURCATION OF A COAXIAL LINE 371

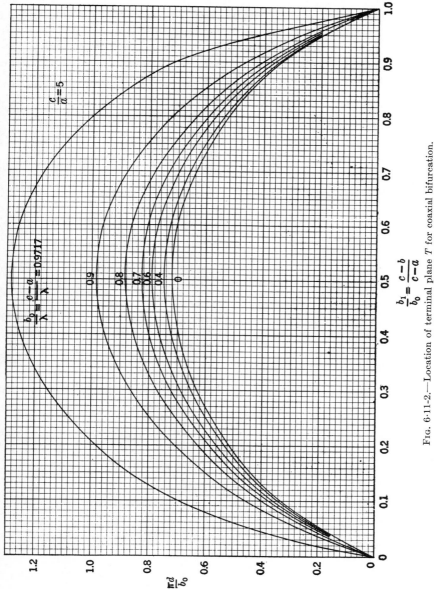

FIG. 6·11-2.—Location of terminal plane T for coaxial bifurcation.

for $c/a = 5$. The corresponding plot for $c/a = 1$ is shown in Fig. 6·4-3 provided the λ_g and b therein is replaced by the λ and b_0 of this section. Corresponding values of $\pi d/b_0$ within the range $c/a = 1$ to 5 differ by at most 10 per cent. The functions $S_1^{7o}(x;0,\alpha)$ are tabulated in the Appendix; the roots $\chi_{0n}(\alpha)$ are tabulated in Table 2·3 (note the α of this section is the ratio c of Table 2·3).

CHAPTER 7

EIGHT-TERMINAL STRUCTURES

An arbitrary junction of four accessible waveguides, each propagating only a single mode, is termed an eight-terminal or four-terminal-pair waveguide structure. So likewise is termed an arbitrary junction of *three* accessible waveguides in one of which two modes can be propagated while in the remaining two of which only one mode can be propagated. The over-all description of the propagating modes in such structures is obtained by representation of the waveguide regions as transmission lines and by representation of the junction region as an eight-terminal lumped-constant equivalent circuit. The four transmission lines and lumped-constant circuit thereby required together comprise an eight-terminal network indicative of the transmission, reflection, standing-wave, etc., properties of the over-all structure. The quantitative description of the transmission lines requires the indication of their characteristic impedances and propagation wavelengths. For the description of the eight-terminal lumped-constant circuit it is, in general, necessary to specify ten circuit parameters and the location of the four associated terminal planes. This number of required circuit parameters may be reduced considerably if the structure possesses geometrical symmetries.

The impedance parameters of the following eight-terminal networks will be referred to the characteristic impedance of one of the transmission lines; this is in accord with the fact that only relative impedances are physically significant in microwave calculations. The equivalent network diagrams of the eight-terminal structures will exhibit explicitly the choice of characteristic impedances for the transmission lines and the locations of the relevant terminal planes. The equivalent networks for the most part will be presented only at those terminal planes which facilitate the derivation of the circuit parameters. The transformation to terminal planes that yield "simple" equivalent networks may be effected by the technique pointed out in Sec. 3·3.

7·1. Slit Coupling of Rectangular Guides, E-plane.—A junction of two contiguous rectangular guides of equal dimensions coupled on their broad sides by a slit in a wall of zero thickness. Sides of slit perpendicular to the electric field (H_{10}-modes in rectangular guides).

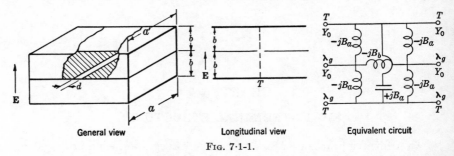

FIG. 7·1-1.

Equivalent-circuit Parameters.—At the common reference plane T

$$\frac{B_a}{Y_0} = \frac{\frac{2b}{\lambda_g}\left[\frac{J_1(\pi d/\lambda_g)}{2b/\lambda_g}\right]^2}{1 - \frac{3}{4}\left(\frac{\pi d}{2\lambda_g}\right)^2 + \left(\frac{\pi d}{2\lambda_g}\right)^2 \ln\frac{4b}{\pi d} + \frac{1}{6}\left(\frac{\pi d}{4b}\right)^2 - 2\left(\frac{\pi d}{4b}\right)^2 \sum_1 \left(\frac{2b}{\lambda_g}\right)}, \quad (1a)$$

$$\frac{B_a}{Y_0} = \frac{4b}{\lambda_g} \ln \cosh \frac{\pi d}{4b}, \qquad \frac{b}{\lambda_g} \ll 1, \quad (1b)$$

$$\frac{B_a}{Y_0} \approx \frac{2b}{\lambda_g}\left(\frac{\pi d}{4b}\right)^2, \qquad \frac{d}{b} \ll 1, \quad (1c)$$

$$\frac{B_b}{Y_0} - \frac{B_a}{Y_0} = \frac{\frac{4b}{\lambda_g}}{[J_0(\pi d/\lambda_g)]^2}\left\{\left[1 - \left(\frac{\pi d}{2\lambda_g}\right)^2\right]\ln\frac{4b}{\pi d} + \sum_0\left(\frac{2b}{\lambda_g}\right) + \frac{1}{2}\left(\frac{\pi d}{2\lambda_g}\right)^2 \right.$$
$$\left. - \frac{1}{6}\left(\frac{\pi d}{4b}\right)^2 + 2\left(\frac{\pi d}{4b}\right)^2 \sum_1\left(\frac{2b}{\lambda_g}\right)\right\}, \quad (2a)$$

$$\frac{B_b}{Y_0} - \frac{B_a}{Y_0} = \frac{4b}{\lambda_g} \ln \operatorname{csch} \frac{\pi d}{4b}, \qquad \frac{b}{\lambda_g} \ll 1, \quad (2b)$$

$$\frac{B_b}{Y_0} \approx \frac{4b}{\lambda_g} \ln \frac{4b}{\pi d}, \qquad \frac{d}{b} \ll 1, \quad (2c)$$

where

$$\sum_0 (x) = \sum_{n=1}^{\infty} \left(\frac{1}{\sqrt{n^2 - x^2}} - \frac{1}{n}\right) \cong \frac{1}{\sqrt{1 - x^2}} - 1 + 0.1010 x^2$$
$$+ 0.0138 x^4 + \cdots,$$

$$\sum_{1}(x) = \sum_{n=1}^{\infty}\left(\sqrt{n^2 - x^2} - n + \frac{x^2}{2n}\right) \cong \sqrt{1-x^2} - 1 + \frac{x^2}{2}$$
$$- 0.0253 x^4 + \cdots.$$

Restrictions.—The equivalent circuit is applicable in the wavelength range $2b/\lambda_g < 1$. Equations (1a) and (2a) have been obtained by a variational method employing static approximations for the electric field in the slit and are estimated to be correct to within a few per cent for $2b/\lambda_g < 1$ and $d/b < 1$. Equations (1b) and (2b) have been obtained by conformal mapping and are exact in the limit $2b/\lambda_g \to 0$. The restriction to zero thickness should be noted.

Experimental Results.—The equivalent circuit parameters have been measured at $\lambda = 3.20$ cm in rectangular guides of dimensions $a = 0.90$ in. and $b = 0.40$ in. For a slit of width $d = 0.04$ in. in a wall of total thickness 0.03 in., the measured values are

$$\frac{B_a}{Y_0} = 0.23, \qquad \frac{B_b}{Y_0} - \frac{B_a}{Y_0} = 0.58.$$

The theoretical values from Eqs. (1a) and (2a) are

$$\frac{B_a}{Y_0} = 0.24, \qquad \frac{B_b}{Y_0} - \frac{B_a}{Y_0} = 0.36.$$

The discrepancy between the theoretical and measured values of B_b may be accounted for by the effect of wall thickness.

7·2. Small-aperture Coupling of Rectangular Guides, E-plane.—A junction of two contiguous rectangular guides coupled on their broad sides by a small circular or elliptical aperture in a wall of zero thickness (H_{10}-modes in rectangular guides).

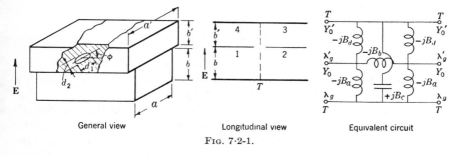

General view Longitudinal view Equivalent circuit

FIG. 7·2-1.

Equivalent-circuit Parameters.—At the common reference plane T the parameters for a centered elliptical aperture of orientation ϕ are

$$\frac{Y_0'}{Y_0} = \frac{\lambda_g ab}{\lambda_g' a' b'}, \tag{1a}$$

$$\frac{B_a}{Y_0} = \frac{\left(\frac{\lambda_g}{\lambda}\right)^2 \frac{2\pi P}{\lambda_g ab}\left(1 + \frac{ab}{a'b'}\right)}{1 - \frac{4\pi P}{\lambda^2 b}\left(1 + \frac{b}{b'}\right)}, \tag{2a}$$

$$\frac{B_a}{Y_0} \approx \left(\frac{\lambda_g}{\lambda}\right)^2 \frac{2\pi P}{\lambda_g ab}\left(1 + \frac{ab}{a'b'}\right), \qquad \frac{d_1}{\lambda} \ll 1, \tag{2b}$$

$$\frac{B_b}{Y_0} = \frac{\lambda_g ab}{4\pi}\left[\frac{1}{M} - \frac{\pi}{a^2 b}\left(1 + \frac{a^2 b}{a'^2 b'}\right)\right], \tag{3a}$$

$$\frac{B_b}{Y_0} \approx \frac{\lambda_g ab}{4\pi M}, \qquad \frac{d_1}{\lambda} \ll 1, \tag{3b}$$

$$\frac{B_c}{Y_0} = \frac{B_a}{Y_0} \frac{2}{1 + \frac{a'b'}{ab}}, \tag{4}$$

$$\frac{B_d}{Y_0} = \frac{B_a}{Y_0} \frac{ab}{a'b'}, \tag{5}$$

where $M = M_1 \cos^2 \phi + M_2 \sin^2 \phi$,

$$\lambda_g = \frac{\lambda}{\sqrt{1 - \left(\frac{\lambda}{2a}\right)^2}}, \qquad \lambda_g' = \frac{\lambda}{\sqrt{1 - \left(\frac{\lambda}{2a'}\right)^2}}.$$

The coefficients P, M_1, and M_2 are defined in Secs. 5·4b and 6·8a. For a centered circular aperture of diameter d

$$P = \frac{d^3}{12}, \qquad M = \frac{d^3}{6}.$$

If for identical guides ($a = a'$, $b = b'$) the aperture dimensions are so chosen that

$$B_a \approx \frac{1}{B_b} \ll 1, \qquad \text{i.e.,} \quad \left(\frac{\lambda_g}{\lambda}\right)^2 P = M,$$

the above junction becomes a directional coupler (Bethe hole). The latter has the property of not coupling power from guide 1 to guide 3 if guides 2 and 4 are matched, and conversely.

Restrictions.—The equivalent circuit is applicable in the wavelength range $2b/\lambda_g < 1$ ($b > b'$). The circuit parameters have been derived by an integral equation method employing small-aperture approximations (*cf.* Sec. 5·4b). It is to be noted that the above formulas apply only to the case of a wall of zero thickness.

Numerical Results.—The coefficients P, M_1, and M_2 are plotted as functions of d_2/d_1 in the graphs of Fig. 5·4-4.

7·3. Aperture Coupling of Coaxial Guides.

A junction of two concentric coaxial guides coupled by a small aperture in a wall of zero thickness (principal modes in coaxial guides).

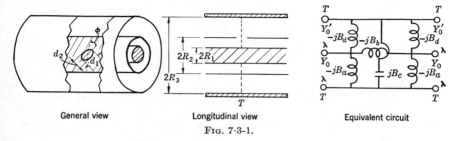

General view   Longitudinal view   Equivalent circuit

Fig. 7·3-1.

Equivalent-circuit Parameters.—At the common central reference plane T

$$\frac{Y_0'}{Y_0} = \frac{\ln \frac{R_2}{R_1}}{\ln \frac{R_3}{R_2}}, \tag{1}$$

$$\frac{B_a}{Y_0} = \frac{P}{2\lambda R_2^2 \ln \frac{R_2}{R_1}} \left(1 + \frac{\ln \frac{R_2}{R_1}}{\ln \frac{R_3}{R_2}}\right), \tag{2}$$

$$\frac{B_b}{Y_0} = \frac{\lambda R_2^2 \ln \frac{R_2}{R_1}}{M}, \tag{3}$$

$$\frac{B_c}{Y_0} = \frac{B_a}{Y_0} \frac{2}{1 + \frac{\ln \frac{R_3}{R_2}}{\ln \frac{R_2}{R_1}}}, \tag{4}$$

$$\frac{B_d}{Y_0} = \frac{B_a}{Y_0} \frac{\ln \frac{R_2}{R_1}}{\ln \frac{R_3}{R_2}}. \tag{5}$$

For a small elliptical aperture of orientation ϕ

$$M = M_1 \cos^2 \phi + M_2 \sin^2 \phi.$$

The coefficients P, M_1, and M_2 are defined in Secs. 5·4b and 6·8a.

For a small circular aperture of diameter d

$$P = \frac{d^3}{12}, \qquad M = \frac{d^3}{6}. \tag{6}$$

For a circumferential slit of width $d_2 = d$ ($\phi = 0$ and $d_1 = 2\pi R_2$)

$$P = \frac{\pi d^2}{16}, \qquad M = -\frac{\lambda^2}{8\pi \ln \dfrac{4\lambda}{\pi d \gamma}}, \tag{7}$$

$$\gamma = 1.781.$$

The susceptance B_b changes sign and hence is capacitive for this case.

Restrictions.—The equivalent circuit is applicable in the wavelength range for which only the principal modes can be propagated. The circuit parameters have been obtained by an integral equation method employing small-aperture approximations (*cf.* Sec. 5·4b). In addition the derivation assumes that the apertures are planar in so far as the higher modes are concerned. It is estimated that the formulas are reasonably accurate for $R_2/R_1 < 2$ and $\pi d/\lambda < 1$. The restriction to zero wall thickness should be emphasized.

Numerical Results.—The coefficients P, M_1, and M_2 for the case of an elliptical aperture are plotted in the graphs of Fig. 5·4-4.

7·4. Slit Coupling of Rectangular Guides, H-plane.—A junction of two contiguous rectangular guides of equal heights coupled on their narrow sides by a slit in a wall of zero thickness. Sides of slit parallel to electric field (H_{10}-modes in rectangular guides).

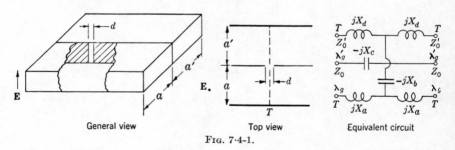

General view   Top view   Equivalent circuit

Fig. 7·4-1.

Equivalent-circuit Parameters.—At the common central reference planes T

$$\frac{Z_0'}{Z_0} = \frac{\lambda_g'}{\lambda_g}\left(\frac{a}{a'}\right)^3, \tag{1}$$

$$\frac{X_a}{Z_0} = \frac{\dfrac{a}{\lambda_g}\left(\dfrac{\pi d}{4a}\right)^4 \left(1 + \dfrac{a^3}{a'^3}\right)}{1 - \dfrac{1}{2}\left(\dfrac{\pi d}{4a}\right)^4 \left(1 + \dfrac{a^4}{a'^4}\right)}, \tag{2a}$$

$$\frac{X_a}{Z_0} \approx \frac{a}{\lambda_g}\left(\frac{\pi d}{4a}\right)^4\left(1 + \frac{a^3}{a'^3}\right), \qquad d \ll a, \tag{2b}$$

$$\frac{X_b}{Z_0} = \frac{a}{\lambda_g}\left[\left(\frac{4a}{\pi d}\right)^2 + \left(1 + \frac{a^2}{a'^2}\right)\right], \tag{3a}$$

$$\frac{X_b}{Z_0} \approx \frac{a}{\lambda_g}\left(\frac{4a}{\pi d}\right)^2, \qquad d \ll a, \tag{3b}$$

$$\frac{X_c}{Z_0} = \frac{X_a}{Z_0}\frac{2}{1 + \left(\frac{a'}{a}\right)^3}, \tag{4},$$

$$\frac{X_d}{Z_0} = \frac{X_a}{Z_0}\left(\frac{a}{a'}\right)^3. \tag{5}$$

Restrictions.—The equivalent circuit is applicable in the wavelength range $a < \lambda < 2a'$ (if $a > a'$). Equations (2a) and (3a) have been derived by an integral equation method employing the small-aperture approximations $\pi d/\lambda \ll 1$, etc. It is to be emphasized that the formulas apply to slits in a wall of zero thickness.

7·5. Aperture Coupling of Rectangular Guides, H-plane.—A junction of two contiguous rectangular guides coupled on their narrow sides by a small elliptical or circular aperture in a wall of zero thickness (H_{10}-modes in rectangular guides).

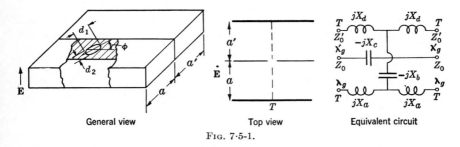

General view Top view Equivalent circuit
Fig. 7·5-1.

Equivalent-circuit Parameters.—At the common reference planes T for the case of an elliptical aperture with orientation ϕ

$$\frac{Z'_0}{Z_0} = \frac{\lambda'_g b}{\lambda_g b'}\left(\frac{a}{a'}\right)^3, \tag{1}$$

$$\frac{X_a}{Z_0} \approx 0, \tag{2}$$

$$\frac{X_b}{Z_0} = \frac{a}{\lambda_g}\left(\frac{a^2 b}{\pi M} + 1 + \frac{a^2 b}{a'^2 b'}\right), \tag{3a}$$

$$\frac{X_b}{Z_0} \approx \frac{a^3 b}{\lambda_g \pi M}, \qquad d_1 \ll a, \tag{3b}$$

$$\frac{X_c}{Z_0} = \frac{X_a}{Z_0} \frac{2}{1 + \left(\frac{a'}{a}\right)^3}, \tag{4}$$

$$\frac{X_d}{Z_0} = \frac{X_a}{Z_0} \left(\frac{a}{a'}\right)^3, \tag{5}$$

where

$$M = M_1 \cos^2 \phi + M_2 \sin^2 \phi,$$

$$\lambda_g = \frac{\lambda}{\sqrt{1 - \left(\frac{\lambda}{2a}\right)^2}}, \quad \lambda'_g = \frac{\lambda}{\sqrt{1 - \left(\frac{\lambda}{2a'}\right)^2}},$$

with the coefficients M_1 and M_2 defined as in Sec. 5·4b.

For a circular aperture of diameter d,

$$M = \frac{d^3}{6}$$

and

$$\frac{X_a}{Z_0} = \frac{4\pi^3}{45} \frac{a^2}{\lambda_g b} \left(\frac{2d}{a}\right)^5 \left(1 + \frac{a^3 b}{a'^3 b'}\right). \tag{6}$$

Restrictions.—The equivalent circuit is applicable in the wavelength range $a < \lambda < 2a$. The circuit parameters have been derived by an integral equation method employing small-aperture approximations (*cf.* Sec. 5·4b). The restriction to zero wall thickness is to be emphasized.

Numerical Results.—The coefficients M_1 and M_2 are plotted vs. d_2/d_1 in the graph of Fig. 5·4-4.

7·6. 0° Y-junction, E-plane.—A symmetrical bifurcation of a rectangular guide by a partition of zero thickness perpendicular to the direction of the electric field (H_{10}-mode in all guides, also H_{11}- and E_{11}-mode in large guide).

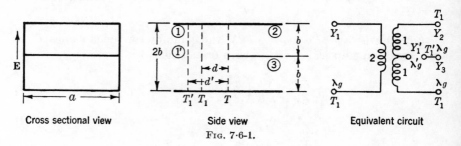

Cross sectional view   Side view   Equivalent circuit

Fig. 7·6-1.

Equivalent-circuit Parameters.—At T_1 the reference plane for the H_{10}-modes in guides 1, 2, and 3; and at T'_1 the reference plane for the H_{11}- and E_{11}-modes in guide 1' the equivalent network can be represented as a hybrid coil junction (*cf.* Fig. 3·18*b* of Sec. 3·3) of four transmission lines whose characteristic admittance ratios are

$$\frac{Y_2}{Y_1} = \frac{Y_3}{Y_1} = 2, \qquad \frac{Y_1'}{Y_1} = \frac{2\lambda_g'}{\lambda_g}.$$

The terminal planes T_1 and T_1' are located at distances d and d' given by

$$\frac{2\pi d}{\lambda_g} = \theta = 2\left(\sin^{-1}\frac{2b}{\lambda_g} - \frac{2b}{\lambda_g}\ln 2\right) - S_2\left(\frac{4b}{\lambda_g};0,0\right) + 2S_2\left(\frac{2b}{\lambda_g};0,0\right), \quad (1a)$$

$$\frac{2\pi d}{\lambda_g} = \theta \approx 2\left(\sin^{-1}\frac{2b}{\lambda_g} - \frac{2b}{\lambda_g}\ln 2\right) - 0.202\left(\frac{2b}{\lambda_g}\right)^3, \quad (1b)$$

$$\frac{2\pi d'}{\lambda_g'} = \theta' = 2\left(\sin^{-1}\frac{4b}{\sqrt{3}\lambda_g'} - \frac{2b}{\lambda_g'}\ln 2\right) - S_2\left(\frac{4b}{\lambda_g'};1,0\right) + 2S_2\left(\frac{2b}{\lambda_g'};\frac{1}{2},0\right), \quad (2a)$$

$$\frac{2\pi d'}{\lambda_g'} = \theta' \approx 2\left(\sin^{-1}\frac{4b}{\sqrt{3}\lambda_g'} - \frac{2b}{\lambda_g'}\ln 2\right) - 0.184\left(\frac{2b}{\lambda_g'}\right) - 0.300\left(\frac{2b}{\lambda_g'}\right)^3, \quad (2b)$$

where

$$\lambda_g = \frac{\lambda}{\sqrt{1 - \left(\frac{\lambda}{2a}\right)^2}}, \qquad \lambda_g' = \frac{\lambda_g}{\sqrt{1 - \left(\frac{\lambda_g}{4b}\right)^2}},$$

and

$$S_2(x;\alpha,0) = \sum_{2}^{\infty}\left(\sin^{-1}\frac{x}{\sqrt{n^2 - \alpha^2}} - \frac{x}{n}\right).$$

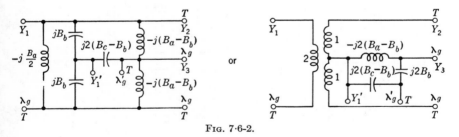

Fig. 7·6-2.

If T is chosen as the reference plane for all guides, the alternative equivalent circuits shown in Fig. 7·6-2 are obtained. At the terminal plane T

$$\frac{B_a}{Y_1} = \frac{1 + \frac{\lambda_g}{\lambda_g'}\tan\theta\tan\theta'}{\tan\theta - \frac{\lambda_g}{\lambda_g'}\tan\theta'}, \quad (3a)$$

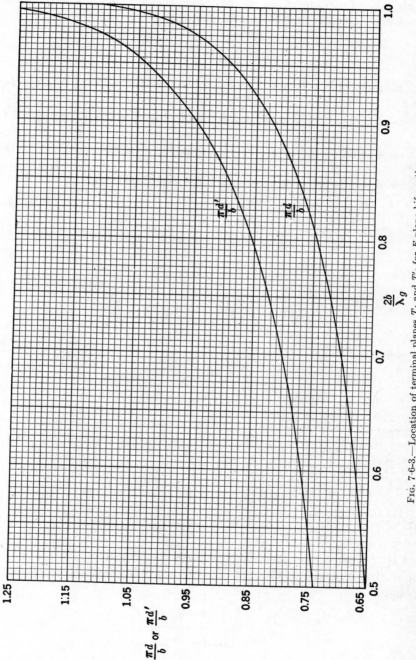

FIG. 7·6-3.—Location of terminal planes T_1 and T'_1 for E-plane bifurcation.

$$\frac{B_b}{Y_1} = \frac{\sec\theta \sec\theta'}{\tan\theta - \frac{\lambda_g}{\lambda_g'}\tan\theta'}, \qquad (3b)$$

$$\frac{B_c}{Y_1} = \frac{1 + \frac{\lambda_g'}{\lambda_g}\tan\theta \tan\theta'}{\tan\theta - \frac{\lambda_g}{\lambda_g'}\tan\theta'}. \qquad (3c)$$

The numbers beside the windings of the ideal transformers in Figs. 7·6-1 and 7·6-2 are indicative of the turns ratios.

Restrictions.—The equivalent circuit is applicable in the wavelength range $2b < \lambda_g < 4b$. The circuit parameters have been obtained by the transform method and are rigorous in the above range. The approximations (1b) and (2b) agree with Eqs. (1a) and (2a), respectively, to within 10 per cent provided $2b/\lambda_g < 0.8$. It should be emphasized that the line of characteristic admittance Y_1' represents both the H_{11}- and E_{11}-mode. Consequently, the indicated equivalent circuits are applicable only if the terminal impedances in this line are the same for both the H_{11}- and E_{11}-modes.

Numerical Results.—In Fig. 7·6-3 the quantities $\pi d/b$ and $\pi d'/b$ are plotted as functions of $2b/\lambda_g$. The $S_2(x;\alpha,0)$ function is tabulated in the Appendix.

7·7. 0° Y-Junction, H-plane.—A symmetrical bifurcation of a rectangular guide by a partition of zero thickness parallel to the electric field (H_{10}-mode in all rectangular guides, in addition, H_{20}-mode in large rectangular guide).

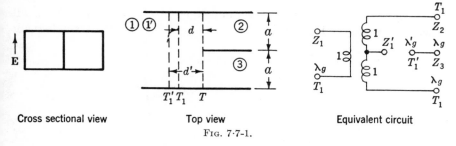

Cross sectional view   Top view   Equivalent circuit

Fig. 7·7-1.

Equivalent-circuit Parameters.—At the reference planes T_1 and T_1' the equivalent network is a hybrid coil junction (*cf.* Fig. 3·18b of Sec. 3·3) of four transmission lines whose characteristic impedance ratios are

$$\frac{Z_2}{Z_1} = \frac{Z_3}{Z_1} = 2, \qquad \frac{Z_1'}{Z_2} = \frac{\lambda_g'}{2\lambda_g}. \qquad (1)$$

The locations of the reference planes T_1 for the H_{20}-mode in guide 1 and the H_{10}-modes in guides 2 and 3, and of the reference plane T'_1 for the H_{10}-mode in guide 1' are given by

$$\theta = \frac{2\pi d}{\lambda_g} = x(2 \ln 2 - 1) - 2S_2(x;1,0) + S_3(2x;2,0), \quad (2)$$

$$\theta' = \frac{2\pi d'}{\lambda'_g} = x'(2 \ln 2 - 1) - 2S_2(x';0.5,0) + S_3(2x';1,0), \quad (3)$$

where

$$x = \frac{2a}{\lambda_g}, \quad x' = \frac{2a}{\lambda'_g},$$

$$\lambda_g = \frac{\lambda}{\sqrt{1 - \left(\frac{\lambda}{2a}\right)^2}}, \quad \lambda'_g = \frac{\lambda}{\sqrt{1 - \left(\frac{\lambda}{4a}\right)^2}},$$

$$S_N(x;\alpha,0) = \sum_{n=N}^{\infty} \left(\sin^{-1} \frac{x}{\sqrt{n^2 - \alpha^2}} - \frac{x}{n}\right).$$

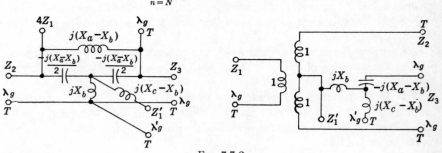

Fig. 7·7-2.

Alternative equivalent circuits at the common reference plane T for all guides are shown in Fig. 7·7-2. The corresponding circuit parameters are

$$\left.\begin{aligned}
\frac{2X_a}{Z_2} &= \frac{1 + \frac{\lambda'_g}{\lambda_g} \tan \theta \tan \theta'}{\frac{\lambda_g}{\lambda'_g} \tan \theta' - \tan \theta}, \\
\frac{2X_b}{Z_2} &= \frac{\sec \theta \sec \theta'}{\frac{\lambda_g}{\lambda'_g} \tan \theta' - \tan \theta}, \\
\frac{2X_c}{Z_2} &= \frac{1 + \frac{\lambda_g}{\lambda'_g} \tan \theta \tan \theta'}{\frac{\lambda_g}{\lambda'_g} \tan \theta' - \tan \theta}.
\end{aligned}\right\} \quad (4)$$

Restrictions.—The equivalent circuit is valid in the range

$$0.5 < a/\lambda < 0.75.$$

SEC. 7·7]   0° Y-JUNCTION, H-PLANE

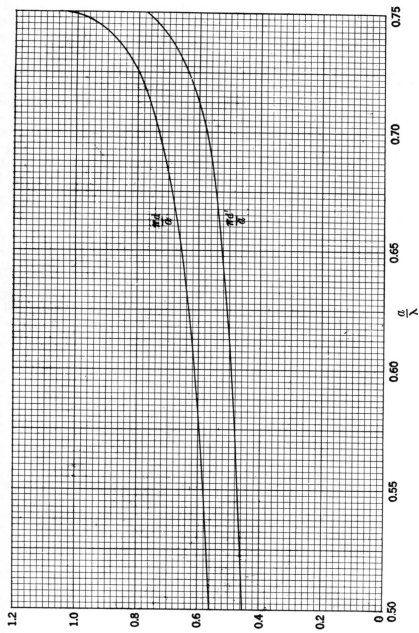

FIG. 7·7-3.—Locations of terminal planes T_1 and T'_1 for H-plane bifurcation.

The circuit parameters have been obtained by the transform method and are rigorous in the above range.

Numerical Results.—Figure 7·7-3 contains graphs of $\pi d/a$ and $\pi d'/a$ as functions of a/λ in the range $0.5 < a/\lambda < 0.75$. A number of the S_N functions are tabulated in the Appendix.

7·8. Magic-Tee (Hybrid) Junction.—A magic-T-junction of four identical guides comprising an E- and H-plane T-junction with a common plane of symmetry (H_{10}-mode in all rectangular guides).

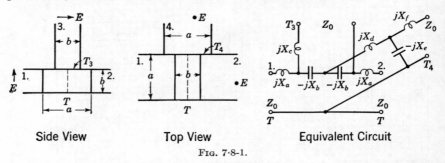

Fig. 7·8-1.

Experimental Results.—Measurements taken at a wavelength of $\lambda = 3.20$ cm with rectangular guides of dimensions $a = 0.90$ in. and $b = 0.40$ in. yield the following data at the indicated terminal planes:

$$\frac{X_a}{Z_0} = 0.34, \quad \frac{X_d}{Z_0} = 0.93,$$

$$\frac{X_b}{Z_0} = 1.30, \quad \frac{X_e}{Z_0} = 0.92,$$

$$\frac{X_c}{Z_0} = 0.18, \quad \frac{X_f}{Z_0} = 1.27.$$

CHAPTER 8

COMPOSITE STRUCTURES

The microwave structures to be described in this chapter are not of the same character as the elementary discontinuities and waveguides considered in previous chapters. They are instead composite structures whose network representations comprise a variety of elementary lumped-constant circuits and transmission lines. In contrast to the case of the elementary structures it is not necessary to solve any field problems to determine the network parameters of composite structures. These parameters may be calculated by straightforward microwave network analysis from the parameters descriptive of the elementary structures. In the following the over-all parameters of a number of composite structures will be derived to illustrate both the use and the results of such network analyses.

Uniform waveguides partially filled with dielectric or metallic slabs represent an important class of composite microwave structures. The determination of the propagation constants of the modes capable of propagating along such waveguides usually constitutes the chief problem of interest. This problem may be phrased alternatively as that of finding the cutoff, or resonant, frequencies of the waveguide cross section; the knowledge of these implies that of the mode propagation constants [$cf.$ Eqs. (1·11b)]. If the waveguide crosssection possesses an appropriate geometrical symmetry, the characteristics of the propagating modes may be determined by regarding not the longitudinal guide axis but rather one of the transverse directions perpendicular thereto as a transmission direction. The composite transmission structure so defined is completely equivalent to the original guide; it is composed in general of elementary discontinuities and waveguides and is terminated by the guide walls, if any. The desired frequencies may be ascertained by simple microwave network analysis of the "transverse equivalent network" representative of the desired mode in the above composite structure. The resonant frequencies are those frequencies for which the total admittance looking in both directions at any point of the transverse equivalent network vanishes. If the transverse network possesses symmetries, such resonance calculations may be simplified considerably by use of bisection theorems.

Although the propagation constants of any mode can be calculated by use of the appropriate transverse network, the results presented below will refer for the most part to the dominant mode. The latter will be designated as an E- or H-type mode depending on whether the dominant mode in the unperturbed guide without metallic or dielectric slabs is an E- or H-mode. No general effort will be made to extend the results to structures containing dissipative dielectrics or metals although this can be done in a manner formally identical with the nondissipative case; in the dissipative case one finds complex rather than real resonant frequencies and hence complex rather than real propagation constants. The determination of the resonant frequencies of a composite waveguide cavity will not be considered explicitly, as it follows readily from that for the corresponding composite waveguide, of which it represents a special case.

Another important type of composite microwave structure to be described below is provided by a junction of two waveguides coupled by a slit or aperture in a wall of finite thickness. The modification in the network representation introduced by the effect of wall thickness requires the representation of the coupling wall as a transmission line whose length is equal to the wall thickness. The over-all network is then a composite of the circuits representative of coupling into and out of an infinitely thick coupling wall and of the transmission line representative of the coupling wall itself. The calculation of the over-all network parameters is effected by a simple network analysis. Although the above composite representation of thickness effects is rigorous only for coupling walls whose thickness is large compared with aperture dimensions, it is nevertheless surprisingly accurate for much smaller wall thickness.

PROPAGATION IN COMPOSITE GUIDES

8·1. Rectangular Guide with Dielectric Slabs Parallel to E.—A uniform rectangular guide containing two dielectric slabs, the sides of which are parallel to the electric field (dominant mode in dielectric-filled guide).

RECTANGULAR GUIDE

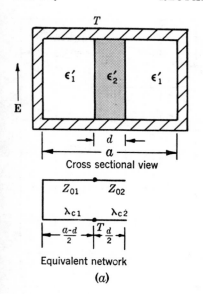
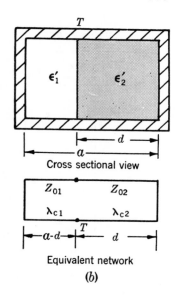

Fig. 8·1-1.

Equivalent Network Results.—At the reference plane T the equivalent transverse network for the dominant mode in the guide of Fig. 8·1-1a consists merely of a junction of a short- and an open-circuited uniform H-mode transmission line. The propagation wavelength λ_g of the dominant mode is determined by the resonant condition

$$Z_{01} \tan \frac{2\pi}{\lambda_{c1}} \left(\frac{a-d}{2} \right) = Z_{02} \cot \frac{2\pi}{\lambda_{c2}} \frac{d}{2}, \tag{1}$$

where

$$\frac{Z_{01}}{Z_{02}} = \frac{\lambda_{c1}}{\lambda_{c2}} = \sqrt{\frac{\epsilon_2' - \left(\dfrac{\lambda_0}{\lambda_g}\right)^2}{\epsilon_1' - \left(\dfrac{\lambda_0}{\lambda_g}\right)^2}}, \tag{2}$$

the quantities ϵ_1' and ϵ_2' being the relative dielectric constants of the two media; λ_0 is the free-space wavelength.

At the reference plane T the equivalent transverse network for the dominant mode in the guide of Fig. 8·1-1b is a junction of two short-

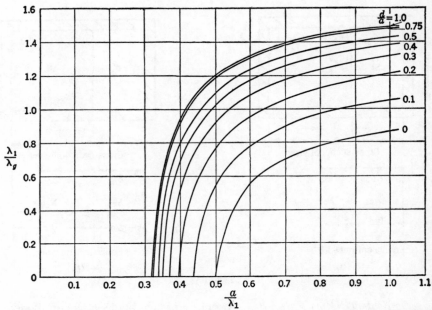

FIG. 8·1-2.—Propagation characteristics of rectangular guide with dielectric slabs ∥ E. Central slab of dielectric constant ϵ_2.

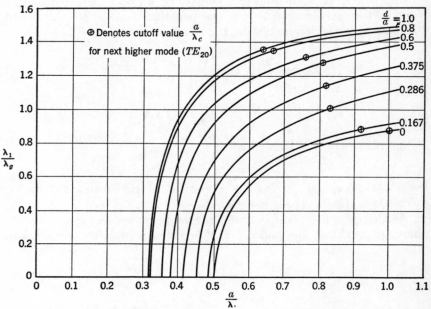

FIG. 8·1-3.—Propagation characteristics of rectangular guide with dielectric slabs ∥ E. Slab of dielectric constant ϵ_2 at one side.

circuited H-mode uniform transmission lines. The propagation wavelength λ_g of the dominant mode is obtained in this case from the resonant condition

$$Z_{01} \tan \frac{2\pi}{\lambda_{c1}} (a - d) = -Z_{02} \tan \frac{2\pi}{\lambda_{c2}} d, \qquad (3)$$

where the various quantities are defined as in Eq. (2).

Restrictions.—Equations (1) and (3) are rigorous provided the conductivity of the guide walls is infinite. These equations are also applicable when the slabs are dissipative and characterized by a complex dielectric constant; in this event λ_g is complex. Equation (1) applies as well to the higher-order modes of the H_{m0} type if m is odd; when m is even, the cotangent function should be replaced by its negative inverse.

Numerical Results.—The ratio λ_1/λ_g ($\lambda_1 = \lambda_0/\sqrt{\epsilon_1'}$), as computed from a numerical solution of Eq. (1) for $\lambda_{c1}/\lambda_{c2}$, is plotted in the graph of Fig. 8·1-2 as a function of a/λ_1 for various values of d/a. A corresponding plot for Eq. (3) is contained in Fig. 8·1-3. In both cases the data apply to nondissipative dielectric slabs with $\epsilon_2'/\epsilon_1' = 2.45$.

8·2. Rectangular Guide with Dielectric Slabs Perpendicular to E.—A uniform rectangular guide containing two parallel dielectric slabs, the sides of which are perpendicular to the electric field (dominant mode in dielectric-filled rectangular guide).

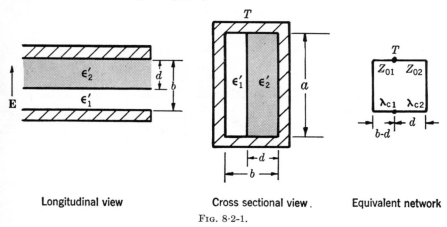

Longitudinal view   Cross sectional view.   Equivalent network
Fig. 8·2-1.

Equivalent Network Results.—At the reference plane T the equivalent transverse network for the dominant mode is a shunt arrangement of two short-circuited uniform E-mode transmission lines. The propagation wavelength λ_g of the dominant mode is determined from the resonance condition

$$Z_{01} \tan \frac{2\pi}{\lambda_{c1}} (b - d) + Z_{02} \tan \frac{2\pi}{\lambda_{c2}} d = 0, \qquad (1)$$

where

$$\frac{Z_{02}}{Z_{01}} = \frac{\epsilon_1' \lambda_{c1}}{\epsilon_2' \lambda_{c2}} = \frac{\epsilon_1'}{\epsilon_2'} \sqrt{\frac{\epsilon_2' - \left(\frac{\lambda_0}{\lambda_g}\right)^2 - \left(\frac{\lambda_0}{2a}\right)^2}{\epsilon_1' - \left(\frac{\lambda_0}{\lambda_g}\right)^2 - \left(\frac{\lambda_0}{2a}\right)^2}}, \qquad (2)$$

the quantities ϵ_1' and ϵ_2' being the relative dielectric constants of the two media. An explicit, but approximate, expression for λ_g is

$$\lambda_g \approx \frac{\lambda_0}{\sqrt{1 - \frac{d}{b}\left(1 - \frac{\epsilon_1'}{\epsilon_2'}\right) - \left(\frac{\lambda_0}{2a}\right)^2}}. \qquad (3)$$

Restrictions.—Equation (1) is rigorous and subject only to the restriction that the guide walls are perfectly conducting. This result applies as well to dissipative dielectrics that are characterized by complex dielectric constants. Equation (3) is an approximate solution for the lowest root of Eq. (1) and is valid in the range for which $2\pi(b - d)/\lambda_{c1}$ and $2\pi d/\lambda_{c2} \ll 1$. It should be noted that the dominant mode in the dielectric-filled guide has components of both **E** and **H** in the direction of propagation.

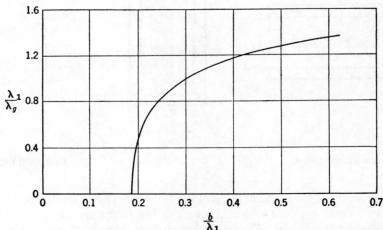

FIG. 8·2-2.—Propagation characteristics of rectangular guide with dielectric slabs $\perp$ **E**.

Numerical Results.—The ratio $\lambda_1/\lambda_g (\lambda_1 = \lambda_0/\sqrt{\epsilon_1'})$, as obtained by numerical solution of Eq. (1) for $\lambda_{c1}/\lambda_{c2}$, is plotted as a function of b/λ_1 in Fig. 8·2-2 for the special case $b/a = 0.45$, $d/b = 0.5$, and $\epsilon_2'/\epsilon_1' = 2.45$. In addition Fig. 8·2-3 contains graphs of λ_1/λ_g vs. d/b for $d/b = 0.31$ and 0.40 with the same values for b/a and ϵ_2'/ϵ_1' as above.

8·3. Circular Guide with Dielectric Cylinders.—A uniform circular guide containing two concentric dielectric cylinders (dominant angularly symmetric E_0- or H_0-mode in dielectric-filled circular guide).

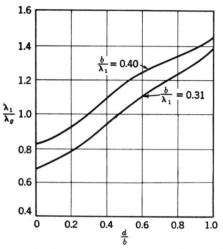

Fig. 8·2-3.—Propagation characteristics of rectangular guide with dielectric slabs $\perp$ E.

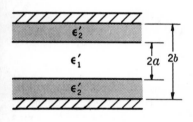

Longitudinal view

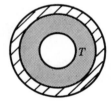

Cross sectional view

Fig. 8·3-1.

Equivalent network

Equivalent Network Results.—At the reference plane T the equivalent transverse network for the dominant E_0-mode consists of a shunt junction of an open-circuited and a short-circuited E-type radial transmission line. The propagation wavelength λ_g of the dominant E_0-mode is determined from the resonance condition

$$Y_{01} \operatorname{Tn}(x_1, 0) = -Y_{02} \operatorname{ct}(x_1', x_2') \tag{1}$$

where

$$\frac{Y_{01}}{Y_{02}} = \frac{\epsilon_1' \lambda_{c1}}{\epsilon_2' \lambda_{c2}} = \frac{\epsilon_1'}{\epsilon_2'} \sqrt{\frac{\epsilon_2' - \left(\dfrac{\lambda_0}{\lambda_g}\right)^2}{\epsilon_1' - \left(\dfrac{\lambda_0}{\lambda_g}\right)^2}}, \tag{2}$$

$$x_1 = \frac{2\pi a}{\lambda_{c1}}, \qquad x_1' = \frac{2\pi a}{\lambda_{c2}}, \qquad x_2' = \frac{2\pi b}{\lambda_{c2}},$$

$$\lambda_g = \frac{\lambda_0}{\sqrt{\epsilon_1' - \left(\frac{\lambda_0}{\lambda_{c1}}\right)^2}} = \frac{\lambda_0}{\sqrt{\epsilon_2' - \left(\frac{\lambda_0}{\lambda_{c2}}\right)^2}},$$

and the radial tangent and cotangent functions Tn and ct are defined in Eqs. (1·70). The quantities ϵ_1' and ϵ_2' are the relative dielectric constants of the two media; λ_0 is the free-space wavelength.

At the reference plane T the equivalent transverse network for the dominant H_0-mode is a shunt junction of an open-circuited and a short-circuited H-type radial transmission line. The propagation wavelength λ_g of this mode is obtained from

$$-Z_{01} \operatorname{ct}(x_1, 0) = Z_{02} \operatorname{Tn}(x_1', x_2'), \tag{3}$$

where

$$\frac{Z_{01}}{Z_{02}} = \frac{\lambda_{c1}}{\lambda_{c2}} = \sqrt{\frac{\epsilon_2' - \left(\frac{\lambda_0}{\lambda_g}\right)^2}{\epsilon_1' - \left(\frac{\lambda_0}{\lambda_g}\right)^2}}, \tag{4}$$

and the remaining quantities are defined as above.

Restrictions.—Equations (1) and (3) are rigorous, subject only to the restriction that the guide walls are perfectly conducting. The roots of these equations determine the propagation wavelengths of all modes with no angular dependence.

Numerical Results.—The values of the radial functions Tn and ct for real arguments are given in Sec. 1·7 (Table 1·3 and Figs. 1·10 and 1·11). With this knowledge the transcendental equations (1) and (3) can be solved graphically.

A convenient procedure is to assume values for λ_1/λ_g ($\lambda_1 = \lambda_0/\sqrt{\epsilon_1'}$), compute $\lambda_{c1}/\lambda_{c2}$ by Eq. (2) or (4), numerically solve Eq. (1) or (3) for, say, b/λ_{c1}, and then compute the corresponding value of $\lambda_1/2b$. E_0-mode propagation curves of λ_1/λ_g ($\lambda_1 = \lambda_0/\sqrt{\epsilon_1'}$) vs. $\lambda_1/2b$ with ϵ_2/ϵ_1 as a parameter are shown in Fig. 8·3-2 for the special case $b/a = 2$. In Fig. 8·3-3 E_0-mode curves of λ_1/λ_g vs. $\lambda_1/2b$ with b/a as a parameter are plotted for $\epsilon_2/\epsilon_1 = 2.54$.[1] It should be noted that, when $\lambda_1/\lambda_g > 1$, imaginary

[1] *Cf.* R. E. Staehler, "Determination of Propagation Constants by Radial Transmission Line Theory," Master's Thesis, Polytechnic Institute of Brooklyn, 1948.

SEC. 8·3] CIRCULAR GUIDE WITH DIELECTRIC CYLINDERS 395

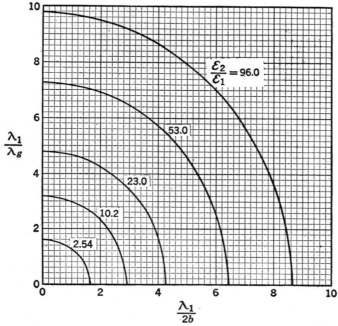

FIG. 8·3-2.—Propagation characteristics of E_0-mode in circular guide with dielectric cylinders ($b/a = 2$).

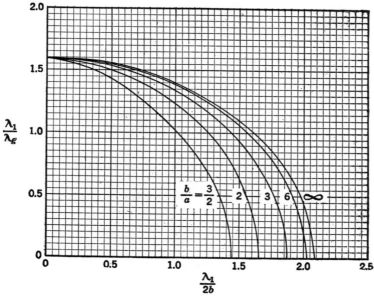

FIG. 8·3-3.—Propagation characteristics of E_0-mode in circular guide with dielectric cylinders ($\epsilon_2/\epsilon_1 = 2.54$).

values of x_1 are possible; this corresponds for $\epsilon_2 > \epsilon_1$, to critical reflection in dielectric ϵ_1 and to the propagation of most of the power in dielectric ϵ_2.

8·4. Coaxial Guide with Dielectric Cylinders.—A uniform coaxial guide containing two concentric dielectric cylinders (principal mode in dielectric-filled coaxial guide).

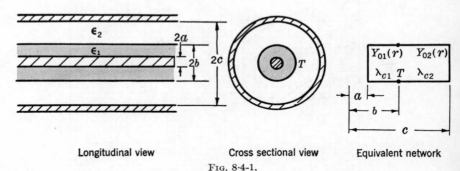

Longitudinal view   Cross sectional view   Equivalent network
FIG. 8·4-1.

Equivalent Network Results.—At the reference plane T the transverse equivalent network is a shunt arrangement of two short-circuited E-type radial transmission lines of different characteristic admittances. The propagation wavelengths λ_g of the principal mode in the dielectric-filled guide is determined from the resonant condition

$$Y_{01} \operatorname{ct}(x_2, x_1) = Y_{02} \operatorname{ct}(x_2', x_3'), \tag{1}$$

where

$$\frac{Y_{02}}{Y_{01}} = \frac{\epsilon_2'}{\epsilon_1'} \frac{\lambda_{c2}}{\lambda_{c1}} = \frac{\epsilon_2'}{\epsilon_1'} \sqrt{\frac{\epsilon_1' - \left(\frac{\lambda_0}{\lambda_g}\right)^2}{\epsilon_2' - \left(\frac{\lambda_0}{\lambda_g}\right)^2}},$$

$$x_1 = \frac{2\pi a}{\lambda_{c1}}, \quad x_2 = \frac{2\pi b}{\lambda_{c1}}, \quad x_2' = \frac{2\pi b}{\lambda_{c2}}, \quad x_3' = \frac{2\pi c}{\lambda_{c2}}.$$

The radial cotangent function $\operatorname{ct}(x, y)$ is defined in Eqs. (1·70). The quantities ϵ_1' and ϵ_2' are the relative dielectric constants of the two regions; λ_0 is the free-space wavelength.

The propagation wavelength of the principal mode is given approximately as

$$\left(\frac{\lambda_g}{\lambda_0}\right)^2 \approx \frac{\dfrac{1}{\epsilon_1'} \ln \dfrac{b}{a} + \dfrac{A}{\epsilon_2'} \ln \dfrac{c}{b}}{\ln \dfrac{b}{a} + A \ln \dfrac{c}{b}}, \tag{2}$$

where

$$A = \frac{1 + \frac{1}{2}\left(\frac{2\pi b}{\lambda_0}\right)^2 \left[\epsilon_1' - \left(\frac{\lambda_0}{\lambda_g}\right)^2\right] \ln \left|\frac{\gamma \pi a}{\lambda_0} \sqrt{\epsilon_1' - \left(\frac{\lambda_0}{\lambda_g}\right)^2}\right|}{1 + \frac{1}{2}\left(\frac{2\pi b}{\lambda_0}\right)^2 \left[\epsilon_2' - \left(\frac{\lambda_0}{\lambda_g}\right)^2\right] \ln \left|\frac{\gamma \pi c}{\lambda_0} \sqrt{\epsilon_2' - \left(\frac{\lambda_0}{\lambda_g}\right)^2}\right|},$$

$$A \approx 1,$$

$$\lambda_g = \frac{\lambda_0}{\sqrt{\epsilon_1' - \left(\frac{\lambda_0}{\lambda_{c1}}\right)^2}} = \frac{\lambda_0}{\sqrt{\epsilon_2' - \left(\frac{\lambda_0}{\lambda_{c2}}\right)^2}}, \qquad \gamma = 1.781.$$

Restrictions.—Equation (1) is rigorous and has been obtained by equating the input admittances of the two short-circuited radial lines that comprise the over-all transverse network. Equation (2) is an approximate solution for the lowest root of Eq. (1) valid in the low-frequency range $2\pi b/\lambda_{c1}$ and $2\pi c/\lambda_{c2} < 1$. It is assumed that no dissipation exists either in the dielectric or in the guide walls, although the results can be readily modified to include the effects of dissipation. Equation (1) applies equally well to all E-modes with no angular dependence.

Numerical Results.—Numerical values as well as a plot of the function $ct(x,y)$ for real values of the arguments are given in Sec. 1·7, Table 1·3. If this function is also computed for imaginary arguments, Eq. (1) can be solved graphically.[1] An approximate solution of Eq. (2) is obtained on placing $A = 1$; the value of λ_0/λ_g so obtained can be inserted into the expression for A to obtain a more accurate value of A, etc.

8·5. Rectangular Guide with "Nonradiating" Slit.—A rectangular guide with a centered longitudinal slit on its broad side (H_{10}-mode in unperturbed rectangular guide).

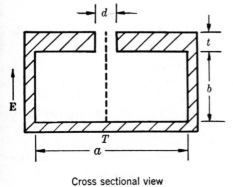
Cross sectional view

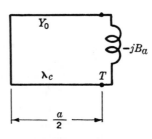
Equivalent network

FIG. 8·5-1.

[1] *Ibid.* (see Sec. 8·3).

Equivalent Network Results.—The relative change $\Delta\lambda_g/\lambda_g$ in guide wavelength produced by the presence of the slit is determined from the resonance condition

$$\frac{B_a}{Y_0} + \cot\frac{\pi a}{\lambda_c} = 0 \qquad (1)$$

for the total transverse admittance at the reference plane T.

For a wall of zero thickness ($t = 0$)

$$\frac{B_a}{Y_0} \approx \frac{b}{\lambda_c}\left(\frac{\pi d}{4b}\right)^2, \qquad \frac{d}{b} \ll 1, \qquad (2)$$

as given in Eq. (2b) of Sec. 6·2. Therefore from Eq. (1) it follows that the fractional increase in guide wavelength is

$$\frac{\Delta\lambda_g}{\lambda_g} \approx \frac{\pi d^2}{16ab}\left(\frac{\lambda_g}{2a}\right)^2, \qquad \frac{d}{b} \ll 1,$$

where

$$\lambda_g = \frac{\lambda}{\sqrt{1 - \left(\frac{\lambda}{2a}\right)^2}}. \qquad (3)$$

For a wall of large thickness ($t \gg d$)

$$\frac{B_a}{Y_0} \approx \frac{2b}{\lambda_c}\left(\frac{d}{2b}\right)^2, \qquad \frac{d}{b} \ll 1, \qquad (4)$$

as obtained from Eq. (2b) of Sec. 6·1 by an appropriate shift in reference plane. Therefore, in this case

$$\frac{\Delta\lambda_g}{\lambda_g} = \frac{d^2}{2\pi ab}\left(\frac{\lambda_g}{2a}\right)^2, \qquad \frac{d}{b} \ll 1. \qquad (5)$$

Restrictions.—The equivalent network is valid in the wavelength range $2b/\lambda_c < 1$. Equation (1) is the rigorous condition for the determination of the cutoff wavelength provided the correct value for the junction admittance is employed. Equations (2) and (4) are approximate expressions for the junction susceptance in the range $d/b \ll 1$; these expressions neglect the radiation conductance and to a first order are independent of the external geometry. Equations (3) and (5) give the first-order change $\Delta\lambda_g$ in the unperturbed guide wavelength λ_g and are applicable in the small slit range $d/b \ll 1$ and λ not too close to the cutoff wavelength of the next higher mode.

8·6. Rectangular Guides with Ridges.

—A uniform rectangular guide with a centered rectangular ridge on one or both of its wide sides (H_{n0}-mode in unperturbed rectangular guide).

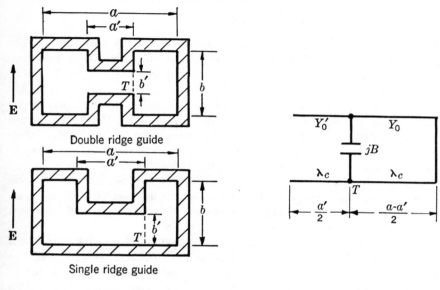

Cross sectional view   Equivalent network

Fig. 8·6-1.

Equivalent Network Results.—On consideration of the cross-sectional symmetry, it is evident that the equivalent transverse network for the dominant mode consists simply of a junction capacitance shunted by two H-mode transmission lines with open- and short-circuit terminations, respectively. The propagation wavelength λ_g of the dominant mode is determined by the resonance condition (at the reference plane T)

$$\frac{Y_0'}{Y_0} \tan \frac{\pi}{\lambda_c} a' + \frac{B}{Y_0} - \cot \frac{\pi}{\lambda_c} (a - a') = 0, \tag{1}$$

where

$$\frac{Y_0'}{Y_0} = \frac{b}{b'}, \qquad \lambda_g = \frac{\lambda}{\sqrt{1 - \left(\frac{\lambda}{\lambda_c}\right)^2}}.$$

For the double-ridge guide with $a' = 0$

$\frac{B}{Y_0}$ = same as $\frac{1}{2}$ the $\frac{B}{Y_0}$ of Eqs. (2) of Sec. 5·1a provided d and λ_g therein are replaced by b' and λ_c of this section. (2)

With finite a' but $(a - a')/2b' \gg 1$

$$\frac{B}{Y_0} = \text{same as the } \frac{B}{Y_0} \text{ of Eqs. (2) of Sec. 5·26a provided } \lambda_g \text{ therein is replaced by } \lambda_c \text{ of this section.} \quad (3)$$

For the single-ridge guide $\lambda_c/2$ should be inserted for λ_c in both Eqs. (2) and (3). Equation (1) yields the same λ_c for both the single- and double-ridge guides if the gap height b' of the former is one-half the gap height b' of the latter.

Restrictions.—The equivalent network is valid in the wavelength range $2b/\lambda_c < 1$ for the single ridge and $b/\lambda_c < 1$ for the double ridge. In the nondissipative case Eq. (1) is the rigorous condition for the determination of the cutoff wavelengths λ_c for the odd H_{n0} modes.* Equations (2) and (3) for the junction susceptance B/Y_0 are subject to the restrictions indicated in Secs. 5·1 and 5·26. In addition these equations are subject to the proviso that no higher-mode interaction exist between the ridge and the narrow side of the guide; this is approximately the case if $(a - a')/2b' \gg 1$. If $a'/b' < 1$, see Sec. 8·8.

Numerical Results.—The junction susceptances B/Y_0 of Eqs. (2) and (3) are plotted in Figs. 5·1-4 and 5·26-3. Numerical solutions of the transcendental equation (1) are tabulated below, wherein the quantity λ_c/a appears as a function of a'/a with b'/b as a parameter. The tables refer both to the case $b/a = 0.25$ for the single-ridge guide and to $b/a = 0.5$ for the double-ridge guide.

CUTOFF WAVELENGTH OF H_{10} MODE

$b'/b = 0.1$		$b'/b = 0.15$		$b'/b = 0.20$		$b'/b = 0.25$	
a'/a	λ_c/a	a'/a	λ_c/a	a'/a	λ_c/a	a'/a	λ_c/a
0.00	3.062	0.00	2.925	0.0	2.747	0.0	2.636
0.05	3.652	0.10	3.623	0.1	3.286	0.1	3.060
0.10	4.111	0.20	4.085	0.2	3.646	0.2	3.349
0.15	4.500	0.25	4.248	0.3	3.869	0.25	3.453
0.20	4.763	0.30	4.370	0.333	3.917	0.3	3.529
0.30	5.164	0.40	4.509	0.4	3.976	0.333	3.567
0.40	5.368	0.50	4.518	0.5	3.977	0.4	3.614
0.50	5.397	0.60	4.396	0.6	3.871	0.5	3.609
0.60	5.255						
0.70	4.927						
0.80	4.369						
0.90	3.485						
1.00	2.000						

* For n even the tangent function in Eq. (1) is to be replaced by the cotangent.

RECTANGULAR GUIDES WITH RIDGES

Cutoff Wavelength of H_{10} Mode (Continued)

$b'/b = 0.3$		$b'/b = 0.35$		$b'/b = 0.4$		$b'/b = 0.5$	
a'/a	λ_c/a	a'/a	λ_c/a	a'/a	λ_c/a	a'/a	λ_c/a
0.0	2.536	0.00	2.505	0.0	2.374	0.00	2.268
0.1	2.881	0.10	2.736	0.10	2.613	0.10	2.435
0.2	3.121	0.20	2.936	0.20	2.785	0.20	2.559
0.25	3.205	0.25	3.008	0.25	2.847	0.25	2.604
0.3	3.270	0.30	3.063	0.30	2.893	0.30	2.638
0.4	3.339	0.40	3.121	0.40	2.942	0.40	2.673
0.5	3.332	0.50	3.114	0.50	2.935	0.50	2.666
		0.60	3.039				

Cutoff Wavelength of H_{20} Mode

$b'/b = 0.1$		$b'/b = 0.15$		$b'/b = 0.2$		$b'/b = 0.25$	
a'/a	λ_c/a	a'/a	λ_c/a	a'/a	λ_c/a	a'/a	λ_c/a
0.0	1.000	0.0	1.000	0.0	1.000	0.0	1.000
0.1	0.911	0.1	0.917	0.1	0.924	0.1	0.929
0.2	0.831	0.2	0.851	0.2	0.869	0.2	0.884
0.275	0.804	0.25	0.841	0.3	0.898	0.25	0.889
0.3	0.815	0.3	0.867	0.333	0.933	0.3	0.920
0.4	0.943	0.4	0.993	0.4	1.019	0.333	0.952
0.5	1.107	0.5	1.139	0.5	1.157	0.4	1.032
0.6	1.266	0.6	1.276	0.6	1.281	0.5	1.157
0.7	1.405						
0.8	1.503						
0.85	1.519						
0.9	1.488						
1.0	1.000						

$b'/b = 0.3$		$b'/b = 0.35$		$b'/b = 0.4$		$b'/b = 0.5$	
a'/a	λ_c/a	a'/a	λ_c/a	a'/a	λ_c/a	a'/a	λ_c/a
0.0	1.000	0.0	1.000	0.0	1.000	0.0	1.000
0.1	0.935	0.1	0.940	0.1	0.946	0.1	0.956
0.2	0.898	0.2	0.908	0.2	0.918	0.2	0.956
0.25	0.909	0.25	0.915	0.25	0.926	0.25	0.942
0.3	0.941	0.3	0.941	0.3	0.950	0.3	0.961
0.4	1.045	0.4	1.031	0.4	1.031	0.4	1.024
0.5	1.159	0.5	1.128	0.5	1.122	0.5	1.095
		0.6	1.210				

CUTOFF WAVELENGTH OF H_{30} MODE

$b'/b = 0.1$		$b'/b = 0.2$		$b'/b = 0.25$	
a'/a	λ_c/a	a'/a	λ_c/a	a'/a	λ_c/a
0.0	0.880	0.0	0.867	0.0	0.858
0.1	0.856	0.1	0.834	0.1	0.825
0.2	0.781	0.2	0.768	0.2	0.762
0.3	0.695	0.3	0.692	0.3	0.690
0.4	0.612	0.333	0.667	0.333	0.667
0.5	0.590	0.4	0.625	0.4	0.631
0.6	0.666	0.5	0.633	0.5	0.647
0.7	0.753	0.6	0.688		
0.8	0.833				
0.9	0.885				
0.95	0.864				
1.0	0.667				

* Computations by W. E. Waller, S. Hopfer, M. Sucher, Report of Polytechnic Research and Development Co.

8·7. Rectangular Guide with Resistive Strip.—A uniform rectangular guide with a central resistive strip of zero thickness parallel to the electric field (H_{10}-mode in unperturbed rectangular guide).

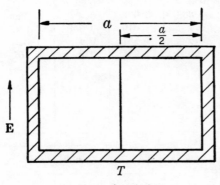

Cross sectional view   Equivalent network

FIG. 8·7-1.

Equivalent Network Results.—The equivalent transverse network indicative of the propagating characteristics of the dominant mode comprises an H-mode transmission line short-circuited at one end and terminated at its other end by twice the resistance R (in ohms per unit square) of the resistive strip. The complex cutoff wave number k_c of the dominant mode is given by the resonant condition

$$j \tan \frac{k_c a}{2} + \frac{2R}{Z_0} = 0, \qquad (i)$$

where

$$Z_0 = \zeta \frac{k}{k_c} = 377 \frac{\lambda_c}{\lambda} \quad \text{ohms.}$$

By decomposition into real and imaginary parts Eq. (1) can be rewritten as

$$-A\xi_2 = \frac{\tan \xi_1 (1 - \tanh^2 \xi_2)}{1 + \tan^2 \xi_1 \tanh^2 \xi_2}, \tag{2a}$$

$$A\xi_1 = \frac{\tanh \xi_2 (1 + \tan^2 \xi_1)}{1 + \tan^2 \xi_1 \tanh^2 \xi_2}, \tag{2b}$$

which implies that

$$\xi_1 \sin 2\xi_1 + \xi_2 \sinh 2\xi_2 = 0, \tag{2c}$$

where

$$\frac{k_c a}{2} = \xi_1 + j\xi_2, \quad A = \frac{R}{\zeta}\left(\frac{2\lambda}{\pi a}\right).$$

The propagation wavelength λ_g and attenuation constant α of the dominant mode follow as

$$\lambda_g = \frac{\lambda}{\sqrt{1 - (\xi_1^2 - \xi_2^2)\left(\frac{\lambda}{\pi a}\right)^2}} \left[\cosh\left(\frac{\sinh^{-1} x}{2}\right)\right]^{-1}, \tag{3a}$$

$$\alpha = \frac{2\pi \sqrt{1 - (\xi_1^2 - \xi_2^2)\left(\frac{\lambda}{\pi a}\right)^2}}{\lambda} \sinh\left(\frac{\sinh^{-1} x}{2}\right), \tag{3b}$$

where

$$x = \frac{2\xi_1 \xi_2 \left(\frac{\lambda}{\pi a}\right)^2}{1 - (\xi_1^2 - \xi_2^2)\left(\frac{\lambda}{\pi a}\right)^2}.$$

Restrictions.—The equivalent transverse network is applicable to the computation of the characteristics of the dominant mode in the entire wavelength range. Equation (1) is the rigorous equation for the determination of the cutoff frequency and has been obtained by equating to zero the total impedance at the central reference plane T. The form of Eqs. (2) and (3) has been adapted for computations in the wavelength range in which only the dominant mode can be propagated ($\pi/2 < \xi_1 < \pi$). It is assumed that the resistive strip has a zero thickness; the effect of thickness can be taken into account as in Sec. 8·1. It is further assumed that there is good electrical contact between the strip and the top and bottom walls of the guide.

Numerical Results.—For purposes of computation, values of ξ_1 are assumed and the corresponding values of ξ_2 and A are computed from

Eqs. (2c), (2a), and (2b), respectively. The quantities λ_g and α may be evaluated with the aid of the graph in Fig. 1·7.

THICKNESS EFFECTS

8·8. Capacitive Obstacles of Large Thickness. *a. Window Formed by Two Obstacles.*—Window formed by obstacles of large thickness with edges perpendicular to the electric field (H_{10}-mode in rectangular guides).

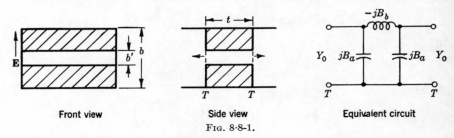

Front view   Side view   Equivalent circuit
FIG. 8·8-1.

Equivalent-circuit Parameters.—At the reference planes T

$$\frac{B_a}{Y_0} = \frac{B_1}{Y_0} + \frac{b}{b'} \tan \frac{\pi t}{\lambda_g}, \tag{1}$$

$$\frac{B_b}{Y_0} = \frac{b}{b'} \csc \frac{2\pi t}{\lambda_g}, \tag{2}$$

where

$$\frac{B_1}{Y_0} = \frac{2b}{\lambda_g} \left[\ln \frac{1-\alpha^2}{4\alpha} \left(\frac{1+\alpha}{1-\alpha}\right)^{\frac{1}{2}\left(\alpha+\frac{1}{\alpha}\right)} + 2\frac{A+A'+2C}{AA'-C^2} \right], \tag{3a}$$

$$\frac{B_1}{Y_0} \approx \frac{2b}{\lambda_g} \left[\ln \frac{e}{4\alpha} + \frac{\alpha^2}{3} + \frac{1}{2}\left(\frac{b}{\lambda_g}\right)^2 (1-\alpha^2)^4 \right], \quad \alpha \ll 1, \quad t \gg b', \tag{3b}$$

and

$$A = \left(\frac{1+\alpha}{1-\alpha}\right)^{2\alpha} \frac{1+\sqrt{1-\left(\dfrac{b}{\lambda_g}\right)^2}}{1-\sqrt{1-\left(\dfrac{b}{\lambda_g}\right)^2}} - \frac{1+3\alpha^2}{1-\alpha^2},$$

$$A' = \left(\frac{1+\alpha}{1-\alpha}\right)^{2/\alpha} \frac{1+\sqrt{1-\left(\dfrac{b'}{\lambda_g}\right)^2} \coth\left(\sqrt{1-\left(\dfrac{b'}{\lambda_g}\right)^2}\dfrac{\pi t}{2b'}\right)}{1-\sqrt{1-\left(\dfrac{b'}{\lambda_g}\right)^2} \coth\left(\sqrt{1-\left(\dfrac{b'}{\lambda_g}\right)^2}\dfrac{\pi t}{2b'}\right)} + \frac{3+\alpha^2}{1-\alpha^2},$$

$$C = \left(\frac{4\alpha}{1-\alpha^2}\right)^2, \quad \alpha = \frac{b'}{b}.$$

Restrictions.—The equivalent circuit is valid in the range $b/\lambda_g < 1$. The above equations have been obtained by the equivalent static method with the two lowest modes in each guide treated accurately. Equation

(2) is in error by less than 2 per cent. Equation (3a) for B_1/Y_0 is in error by less than 5 per cent for $\alpha < 0.5$, $b/\lambda_g < 1$, and for all values of t; for $\alpha < 0.7$ the error is less than 10 per cent. For $\alpha > 0.5$ Eq. (3a) is in error by less than 5 per cent when $t/b' > 0.3$; for $t/b' \geqq 1$ the error is less than 1 per cent for all values of α. Equation (3b) is an asymptotic expansion of Eq. (3a) that agrees with the latter to within 5 per cent if $\alpha < 0.6$, $b/\lambda_g < 0.5$, and $t/b' > 1$. For $\alpha < 0.5$ it is to be noted that B_1/Y_0 ranges from its value at $t = 0$ of one-half the susceptance of a window of zero thickness (Sec. 5·1) to a value for $t/b' > 1$ equal to the junction susceptance for a change of height (Sec. 5·26), the latter value being less than 10 per cent greater than the former.

Numerical Results.—Curves of $B_1\lambda_g/Y_0 b$ as a function of t/b have been plotted for several values of α and b/λ_g in Fig. 8·8-4. Curves for other values of α and b/λ_g may be approximated by employing the results of the known limiting cases $t = 0$ (see Sec. 5·1) and $t = \infty$ (see Sec. 5·26).

b. *Window Formed by One Obstacle.*—Window formed by an obstacle of large thickness with edges perpendicular to the electric field (H_{10}-mode in rectangular guide).

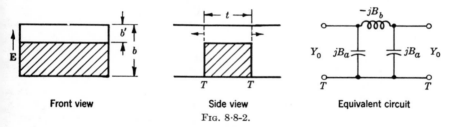

Front view Side view Equivalent circuit
Fig. 8·8-2.

Equivalent-circuit Parameters.—Same as in Sec. 8·8a except that λ_g is replaced by $\lambda_g/2$.

Restrictions.—Same as in Sec. 8·8a except that λ_g is replaced by $\lambda_g/2$.

Numerical Results.—Same as in Sec. 8·8a except that λ_g is replaced by $\lambda_g/2$.

c. *Symmetrical Obstacle.*—A symmetrical obstacle of large thickness with edges perpendicular to the electric field (H_{10}-mode in rectangular guide).

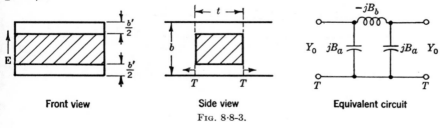

Front view Side view Equivalent circuit
Fig. 8·8-3.

Equivalent-circuit Parameters.—Same as in Sec. 8·8a.
Restrictions.—Same as in Sec. 8·8a.
Numerical Results.—Same as in Sec. 8·8a.

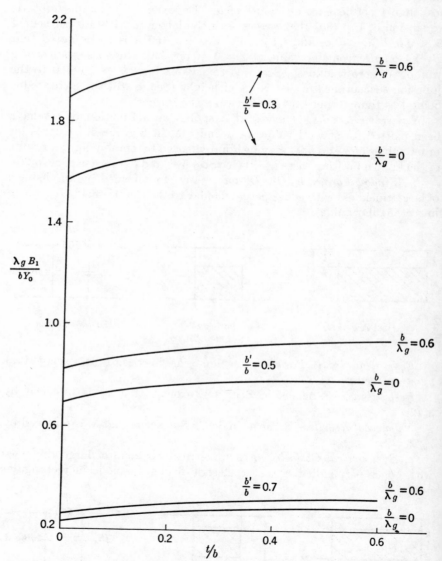

FIG. 8·8-4.—Zero to infinite thickness variation of junction susceptance B_1.

8·9. Inductive Obstacles of Large Thickness.

a. Window Formed by Two Obstacles.—A window formed by two rectangular obstacles of large thickness with edges parallel to the electric field (H_{10}-mode in rectangular guides).

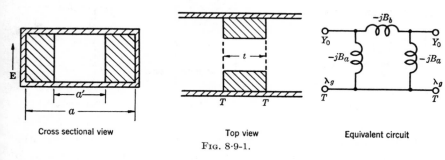

Cross sectional view   Top view   Equivalent circuit

FIG. 8·9-1.

Equivalent-circuit Parameters.—At the reference planes T

$$\frac{B_a}{Y_0} = \frac{B}{Y_0} - \frac{Y_0'}{Y_0} \tan \frac{\pi(t + 2l)}{\lambda_g'}, \qquad (1)$$

$$\frac{B_b}{Y_0} = \frac{Y_0'}{Y_0} \csc \frac{2\pi(t + 2l)}{\lambda_g'}, \qquad (2)$$

where

$$\frac{Y_0}{Y_0'} = \frac{Z_0'}{Z_0} = \text{same as in Eqs. (1) of Sec. 5·24}a,$$

$$\frac{Y_0}{B} = \frac{X}{Z_0} = \text{same as in Eqs. (2) of Sec. 5·24}a,$$

$$\frac{l}{a} = \text{same as in Eqs. (3) of Sec. 5·24}a,$$

$$\lambda_g' = \frac{\lambda}{\sqrt{1 - \left(\frac{\lambda}{2a'}\right)^2}}, \qquad \lambda_g = \frac{\lambda}{\sqrt{1 - \left(\frac{\lambda}{2a}\right)^2}}.$$

Restrictions.—The equivalent circuit is valid in the wavelength range $0.5 < a/\lambda < 1.5$. Equations (1) and (2) have been obtained from the results of Sec. 5·24 by standard transmission-line analysis and are subject to the restriction that $t \gg a'$. This restriction is not serious; for when $t < a'$, the relative reactance X/Z_0 varies between the value given and twice the value for $t = 0$ given in Sec. 5·2; the two values are within about 20 per cent of each other. For further restrictions see Sec. 5·24a. If $\lambda > 2a'$, λ_g' becomes imaginary but the circuit is still applicable.

Numerical Results.—The quantities Z_0'/Z_0, X/Z_0, and l/a may be obtained from the graphs in Figs. 5·24-2 and 5·24-3.

b. Window Formed by One Obstacle.—A window formed by a single rectangular obstacle of large thickness with edges parallel to the electric field (H_{10}-mode in rectangular guides).

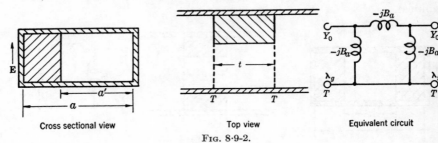

Cross sectional view   Top view   Equivalent circuit
FIG. 8·9-2.

Equivalent-circuit Parameters.—Same as in Sec. 8·9a except that all parameters are to be determined from Eqs. (4) to (6) of Sec. 5·24b rather than Eqs. (1) to (3) of Sec. 5·24a.

Restrictions.—The equivalent circuit is valid in the wavelength range $0.5 < a/\lambda < 1.0$. See Restrictions under Secs. 8·9a and 5·24b.

Numerical Results.—The quantities Z_0'/Z_0, X/Z_0, and l/a of this section may be obtained from the graphs in Figs. 5·24-5 and 5·24-6.

8·10. Thick Circular Window.—A centered circular window in a thick metallic plate transverse to the axis of a rectangular guide (H_{10}-mode in rectangular guide).

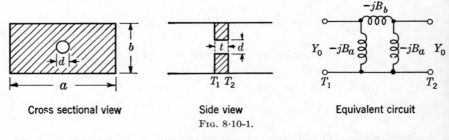

Cross sectional view   Side view   Equivalent circuit
FIG. 8·10-1.

Equivalent-circuit Parameters.—At the reference planes T_1 and T_2

$$\frac{B_a}{Y_0} = \frac{Z_0}{X_a} = \frac{B}{2Y_0} + \frac{|Y_0'|}{Y_0} \tanh \frac{\pi t}{|\lambda_g'|} \tag{1}$$

$$\frac{B_b}{Y_0} = \frac{Z_0}{X_b} = \frac{|Y_0'|}{Y_0} \operatorname{csch} \frac{2\pi t}{|\lambda_g'|}, \tag{2}$$

where

$\dfrac{Y_0'}{Y_0}$ = same as in Eqs. (1) of Sec. 5·32,

$\dfrac{B}{2Y_0}$ = same as in Eqs. (2) of Sec. 5·32,

$$|\lambda_g| = \frac{1.706d}{\sqrt{1 - \left(\frac{1.706d}{\lambda}\right)^2}}.$$

The power-transmission coefficient, in decibels, into a matched output line is

$$T = -10 \log_{10}\left[\left(\frac{B_a}{Y_0}\right)^2\left(1 + \frac{B_a}{2B_b} - \frac{Y_0^2}{B_aB_b}\right)^2 + \left(1 + \frac{B_a}{B_b}\right)^2\right], \quad (3a)$$

$$T \approx A_1 + A_2 \frac{t}{d}, \quad t > d, \quad (3b)$$

where

$$A_1 = -20 \log_{10} \frac{\left(\frac{B}{2Y_0} + \frac{|Y_0'|}{Y_0}\right)^2 + 1}{\frac{4|Y_0'|}{Y_0}},$$

$$A_1 \approx -20 \log_{10} \frac{0.263ab\lambda_g}{d^3}, \quad \frac{d}{\lambda} \ll 1,$$

$$A_2 = -32.0 \sqrt{1 - \left(\frac{1.706d}{\lambda}\right)^2},$$

$$A_2 \approx -32.0, \quad \frac{d}{\lambda} \ll 1.$$

The approximate expression for A_1, valid for $d \ll \lambda$, is almost equal to the transmission coefficient of a hole in an infinitely thin plate (*cf.* Sec. 5·4). Hence the approximate rule: the power-transmission coefficient of a hole in a thick plate is equal to that of a hole in an infinitely thin plate plus a thickness correction of -32.0 db per diameter of the hole.

Restrictions.—The equivalent circuit is valid in the wavelength range $b/\lambda_g < 1$. The equivalent-circuit parameters have been obtained from Secs. 5·4 and 5·32 and are subject to the restrictions listed therein. In addition, since the effect of thickness is taken into account in transmission-line fashion, it is assumed that there is only a lowest mode interaction between the opposite faces of the hole. This is rigorously valid only for a hole of thickness large compared with the hole diameter; however, it is not too inaccurate even for relatively small thickness ($t < d$) because of the weak excitation of the higher modes within the hole. Equation (3b), though strictly valid for holes in very thick plates, is a fairly good approximation even for small thickness (*cf.* Figs. 8·10-3,4).

Numerical Results.—The relative reactances X_a/Z_0 and X_b/Z_0 are plotted as a function of thickness in Fig. 8·10-2 for a 0.375-in. hole in a rectangular guide of inner dimensions 0.400 by 0.900 in. and at a wavelength $\lambda = 3.20$ cm; the points are measured values and the solid curves are computed from Eqs. (1) and (2). The transmission coefficient T,

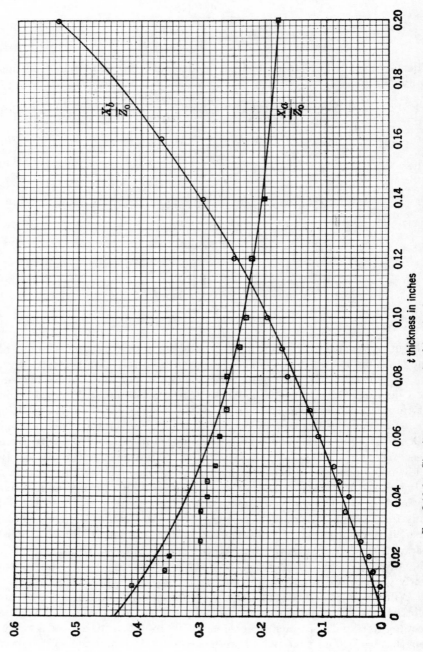

Fig. 8·10-2.—Circuit parameters of a ⅜-in. circular aperture as a function of thickness.

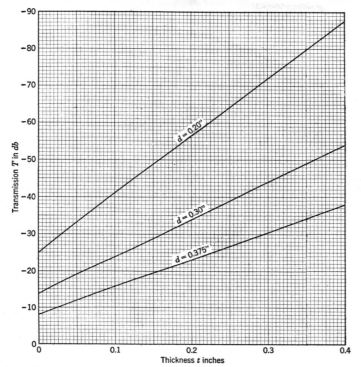

Fig. 8·10-3.—Transmission curves of circular apertures in rectangular guide ($a = 0.90$ in., $b = 0.40$ in., $\lambda = 3.20$ cm).

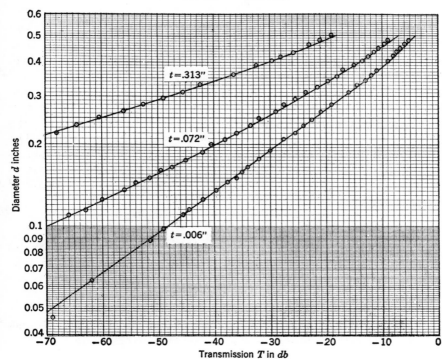

Fig. 8·10-4.—Transmission curves for circular apertues in rectangular guide ($a = 1.122$ in., $b = 0.497$ in., $\lambda = 3.21$ cm).

as obtained from Eq. (3a), is plotted vs. t in Fig. 8·10-3. In Fig. 8·10-4 there are indicated the transmission coefficients for a number of variable diameter holes in plates of variable thickness. These data were measured in a rectangular guide of dimensions 0.497 by 1.122 in. at $\lambda = 3.21$ cm; the circled points are experimental (by W. A. Tyrrell—Bell Telephone Laboratories) and the solid curves are computed from Eq. (3a).

8·11. E-plane T with Slit Coupling.—A symmetrical E-plane T-junction of three rectangular guides of equal widths but unequal heights coupled on their broad sides by a capacitive slit in a wall of finite thickness (H_{10}-modes in rectangular guides).

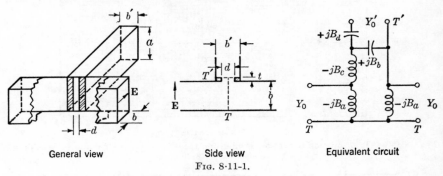

General view   Side view   Equivalent circuit
Fig. 8·11-1.

Equivalent-circuit Parameters.—For large thickness, t, the over-all junction may be regarded as a composite structure consisting of a main guide of height b joined to a stub guide of height b' by a coupling guide of height d. The three guides are described by transmission lines of equal propagation wavelengths λ_g but with characteristic impedances proportional to the guide heights. The circuit description at the junction of the main and coupling guides is given in Sec. 6·1, whereas that at the junction of the coupling and stub guides is shown in Sec. 5·26a. The circuit description of the length t of the coupling guide may be obtained from Fig. 1·4. The over-all network can then be reduced to the form indicated in Fig. 8·11-1. Because of their complexity explicit values of the corresponding circuit parameters will not be presented.

For a slit of thickness $t = 0$ see Sec. 6·2.

Restrictions.—See Secs. 6·1 and 5·26a. The transmission-line treatment of the effect of wall thickness neglects higher-mode interactions and hence is strictly valid only for $t \gg d$. However, it is quite accurate even for $t < d$ because of the weak excitation of the higher modes in the coupling guide.

Experimental and Numerical Results.—The equivalent-circuit parameters of the representation in Fig. 8·11-1 were measured at $\lambda = 3.20$

cm in guides of dimensions $b = b' = 0.40$ in. and $a = 0.90$ in. for a few slits of width d in a wall of thickness $t = 0.02$ in. The resulting data together with the corresponding theoretical values, computed as above, are shown in the following table:

$\dfrac{d}{b}$	$\dfrac{B_a}{Y_0}$		$\dfrac{Y_0}{B_b}$		$\dfrac{Y_0}{B_c}$		$\dfrac{Y_0}{B_d}$	
	Exp.	Theor.	Exp.	Theor.	Exp.	Theor.	Exp.	Theor.
0.50	0.03	0.028	1.31	1.20	0.05	0.042	0.08	0.008
0.625	0.05	0.043	1.76	1.60	0.10	0.076	0.11	0.036
0.75	0.08	0.059	2.20	2.10	0.13	0.133	0.19	0.089

The agreement between the experimental results, which are not too reliable, and the theoretical values is fair even though the latter have been computed in the range $t \ll d$ in which the theory is not strictly applicable.

APPENDIX

ARC SINE SUMS

A. The Function $S_N(x;\alpha,\beta)$. The solution of rectangular-guide problems by the transform method requires the evaluation of the sum function

$$S_N(x;\alpha,\beta) = \sum_{n=N}^{\infty}\left[\sin^{-1}\frac{x}{\sqrt{(n-\beta)^2-\alpha^2}}-\frac{x}{n}\right] \quad (1)$$

This function may be expressed in a form more suitable for numerical calculation as

$$S_N(x;\alpha,\beta) = \sum_{p=0}^{\infty} C_p{}^{p-\frac{1}{2}}\zeta^{-\beta}{}_{2p+1,N}\, x(x^2+\alpha^2)^p W_p\left(\frac{x}{\alpha}\right), \quad (2)$$

where

$$C_p{}^{p-\frac{1}{2}} = \frac{1}{2^p}\cdot\frac{1\cdot 3\cdot 5\cdots(2p-1)}{1\cdot 2\cdot 3\cdots p} = \frac{(2p)!}{2^{2p}(p!)^2} \quad (2a)$$

$$\zeta^{-\beta}{}_{1,N} = \Psi(N) - \Psi(N-\beta) = \sum_{n=N}^{\infty}\left(\frac{1}{n-\beta}-\frac{1}{n}\right) \quad (2b)$$

$$\zeta^{-\beta}{}_{2p+1,N} = -\frac{\Psi^{(2p)}(N-\beta)}{(2p)!} = \sum_{n=N}^{\infty}\frac{1}{(n-\beta)^{2p+1}}, \quad p > 0 \quad (2c)$$

$$\Psi(x) = \frac{d}{dx}\ln\Gamma(x) = \frac{d}{dx}\ln(x-1)! \quad (2d)$$

$$\zeta^0{}_{2p+1,N} = \zeta(2p+1) - \sum_{n=1}^{N-1}\frac{1}{n^{2p+1}} \quad (2e)$$

$$W_p\left(\frac{x}{\alpha}\right) = \frac{1}{2p+1}\left[1+\frac{2p}{1+(x/\alpha)^2}W_{p-1}\left(\frac{x}{\alpha}\right)\right], \quad W_0 = 1 \quad (2f)$$

The polygamma function $\Psi^{(2p)}(x)$ and the zeta function $\zeta(x)$ are tabulated, for example, in "Tables of the Higher Mathematical Functions," Vols. I and II, by H. T. Davis.

Special cases of Eq. (2) that are of importance obtain for $N = 2$, $\beta = 0$:

$$S_2(x;\alpha,0) = \sum_{p=1}^{\infty} C_p{}^{p-\frac{1}{2}}[\zeta(2p+1)-1]x(x^2+\alpha^2)^p W_p\left(\frac{x}{\alpha}\right) \quad (3)$$

and for $N = 2$, $\beta = 0$, $\sigma = 0$:

$$S_2(x;0,0) = \sum_{p=1}^{\infty} C_p p^{-\frac{1}{2}}[\zeta(2p+1) - 1]\frac{x^{2p+1}}{2p+1} \tag{4}$$

Values of the function $S_2(x;\alpha,0)$ are given in Table I for $0 < x < 2$ and $0 < \alpha < 1.9$. The values of the function $S_N(x;\alpha,0)$ may be readily obtained from $S_2(x;\alpha,0)$ by means of Eq. (1).

B. The Function $S_N^{J_0}(x;\alpha)$. The function $S_N^{J_0}(x;\alpha)$ which arises in the solution of circular-guide problems by the transform method is defined, in a manner analogous to that in (1), by

$$S_N^{J_0}(x;\alpha) = \sum_{n=N}^{\infty}\left[\sin^{-1}\frac{x}{\sqrt{\beta_n^2 - \alpha^2}} - \frac{x}{n}\right], \tag{5}$$

where
$$J_0(\pi\beta_n) = 0,$$

the roots $\chi_{0n} = \pi\beta_n$ being obtained by the formula given in Table 2.1, page 66. Equation (5) can be expressed in a form suitable for numerical evaluation as,

$$S_N^{J_0}(x;\alpha) = \sum_{p=0}^{\infty} C_p p^{-\frac{1}{2}} s_{2p+1,N}^0 x(x^2 + \alpha^2)^p W_p\left(\frac{x}{\alpha}\right) \tag{6}$$

where

$$s_{1,N}^0 = \sum_{n=N}^{\infty}\left(\frac{1}{\beta_n} - \frac{1}{n}\right) \tag{6a}$$

$$s_{2p+1,N}^0 = \sum_{n=N}^{\infty}\frac{1}{\beta_n^{2p+1}}, \quad p > 0. \tag{6b}$$

The quantities $C_p p^{-\frac{1}{2}}$ and $W_p(x/\alpha)$ are defined as in Eqs. (2a) and (2f). A number of typical values for the sums $s_{2p+1,3}^0$ are:

$s_{1,3}^0 = 0.1026964$ $s_{11,3}^0 = 0.000014956$
$s_{3,3}^0 = 0.0940399$ $s_{13,3}^0 = 0.000001938$
$s_{5,3}^0 = 0.00837962$ $s_{15,3}^0 = 0.000000253$
$s_{7,3}^0 = 0.00095207$ $s_{17,3}^0 = 0.000000033$
$s_{9,3}^0 = 0.000117294$ $s_{19,3}^0 = 0.000000004$

Explicit values of the function $S_2^{J_0}(x;\alpha)$ are given in Table II for $0 < x < 1.8$ and $0 < \alpha < 0.7655$.

It should be noted that a function $S_N^{J_m}(x;\alpha)$ can be defined by a simple generalization of that in Eq. (5).

C. The Function $S_N^{Z_0}(x;\alpha,c)$. The function $S_N^{Z_0}(x;\alpha,c)$ which arises from the transform method of solution of coaxial-guide problems is

APPENDIX

defined by

$$S_N^{Z_0}(x;\alpha,c) = \sum_{n=N}^{\infty} \left[\sin^{-1} \frac{n}{\sqrt{\gamma_n^2 - \alpha^2}} - \frac{x}{n} \right], \quad (7)$$

where

$$\pi\gamma_n(c) = (c-1)\chi_{0n},$$

χ_{0n} being the nth root (*cf.* Table 2.3, p. 74) of the equation

$$Z_0 = J_0(cx)N_0(x) - N_0(cx)J_0(x) = 0.$$

A convenient procedure for the numerical evaluation of $S_N^{Z_0}(x;\alpha,c)$ utilizes the $S_N(x;\alpha,0)$ function tabulated in A above. Thus

$$S_N^{Z_0}(x;\alpha,c) = S_N(x;\alpha,0) + \sum_{p=0}^{\infty} C_p p^{-\frac{1}{2}} D'_{2p+1,N}(c) \, x \, (x^2 + \alpha^2)^p W_p\left(\frac{x}{\alpha}\right) \quad (8)$$

where

$$D_{2p+1,N}(c) = \sum_{n=N}^{\infty} \left(\frac{1}{\gamma_n^{2p+1}} - \frac{1}{n^{2p+1}} \right). \quad (8a)$$

With the aid of the asymptotic series*

$$\gamma_n = n\left[1 - 0.012665\frac{\nu}{n^2} + (0.001845 + 0.000668\nu)\left(\frac{\nu}{n^2}\right)^2 \right.$$
$$\left. - (0.000992 + 0.001056\nu + 0.000218\nu^2)\left(\frac{\nu}{n^2}\right)^3 \cdots \right], \quad (9)$$

with $\nu = c + \frac{1}{c} - 2$, the following explicit expressions for $D_{2p+1,N}$ can be obtained:

$$D_{1,N}(c) = 0.012665\nu\zeta_{3,N}^0 - (0.001684 + 0.000668\nu)\nu^2\zeta_{5,N}^0$$
$$+ (0.000948 + 0.001039\nu + 0.000218\nu^2)\nu^3\zeta_{7,N}^0 \quad (10a)$$
$$D_{3,N}(c) = 0.037995\nu\zeta_{5,N}^0 - (0.004572 + 0.002005\nu)\nu^2\zeta_{7,N}^0 \quad (10b)$$
$$D_{5,N}(c) = 0.063326\nu\zeta_{7,N}^0 \quad (10c)$$

The cutoff zeta functions $\zeta_{m,N}^0$ are defined in (2e). For $\sqrt{x^2 + \alpha^2} < 2$ the first three terms of the series in (8), utilizing the approximations (10a − c), provide values of $S_N^{Z_0}$ accurate to four decimal places if $N = 3$, $c \leq 5$ or if $N = 4$, $c \leq 7$. Values of the function $S_N^{Z_0}(x;0,c)$ are shown in Table III for the range $0 < x < 1$ and $c = 2, 3, 4, 5$ with $N = 1, 2, 3, 4$.

It is evident that a function $S_N^{Z_m}(x;\alpha,c)$ can be defined by a natural generalization of the $S_N^{Z_0}$ function in (7); however, no tables are available for $S_N^{Z_m}$.

* *Cf.* "Bessel Functions," Gray, Mathews, and Macrobert, p. 261.

TABLE I.—$S_2(x;\alpha,0)$

x \ α	0	0.1	0.2	0.3	0.4	0.5	0.6	0.7	0.8	0.9
0.1	.00003	.00014	.00044	.00096	.00169	.00267	.00387	.00536	.00715	.00930
0.2	.00027	.00047	.00109	.00212	.00359	.00552	.00796	.01095	.01455	.01887
0.3	.00092	.00122	.00215	.00370	.00592	.00883	.01251	.01701	.02245	.02896
0.4	.00218	.00259	.00384	.00592	.00890	.01281	.01775	.02380	.03111	.03986
0.5	.00430	.00482	.00638	.00901	.01277	.01771	.02393	.03149	.04081	.05187
0.6	.00750	.00813	.01002	.01322	.01778	.02377	.03133	.04062	.05185	.06531
0.7	.01205	.01279	.01503	.01881	.02420	.03130	.04025	.05125	.06456	.08055
0.8	.01824	.01910	.02170	.02609	.03235	.04061	.05102	.06383	.07934	.09800
0.9	.02639	.02729	.03036	.03540	.04259	.05205	.06403	.07876	.09663	.11816
1.0	.03689	.03802	.04141	.04714	.05531	.06610	.07973	.09654	.11697	.14163
1.1	.05020	.05146	.05530	.06177	.07102	.08323	.09869	.11778	.14103	.16917
1.2	.06685	.06828	.07260	.07989	.09032	.10411	.12160	.14324	.16967	.20179
1.3	.08755	.08916	.09402	.10224	.11399	.12957	.14936	.17394	.20407	.24089
1.4	.11322	.11504	.12052	.12979	.14309	.16075	.18326	.21132	.24593	.28856
1.5	.14513	.14719	.15341	.16395	.17910	.19928	.22513	.25756	.29802	.34831
1.6	.18515	.18751	.19466	.20680	.22430	.24774	.27798	.31634	.36486	.42699
1.7	.23637	.23914	.24754	.26186	.28263	.31070	.34745	.39509	.45757	.54318
1.8	.30463	.30803	.31837	.33613	.36227	.39838	.44740	.51536	.61926	
1.9	.40478	.40947	.42395	.44951	.48910	.54993	.66340			
2.0	.69151									

x \ α	1.0	1.1	1.2	1.3	1.4	1.5	1.6	1.7	1.8	1.9
0.1	.01187	.01493	.01860	.02307	.02861	.03567	.04508	.05854	.08051	.12871
0.2	.02401	.03016	.03755	.04653	.05767	.07188	.09085	.11807	.16271	.26191
0.3	.03673	.04601	.05717	.07077	.08764	.10920	.13805	.17965	.24846	.40529
0.4	.05031	.06281	.07786	.09620	.11901	.14824	.18751	.24452	.34012	.56805
0.5	.06509	.08091	.09999	.12330	.15234	.18971	.24019	.31423	.44112	.77048
0.6	.08141	.10072	.12404	.15259	.18829	.23446	.29733	.39088	.55675	1.10180
0.7	.09969	.12267	.15051	.18469	.22764	.28355	.36056	.47768	.69813	
0.8	.12038	.14731	.18003	.22038	.27139	.33842	.43229	.58011	.89722	
0.9	.14404	.17528	.21338	.26064	.32089	.40115	.51643	.70982		
1.0	.17136	.20740	.25158	.30680	.37808	.47501	.62040	.90537		
1.1	.20324	.24475	.29603	.36085	.44604	.56601	.76317			
1.2	.24089	.28887	.34880	.42587	.53035	.68818	1.14402			
1.3	.28604	.34209	.41328	.50763	.64379	.90559				
1.4	.34147	.40834	.49597	.61965	.84055					
1.5	.41214	.49570	.61335	.82670						
1.6	.50927	.62791	.93821							
1.7	.67824									

TABLE II.—$S_2^{J_0}(x;\alpha)$

x \ α	0	0.1	0.2	0.3	0.4	0.5	0.6	0.7	0.76548
0.1	.01723	.01736	.01778	.01851	.01954	.02089	.02261	.02476	.02642
0.2	.03474	.03502	.03586	.03730	.03935	.04207	.04554	.04986	.05320
0.3	.05281	.05323	.05451	.05668	.05973	.06390	.06913	.07566	.08069
0.4	.07175	.07232	.07405	.07695	.08112	.08668	.09372	.10252	.10934
0.5	.09188	.09260	.09476	.09845	.10373	.11075	.11969	.13082	.13946
0.6	.11354	.11441	.11706	.12155	.12799	.13655	.14746	.16108	.17162
0.7	.13713	.13816	.14132	.14666	.15432	.16451	.17751	.19375	.20636
0.8	.16311	.16431	.16800	.17425	.18321	.19515	.21041	.22949	.24433
0.9	.19199	.19340	.19766	.20490	.21527	.22913	.24684	.26906	.28637
1.0	.22449	.22610	.23100	.23930	.25125	.26723	.28769	.31346	.33358
1.1	.26138	.26323	.26885	.27838	.29211	.31049	.33413	.36403	.38748
1.2	.30385	.30596	.31241	.32337	.33916	.36039	.38785	.42277	.45039
1.3	.35346	.35588	.36335	.37601	.41437	.41913	.45141	.49288	.52601
1.4	.41263	.41546	.42422	.43913	.46086	.49040	.52932	.58030	.62198
1.5	.48564	.48907	.49879	.51776	.54443	.58130	.63116	.69946	.75929
1.58	.....	.....	.....	.....	.....	.....	.....	.....	1.01924
1.6	.58127	.58573	.59941	.62328	.65927	.71140	.78909	.93124	
1.7	.72799	.73524	.75821	.80177	.88556				

TABLE III.—$S_{N^0}^{Z^0}(x,0,c)$

		$c = 2$				$c = 3$		
x \ N	4	3	2	1	4	3	2	1
0.1	.00003	.00006	.00016	.00092	.00007	.00014	.00036	.00198
0.2	.00010	.00019	.00051	.00308	.00019	.00036	.00093	.00523
0.3	.00026	.00050	.00130	.00786	.00038	.00073	.00190	.01113
0.4	.00053	.00102	.00269	.01680	.00070	.00134	.00351	.02133
0.5	.00096	.00186	.00494	.03197	.00117	.00226	.00597	.03792
0.6	.00160	.00310	.00828	.05625	.00185	.00358	.00953	.06389
0.7	.00249	.00483	.01298	.09422	.00278	.00539	.01445	.10409
0.8	.00366	.00712	.01931	.15457	.00400	.00777	.02103	.16781
0.9	.00516	.01007	.02762	.25982	.00554	.01081	.02959	.28013
1.0	.00705	.01380	.03830		.00747	.01462	.04053	

		$c = 4$				$c = 5$		
x \ N	4	3	2	1	4	3	2	1
0.1	.00012	.00023	.00057	.00297	.00016	.00031	.00077	.00386
0.2	.00028	.00053	.00135	.00725	.00037	.00070	.00175	.00905
0.3	.00052	.00099	.00254	.01423	.00065	.00124	.00314	.01699
0.4	.00088	.00168	.00435	.02559	.00106	.00202	.00517	.02941
0.5	.00140	.00269	.00704	.04352	.00163	.00312	.00807	.04854
0.6	.00213	.00410	.01083	.07111	.00240	.00461	.01207	.07757
0.7	.00310	.00599	.01598	.11341	.00342	.00659	.01745	.12177
0.8	.00436	.00846	.02280	.18035	.00473	.00915	.02451	.19165
0.9	.00596	.01160	.03162	.29973	.00638	.01238	.03358	.31772
1.0	.00793	.01550	.04283		.00840	.01638	.04507	

GLOSSARY

B = susceptance.

$Ci(x) = -\int_x^\infty \dfrac{\cos y}{y}\, dy$ = cosine integral function.

$E(x) = \int_0^{\pi/2} \sqrt{1 - x^2 \sin^2 \phi}\, d\phi$ = complete elliptic integral of second kind.

E_{mn} = mnth transverse magnetic mode.[1]

$F(x) = \int_0^{\pi/2} \dfrac{1}{\sqrt{1 - x^2 \sin^2 \phi}}\, d\phi$ = complete elliptic integral of first kind.

G = conductance.

$H_n^{(1)}(x), H_n^{(2)}(x)$ = Hankel function of $\begin{cases} \text{first} \\ \text{second} \end{cases}$ kind.

$\hat{H}_n^{(1)}(x), \hat{H}_n^{(2)}(x) = \sqrt{\dfrac{\pi x}{2}}\, H_{n+\frac{1}{2}}^{(1),(2)}(x)$ = spherical Hankel function of $\begin{cases} \text{first} \\ \text{second} \end{cases}$ kind.

H_{mn} = mnth transverse electric mode.[1]

I = rms current amplitude.

Im = imaginary part of.

$I_n(x) = j^{-n} J_n(jx)$ = modified Bessel function of first kind.

$j = \sqrt{-1}$.

$J_n(x)$ = Bessel function of order n.

$\hat{J}_n(x) = \sqrt{\dfrac{\pi x}{2}}\, J_{n+\frac{1}{2}}(x)$ = spherical Bessel function.

$k = \dfrac{2\pi}{\lambda} = \omega\sqrt{\mu\epsilon}$ = propagation wavenumber in medium.

$k_c = \dfrac{2\pi}{\lambda_c}$ = cutoff wavenumber of guide.

$K_n(x) = j^{n+1} \dfrac{\pi}{2} H_n^{(1)}(jx) = (-j)^{n+1} \dfrac{\pi}{2} H_n^{(2)}(-jx)$ = modified Bessel function of second kind.

$N_n(x)$ = Neumann function of order n.

$\hat{N}_n(x) = \sqrt{\dfrac{\pi x}{2}}\, N_{n+\frac{1}{2}}(x)$ = spherical Neumann function.

R = resistance.

$\mathcal{R}$ = characteristic resistance of metallic medium.

Re = real part of.

S = scattering coefficient.

$Si(x) = \int_0^x \dfrac{\sin y}{y}\, dy$ = sine integral function.

V = rms voltage amplitude.

X = reactance.

Y = admittance.

Z = impedance.

α = attenuation constant.

---

[1] In circular and coaxial waveguides m denotes the periodicity in the angular direction while n denotes the periodicity in the radial direction.

GLOSSARY

β = phase constant.
$\gamma = \alpha + j\beta$ = propagation constant.
Γ = reflection coefficient.
δ = skin depth of medium.
ϵ = absolute dielectric constant of medium.
$\dfrac{\epsilon}{\epsilon_0} = \epsilon' - j\epsilon''$ = relative dielectric constant of medium.
$\kappa = \dfrac{2\pi}{\lambda_g} = -j\gamma$ = propagation wavenumber in guide.
λ = free space wavelength.
λ_c = cutoff wavelength.
$\lambda_g = \dfrac{\lambda}{\sqrt{1 - (\lambda/\lambda_c)^2}}$ = guide wavelength.
μ = absolute magnetic permeability of medium.
σ = conductivity of medium.
ω = angular frequency.
$\eta = \dfrac{1}{\zeta} = \sqrt{\dfrac{\epsilon}{\mu}}$ = intrinsic admittance of medium.
$\zeta = \dfrac{1}{\eta} = \sqrt{\dfrac{\mu}{\epsilon}}$ = intrinsic impedance of medium.

Index

A

Admittance, relative or normalized, 10
   frequency derivatives of, 12
Admittance matrix, 106
Aperture coupling, of concentric coaxial guides, 377–378
   of contiguous rectangular guides, E-plane, 375–376
   H-plane, 379–380
Apertures, circular, in rectangular guide, 194
   rectangular, in rectangular guide, 193
Arrays, in free space, capacitive posts in, 285–286
   capacitive strips in, 280–284
   inductive posts in, 286–289
   inductive strips in, 284–285
   of semi-infinite planes, E-plane, 289–292
   H-plane, 195, 292–295
Attenuation constant, 18
   of circular waveguides, E-modes, 67
   of coaxial waveguides, H-modes, 79
      higher E-modes, 75
      lowest E-mode, 73
   of conical waveguides, dominant E-mode, 99
   due to losses in metallic guide walls, 24
   of E-mode in arbitrary uniform guide with dissipative metallic walls, 24
   of parallel plate guide, E-modes, 64
      H-modes, 65
   of plane waves in free space, 87
   of rectangular waveguides, E-modes, 57
      H-modes, 60
      H_{10}-mode, 61

B

Bennett, H. S., 46
Bessel functions, 32
Bessel functions, half-order, 52
   mth-order, 66
Bifurcation, of coaxial line, 369–372
   E-plane, 112, 160
   of rectangular guides, E-plane, 353–355, 380–383
   H-plane, 172, 302–307, 383–386
Bisection theorem, 128
Boundary-value problem, electromagnetic-, 101
Bridge circuits, 128

C

Capacitive gap, termination of coaxial line by, 178
Characteristic impedance, 6
   of mode, 8
Chu, L. J., 79
Circle diagram, 11
Circuit parameters (*see* Network parameters)
Circular guides, annular obstacles in, of zero thickness, 249
   annular window in, 247–249
   aperture coupling of, 330
      to rectangular guide, 332
   circular obstacle in, 273–275
   coupling of, to coaxial guide, 323–324
   with dielectric cylinders, 393–396
   E-modes, 66
      attenuation constant of, 67
      cutoff wavelength of, 67
      field components of, 66
      total power of, 67
   elliptical and circular apertures in, 243–246
   H-modes, 69
      attenuation constant of, 70
      cutoff wavelength of, 70
      field components of, 69
      total power of, 70

Circular guides, radiation from, E_{01}-mode, 196
  H_{01}-mode, 201
  H_{11}-mode, 206
  and rectangular guide, junction of, 324–329
  resonant ring in, 275–280
Coaxial guides, 72
  aperture coupling of, 331
  capacitive windows in, with disk on inner conductor, 229–234
    with disk on outer conductor, 234–238
  coupling of, to circular guide, 323–324
  with dielectric cylinders, 396–397
  H-modes, 77
    attenuation constant of, 79
    cutoff wavelength of, 77
    field components of, 78
    total power of, 79
  higher E-modes, 73
    attenuation constant of, 75
    cutoff wavelength of, 74
    field components of, 75
  junction of, 310–312
  lowest E-mode, attenuation constant of, 73
    field components of, 72
  small elliptical and circular apertures in, 246–247
Coaxial line, bifurcation of, 369–372
  coupling of, to circular guide, 174
  with infinite center conductor, 208
  radiating into half space, 213
  termination of, by capacitive gap, 178
Components of conical waveguides, dominant E-mode, 99
Composite structures, 387–413
Conformal mapping, 154
Conical regions, 53
Conical waveguides, 98
  dominant E-mode, attenuation constant of, 99
  components of, 99
  cutoff wavelength of, 99
Cutoff wave number, 8
Cutoff wavelength, of circular waveguides, E-modes, 67
  H-modes, 70
  of coaxial waveguides, H-modes, 77
    higher E-modes, 74
  of conical waveguides, dominant E-mode, 99

Cutoff wavelength, of E_i and H_i-modes, 88
  of elliptical waveguides, $_eH_{mn}$-mode, 83
  of parallel plate guide, E-modes, 64
  H-modes, 65
  of radial waveguides, E-type modes, 91, 94
  H-type modes, 92, 95
  of rectangular waveguides, E-modes, 57
  H-modes, 60

D

Delta function, 162
Derivatives, frequency (*see* Frequency derivatives)
Desikachar, P. R., 53
Dielectric constant, complex, 18
Dielectric materials, properties of, 19
Discontinuities, equivalent circuits for, 108–117
  representation of, 101–108
Dissipation, electric-type, 18
  magnetic-type, 23
  waveguides with, 17
Dissipative case, standing waves for, 26
Dissipative guide, scattering description in, 27
Dissipative transmission lines, perturbation method of calculation for, 26
Dominant mode, 2
Duality principle, 10

E

E-mode functions, 4
Eight-terminal structures, 373–386
Electrical length, 12
Elliptical waveguides, 79
  $_eE_{mn}$-mode, field components of, 81
  $_eH_{mn}$-mode, attenuation in, 83
    cutoff wavelengths of, 83
    degenerate modes of, 83
    field components of, 82
    power flow in, 83
Equivalent circuit, 12, 104
  for discontinuities, 108–117
Error curve, 136

F

Feshbach, H., 54
Field, longitudinal, 5

INDEX

Field, magnetic, 5
  transverse electric, 5
Field distribution, of circular waveguides, E-modes, 66
  H-modes, 69
  of coaxial waveguides, H-modes, 78
    higher E-modes, 75
    lowest E-mode, 72
  of elliptical waveguides, $_eE_{mn}$-mode, 81
    $_eH_{mn}$-mode, 82
  of parallel plate guide, H-modes, 65
  of radial waveguides, E-type modes, 90, 93
    H-type modes, 92, 94
  of rectangular waveguides, E-modes, 57
    H-modes, 60
    H_{10}-mode, 61
  of spherical waveguides, E_{mn}-mode, 97
    H_{mn}-mode, 97
Field equations, invariant transverse vector formulation of, 3
  for radial waveguides, 29
Four-terminal structures, 217–335
Fourier transform, 164
Free space, gratings in, 88
  plane waves in, attenuation constant of, 87
  as uniform waveguide, fields in, 84
    E_i-mode, 85, 88
      cutoff wavelengths of, 88
    H_i-mode, 86, 88
      cutoff wavelengths of, 88
    power flow in, 86
Frequency derivatives, of reflection coefficients, 15
  of relative admittances, 12
Frequency sensitivity, on radial line, 46
  in radial structures, 43

G

Gradient operator, 4
Gratings in free space, 88
Green's function, 162
Guide wavelength of rectangular waveguides, H_{10}-mode, 62

H

H-mode functions, 4
Hankel function, 43
  amplitude of, 44

Hankel function, phase of, 44
  spherical, 54
Hutner, R. A., 79
Hybrid coil, 128
Hybrid junction, 386

I

Ideal transformer, three-winding, 127
Impedance descriptions of uniform transmission lines, 9
Impedance matrix, 103
Impedance measurements, input-output, 131
Integral equation, homogeneous, 163
  Wiener-Hopf, 164
Integral-equation method, 146

J

Junction, coaxial to waveguide, 116
  coaxial T-, 114
  cross, of two coaxial guides, 116
  with E-plane symmetry, 114
  with H-plane symmetry, 116
  Magic-T-, 116
  of N waveguides, 102
  probe coupled, 114
  T- (see T-junction)

L

Lax, 54
Legendre function, associated, 96
Levine, H., 143
Lowan, 54

M

Magic-T-junctions, 116, 386
Mathieu function, angular, 81
  radial, 81
Matrix, admittance, 106
  impedance, 103
  scattering, 107
Maxwell, E., 21
Measurement, of network parameters, 130–138
  precision method of, for nondissipative two-terminal-pair structures, 132
Microwave networks, 101–167

Minimum, position of, 132
Mode, 1
  characteristic impedance of, 8
  dominant, 2
Mode characteristics, 55–56
Mode patterns, 56
Morse, P. M., 54, 79

N

N-terminal-pair structures, 126
Network equations, 103
Network parameters, measurement of, 130–138
  sensitivity of, 137
  theoretical determination of, 138–167
Networks, four-terminal-pair, 114
  "natural," 105
  three-terminal-pair, 111
  two-terminal-pair, 110
Nonuniform regions, 1

O

Orthogonality properties, 5

P

Parallel plate guide, E-modes, 62
    attenuation constant of, 64
    cutoff wavelength of, 64
  E-plane, gain pattern for, 182
  H-modes, 65
    attenuation constant of, 65
    cutoff wavelength of, 65
    field components of, 65
  radiating into half space, E-plane, 183
    H-plane, 187
  radiating into space, E-plane, 179
    H-plane, 186
Power flow, average, 5
Propagation constant, 6

R

Radial functions, cotangent, 33
  tangent, 33
Radial line, dominant E-type mode in, scattering description of, 43
  dominant H-type mode in, 46
  E-type, circuit representation of, 42
  frequency sensitivity on, 46

Radial line, H-type, T-circuit representation of, 42
  voltage reflection coefficient of, 44
Radial structures, frequency sensitivity in, 43
Radial waveguides, 89
  cylindrical cross sections of, 89
  cylindrical sector cross sections of, 93
  dominant E-type mode, impedance description of, 31
  E-type modes, cutoff wavelength of, 91, 94
    field components of, 90, 93
    total outward power of, 91
    field representation in, 29–47
  H-type modes, cutoff wavelength of, 92, 95
    field components of, 92, 94
Radiation, into bounded half space, rectangular guide, E-plane, 184
  into bounded space, rectangular guide, E-plane, 183
  from circular guide, E_{01}-mode, 196
    H_{01}-mode, 201
    H_{11}-mode, 206
  into half space, coaxial line, 213
    parallel-plate guide, E-plane, 183
    H-plane, 187
  into space, parallel-plate guide, E-plane, 179
    H-plane, 186
Ragan, G. L., 10
Rayleigh-Ritz procedure, 145
Reciprocity relations, 104
Rectangular guides, 405
  aperture coupling of, 329–333
    to circular guide, 332
  apertures in, 193
  bifurcation of (see Bifurcation, of rectangular guides)
  capacitive obstacles in, of finite thickness, 249–255
    of large thickness, 404–406
  capacitive obstacles and windows in, of zero thickness, 218–221
  capacitive post in, 268–271
  change of, in cross section, to circular guide, 176
  change in height of, 307–310
  circular aperture in, 410
  circular bends in, E-plane, 333–334
    H-plane, 334–335

INDEX 427

Rectangular guides, circular and elliptical apertures in, of zero thickness, 238–243
  dielectric posts in, 266–267
  with dielectric slabs, parallel to E, 388–391
    perpendicular to E, 391–393
  E-modes, 56
    attenuation constant of, 57
    cutoff wavelength of, 57
    field distribution of, 57
    maximum electric-field intensity for, 58
    total power flow of, 58
  E-plane corners of, arbitrary angle bends of, 316–318
    right-angle bends of, 312–316
  H-modes, 58
    attenuation constant of, 60
    cutoff wavelength of, 60
    field components of, 60
    maximum field-intensity for, 60
    total power of, 61
  H-plane corners of, arbitrary angle, 319–322
    right-angle bends of, 318–319
  H_{10}-mode, 61
    attenuation constant of, 61
    field components of, 61
    guide wavelength of, 62
  inductive obstacles in, of finite thickness, 255–257
    of large thickness, 407–408
  inductive obstacles and windows in, of zero thickness, 221–229
  junction of, and circular guide, 324–329
    H-plane, 168, 296–302
    and radial guide, E-plane, 322–323
    with "nonradiating" slit, 397–398
  120° Y-junction of, E-plane, 352
    H-plane, 362
  post of variable height in, 271–273
  radiating into bounded half space, E-plane, 184
  radiating into bounded space, E-plane, 183
  with rectangular ridge, 399–402
  with resistive strip, 402–404
  solid inductive post in, centered, 258–263
    noncircular, 263–266
    off-centered, 257–258

Rectangular guides, spherical dent in, 273
  T-junction in (see T-junction, in rectangular guide)
  thick circular window of, 408–412
Reference planes, transformations of, 120
Reflection coefficient, 15
  current, 13, 44
  frequency derivatives of, 15
  voltage, 13
    radial line, 44
    in uniform transmission line, 13
Representations, circuit, of two-terminal-pair structure, 118
  of discontinuities, 101–108
  distance invariant, 137
  equivalent, of microwave networks, 117–130
  ideal transformer, of two-terminal-pair structure, 120
  series, of two-terminal-pair network, 124
  shunt, of two-terminal-pair network, 124
Resonant cavities, 89

S

Scattering coefficient, 107
Scattering description in dissipative guide, 27
Scattering matrix, 107
Schelkunoff, S. A., 96
Schwartz-Christoffel transformation, 155
Schwinger, J., 138
Six-terminal structures, 336–372
Slit coupling of contiguous rectangular guides, E-plane, 373–375
  H-plane, 378–379
Smith chart, 10
Spherical cavities, 96
Spherical functions, 53
Spherical Hankel functions, amplitude of, 54
  phase of, 54
Spherical transmission line, scattering description of, 53
Spherical waveguides, 96
  cotangents of, 53
  E-mode functions of, 49
  E_{mn}-mode, cutoff wavelength of, 98
    field components of, 97
    total power of, 97

Spherical waveguides, equations of, for electric and magnetic fields transverse, 47
   field representation in, 47–54
   H-mode functions of, 49
   H_{mn}-mode, field components of, 97
   longitudinal components of, 47
   tangents of, 53
   wave equations for, 51
Staehler, R. E., 394
Standing-wave ratio, 15, 132
Static method, equivalent, 153
Stratton, J. A., 79
Stream function, 156
Stub guide, 111
Symmetrical structures, 109
Symmetry, E-plane, 111
   H-plane, 111

T

T-junction, coaxial, 114
   in coaxial guide, aperture-coupled, 368
   in rectangular guide, aperture-coupled, E-plane, 363–366
      H-plane, 366
   aperture-coupled circular stub, 364
      H-plane, 367
   aperture-coupled rectangular stub, 363, 366
      H-plane, 366
   open, E-plane, 337–339
      H-plane, 355–360
   slit-coupled, E-plane, 339–352, 412–413
      H-plane, 360–362
Tangent relation, 121
Terminal planes, 104
Theoretical determination of circuit parameters, 138–167
Transform method, 160
Transmission line, π-network of, 12
   dissipative, perturbation method of calculation for, 26
   spherical, scattering description of, 53
   T network of, 12
   uniform (see Uniform transmission lines)
Transmission-line equations, 6
Traveling wave, 13

Trial field, 143
Two-terminal-pair structure, circuit representations of, 118

U

Uniform regions, 1
Uniform transmission-line descriptions, interrelations among, 16–17
Uniform transmission lines, 7–13
   with complex parameters, 17–29
   impedance descriptions of, 9
   scattering descriptions of, 13–16
   voltage reflection coefficient in, 13
Uniform waveguides, field representation in, 3–7
Uniqueness theorem, 103
Unitary relations, 108

V

Variational expression, 150
Variational method, 143
von Hippel, A. R., 20

W

Wave equations, 9
Waveguides, circular (see Circular guides)
   coaxial (see Coaxial guides)
   conical (see Conical waveguides)
   beyond cutoff, 27
      lines terminating in, 168–178
   with dissipation, 17
   elliptical (see Elliptical waveguides)
   parallel plate (see Parallel plate guide)
   radial (see Radial waveguides)
   rectangular (see Rectangular guides)
   spherical (see Spherical waveguides)
   as transmission lines, 1–3
   uniform, field representation in, 3–7
Weissfloch, A., 122, 133
Wiener-Hopf integral equation, 164
Window, asymmetrical capacitive, field problems of, 140

Y

Y-junction, 112
   H-plane, 113